METHODS IN MOLECULAR BIOLOGY

Series Editor
John M. Walker
School of Life and Medical Sciences
University of Hertfordshire
Hatfield, Hertfordshire, UK

For further volumes:
http://www.springer.com/series/7651

For over 35 years, biological scientists have come to rely on the research protocols and methodologies in the critically acclaimed *Methods in Molecular Biology* series. The series was the first to introduce the step-by-step protocols approach that has become the standard in all biomedical protocol publishing. Each protocol is provided in readily-reproducible step-by-step fashion, opening with an introductory overview, a list of the materials and reagents needed to complete the experiment, and followed by a detailed procedure that is supported with a helpful notes section offering tips and tricks of the trade as well as troubleshooting advice. These hallmark features were introduced by series editor Dr. John Walker and constitute the key ingredient in each and every volume of the *Methods in Molecular Biology* series. Tested and trusted, comprehensive and reliable, all protocols from the series are indexed in PubMed.

Base Excision Repair Pathway

Methods and Protocols

Edited by

Kishor K. Bhakat

Department of Genetics, Cell Biology and Anatomy, University of Nebraska Medical Center, Omaha, NE, USA

Tapas K. Hazra

Department of Internal Medicine, The University of Texas Medical Branch, Galveston, TX, USA

Humana Press

Editors
Kishor K. Bhakat
Department of Genetics, Cell Biology and Anatomy
University of Nebraska Medical Center
Omaha, NE, USA

Tapas K. Hazra
Department of Internal Medicine
The University of Texas Medical Branch
Galveston, TX, USA

ISSN 1064-3745 ISSN 1940-6029 (electronic)
Methods in Molecular Biology
ISBN 978-1-0716-3372-4 ISBN 978-1-0716-3373-1 (eBook)
https://doi.org/10.1007/978-1-0716-3373-1

This Humana imprint is published by the registered company Springer Science+Business Media, LLC, part of Springer Nature.
The registered company address is: 1 New York Plaza, New York, NY 10004, U.S.A.

Paper in this product is recyclable.

Preface

Cellular DNA is constantly threatened by various endogenous and environmental agents that can damage DNA through various chemical reactions and physical interactions. Reactive oxygen species (ROS) generated endogenously from cellular respiration or chronic inflammatory conditions also cause variety of modified or damaged DNA bases and/or DNA strand breaks in the genome. Cells have developed multiple DNA repair pathways to repair such DNA lesions, and failure in these repair processes can lead to a variety of human diseases including neurodegeneration, cancer, and premature aging.

The base excision repair (BER) pathway is the primary mechanism for repairing small, non-bulky DNA lesions, including oxidized, alkylated, and deaminated bases in the genome. Over the past several decades, scientists around the world have been examining repair of DNA lesions in vitro with purified proteins and/or cell extracts. However, there is no systematic and more elaborated easy-to-follow protocols/assays to examine excision and repair of damaged bases in DNA through BER mechanism in vitro and in cells. This book is dedicated to BER assays that describe multiple assay systems to monitor and examine BER of damaged DNA base in naked DNA or DNA wrapped in a nucleosome with recombinant purified BER proteins in vitro and in cells. Detailed protocols for examining each step of the BER pathway and complete in vitro reconstitution of a damaged base with purified proteins have been described. Further, we present methods for detection and quantification of endogenously generated oxidized DNA bases, DNA double-strand breaks (DSBs), and protein-DNA crosslinks, and for the identification of repair proteins interactome and their genome-wide binding in response to DNA damage. The leading experts in the field of BER were recruited for this book to contribute useful laboratory protocols that represent a comprehensive collection of great utility in the field of BER and genome maintenance. We anticipate that this *Methods in Molecular Biology* book on BER will be widely read and applied in almost every biological setting of experimentation to examine the crucial roles of endogenous damage repair pathways to maintain genomic integrity.

This book begins with six chapters describing multiple molecular and cellular techniques to examine the excision of damaged bases from double-stranded DNA or DNA wrapped in a nucleosome in vitro as well as in live cells. Historically, most of the BER mechanisms were studied in vitro using purified proteins and/or cell lysates, but those may not truly reflect DNA repair conditions in cells. Dr. Roy describes a new method to study BER of a single lesion in live cells using a plaque-based host cell reactivation assay (PL-HCR). This assay system not only measures repair of a damaged base in a plasmid DNA in cells but also provides the information on a BER subpathway that is utilized to repair this type of damage (Chap. 1). This is followed by a contribution from Dr. Tell's lab that describes an experimental procedure to examine excision of a ribose monophosphate abasic site from DNA and provides a pipeline of protocols to quantify endodeoxyribonuclease activity of one of the key BER enzyme, APE1, on these substrates by using recombinant protein and whole cell extracts. The repair capacity of APE1 was measured using fluorescent oligonucleotide substrates, which are then separated by polyacrylamide gel electrophoresis and detected by a laser scanning imaging system (Chap. 2). Non-ligatable DNA ends, such as 3′-phosphates, are generated either by DNA glycosylase-mediated removal of damaged bases or at the site of ROS-mediated single-strand breaks (SSBs) in the genome. These "dirty" DNA ends, if not rapidly processed, will impede transcription and replication leading to various cellular pathologies. Dr. Hazra's group describes two novel assay systems of repair of 3′-phosphate by polynucleotide kinase 3′-phosphatase (PNKP) in vitro or cell free nuclear extracts from mammalian cells/tissues. These assays could be useful for comparison of PNKP-mediated SSBR activity in diseased mammalian cells/tissues versus normal healthy controls (Chap. 3). Genomic DNA in the nucleus is wrapped around nucleosomes, a repeating unit of chromatin, which can pose a barrier for DNA repair enzymes to locate, access, and process DNA damage in the cell. Dr. Freudenthal's group describes a protocol for generating nucleosome containing site-specific DNA damage in vitro. This protocol describes several key steps for expression and purification of recombinant histones and the reconstitution of nucleosomes with DNA oligo containing site-specific damaged DNA base. These methods will enable researchers to generate nucleosomes containing site-specific DNA damage for extensive biochemical and structural studies of DNA repair in the context of nucleosome (Chap. 4). Apart from BER, there are several major repair pathways including nucleotide incision repair (NIR) involved in excising mutagenic or replication blocking modified bases or a modifying group from DNA. Dr. Hang has developed and optimized a DNA cleavage assay that can be used to explore whether a lesion is processed by NIR or nucleotide excision repair (NER) activity, as well as to study its miscoding properties in translesion DNA synthesis (Chap. 5). Finally, Jaiswal et al. provide a detailed assay procedure to reconstitute in vitro complete BER assay using recombinant purified BER proteins (UDG, APE1, Pol β, and Ligase III), providing an opportunity to monitor the sequential enzymatic steps involved in BER pathway. This aids our understanding of the biological activities of these proteins and their coordinated interactions during the repair process (Chap. 6).

Experimental procedures to detect and quantify the damaged bases, protein DNA crosslinks, and DSBs are presented in five modules in Part II of the book. 8-oxoguanine (8-oxoG) base damage has been considered one of the oxidative stress markers in genomic DNA. Lang et al. describe a new and sensitive method that can identify 8-xoG in specific genomic regions with high accuracy in a laboratory setting by fragment length analysis with repair enzyme (FLARE)-coupled to quantitative (q)-PCR or FLARE-qPCR. This assay can be performed using nanogram quantities of DNA samples compared to conventional gas chromatography-mass spectrometry assays or tandem liquid chromatography-mass spectrometry, which require microgram quantities of DNA for analysis and highly trained personnel (Chap. 7). DNA-protein crosslinks (DPCs) are irreversible covalent crosslinks of proteins to the DNA and are steric blockades to virtually all DNA metabolic processes, namely repair, replication, and transcription. Direct visualization or detection of specific protein-DNA crosslinks has been challenging, and it limits determining DPC in a stoichiometry and quantitative manner. Dr. Ghosal's lab describes a modified protocol for direct detection of specific enzymatic DPCs by the rapid approach to DNA adduct recovery (RADAR) assay (Chap. 8). R loops (DNA-RNA hybrid), 3-stranded nucleic acid structures that are comprised of template DNA strand hybridized with the nascent RNA leaving the displaced non-template strand, play an important role in inducing genomic instability. Drs. Sarker and Cooper describe an improved slot-blot protocol to detect and estimate R loops using S9.6 antibody. Since specific factors protecting cells from harmful R loops accumulation are expanding, this protocol can be used to determine R loop accumulation in cells under normal conditions and clinical settings (Chap. 9). Infection and chronic inflammation can cause DNA damage, and the accumulation of mutations leads to cancer development. Well-known examples of cancer-associated microbes are *Helicobacter pylori* in gastric cancer and *Fusobacterium nucleatum in* colorectal carcinoma. Dr. Das's group utilized stem cell-based murine or human colon organoid models to assess the impact of infection on the expression of BER enzymes on the transcriptional and translational levels and develop other functional assays (Chap. 10). Unrepaired SSBs can lead to DSBs during replication, and DSBs are primarily repaired by homologous recombination (HR) and the non-homologous end joining (NHEJ) pathway. A third pathway, microhomology-mediated alternative end-joining (MM/Alt-EJ), has emerged as a specialized back-up pathway for DSB repair that also occurs in non-proliferating cells. Drs. Mitra's and Hegde's groups describe reporter plasmid-based assays to estimate the relative contribution of various DSB repair mechanisms and compare the advantages, limitations, and challenges of prevailing methods (Chap. 11).

Efficient repair of small base damages in the genome in cells via BER not only require the action of individual proteins but their coordinated recruitments and actions through protein-protein interactions. Increasing evidence suggests that DNA damage-induced interactions among repair proteins facilitate their recruitment to damage sites and repair efficiency. Part III of this book describes multiple experimental procedures to identify DNA repair protein interactome by conventional tandem affinity purification followed by mass spectroscopy analysis. Dr. Wood's group describes an immunoprecipitation (IP) procedure that is compatible with mass spectrometry proteomic analysis to examine protein-protein interactions in response to DNA damage. The in-gel digestion protocol is optimized to profile the interactome when there is a limited amount of input lysate, or the study's focus is on posttranslational modifications of target proteins (Chap. 12). Affinity pulldown is also a powerful technique to discover novel interaction partners and verify a predicted physical association between two or more proteins. Dr. Tsai describes a detailed method of pulldown assays for two high-affinity peptide fusion tags, Flag tag and histidine tag, to study protein-protein interactions of human DNA glycosylase NEIL1 and the checkpoint protein complex RAD9-RAD1-HUS1 (9-1-1) (Chap. 13). Repair of damaged bases in the nucleosome context in cells requires the help of nucleosome remodeling complexes which facilities repair in chromatin. The histone chaperone facilitates chromatin transcription (FACT) complex interacts with many BER proteins to facilitate BER. Kaja et al. provide a tandem affinity purification strategy that allows rapid isolation of endogenous FACT. This method of purification of FACT will allow us to know how it works in vitro to facilitate the repair process, and TAP-mediated isolation strategy can be combined with mass spectrometry to identify the protein interaction partners of FACT (Chap. 14).

In addition to specific DNA repair proteins or the interactions with downstream proteins, the influence of copy number variation (CNV) of DNA repair genes can affect cellular response to stress and chemotherapy resistance and/or response. Dr. Izumi presents an experimental protocol to examine the copy number variation of BER genes in tumor tissues using the expression profile of BER protein XRCC1 with its neighbor genes *LIG1*, PNKP, and *POLD1* as an example. Methods for CNV assay at the individual gene level on formalin-fixed, paraffin-embedded (FFPE) tissues using the NanoString Counter technology has been described (Chap. 15). Dr. Bhakat's group describes genome-wide binding analysis of the key BER protein APE1 in both cancer cells and transformed primary cells by chromatin-immunoprecipitation followed by next-generation sequencing (ChIP-seq) methods. Further, simple biomimetics tools that are commonly being used to analyze the data have also been briefly discussed (Chap. 16). Finally, Dr. Ray describes a tumor sphere formation assay to examine the role of STAT3 in promoting tumor growth in vitro

(Chap. 17). These final chapters emphasize the translational values of these assays to test the role of different DNA BER proteins in promoting therapy resistance or tumorigenesis.

Altogether, the compiled 17 chapters in this book provide a comprehensive set of experimental protocols/techniques and useful strategies to examine repair of damaged bases via BER pathway in vitro and in cells. We sincerely believe that this unique collection of chapters provides a valuable resource for novices and experts to examine the repair of various types of DNA lesions in vitro and in cell by the distinct set of proteins in the BER pathway.

We wish to thank all the authors for their outstanding contributions in this book. Their sincere efforts are very much appreciated, and it was a great pleasure to work with them in producing the book. We also acknowledge John Walker, the series editor, for his helpful suggestion, advice, editing of the chapters, and guidance in making/organizing/preparing this book. We also thank our postdoctoral mentor Dr. Sankar Mitra, who is one of the pioneers in developing many BER assays. Throughout our scientific career, Dr. Mitra has taught us the usefulness and mechanics of the scientific method, and the benefits of creative thinking and of developing new assays to understand how cells have evolved multiple repair pathways to maintain the genomic integrity.

Omaha, NE, USA *Kishor K. Bhakat*
Galveston, TX, USA *Tapas K. Hazra*

Contents

Contributors

GIULIA ANTONIALI • *Laboratory of Molecular Biology and DNA Repair, Department of Medicine, University of Udine, Udine, Italy*

PRIYANKA BARMAN • *Department of Biochemistry and Molecular Biology, Southern Illinois University School of Medicine, Carbondale, IL, USA*

KISHOR K. BHAKAT • *Department of Genetics, Cell Biology and Anatomy, Fred and Pamela Buffett Cancer Center, University of Nebraska Medical Center, Omaha, NE, USA*

SUKESH R. BHAUMIK • *Department of Biochemistry and Molecular Biology, Southern Illinois University School of Medicine, Carbondale, IL, USA*

ISTVAN BOLDOGH • *Department of Microbiology and Immunology, University of Texas Medical Branch at Galveston, Galveston, TX, USA*

ANIRBAN CHAKRABORTY • *Department of Internal Medicine, Division of Pulmonary, Critical Care and Sleep Medicine, University of Texas Medical Branch, Galveston, TX, USA*

YINGLING CHEN • *Department of Genetics, Cell Biology and Anatomy, Fred and Pamela Buffett Cancer Center, University of Nebraska Medical Center, Omaha, NE, USA*

PRISCILLA K. COOPER • *Department of BioEngineering and BioMedical Sciences, Biological Systems and Engineering Division, Lawrence Berkeley National Laboratory, Berkeley, CA, USA*

SOUMITA DAS • *Department of Pathology, School of Medicine, University of California San Diego, San Diego, CA, USA*

ARIJIT DUTTA • *Department of Biochemistry and Structural Biology, University of Texas Health Science Center at San Antonio, San Antonio, TX, USA*

BRET D. FREUDENTHAL • *Department of Biochemistry and Molecular Biology, Department of Cancer Biology, University of Kansas Medical Center, Kansas City, KS, USA*

GARGI GHOSAL • *Department of Genetics, Cell Biology and Anatomy, University of Nebraska Medical Center, Omaha, NE, USA; Fred and Pamela Buffett Cancer Center, Omaha, NE, USA*

SHALINI GUHA • *Department of Biochemistry and Molecular Biology, Southern Illinois University School of Medicine, Carbondale, IL, USA*

BO HANG • *Division of Biological Systems and Engineering, Lawrence Berkeley National Laboratory, Berkeley, CA, USA*

TAPAS K. HAZRA • *Department of Internal Medicine, Division of Pulmonary, Critical Care and Sleep Medicine, University of Texas Medical Branch, Galveston, TX, USA*

MURALIDHAR L. HEGDE • *Division of DNA Repair Research, Center for Neuroregeneration, Department of Neurosurgery, Houston Methodist Research Institute, Houston, TX, USA*

PAVANA M. HEGDE • *Division of DNA Repair Research, Center for Neuroregeneration, Department of Neurosurgery, Houston Methodist Research Institute, Houston, TX, USA*

ROBERT A. HROMAS • *Division of Hematology and Medical Oncology, Department of Medicine and the Mays Cancer Center, University of Texas Health Science Center, San Antonio, TX, USA*

TADAHIDE IZUMI • *Department of Toxicology and Cancer Biology, University of Kentucky, Lexington, KY, USA*

ARUNA S. JAISWAL • *Department of Medicine, University of Texas Health Science Center, San Antonio, TX, USA*

ARUNIMA S. JAISWAL • *Division of Hematology and Medical Oncology, Department of Medicine and the Mays Cancer Center, University of Texas Health Science Center, San Antonio, TX, USA*
JENNIFER MOITOZA JOHNSON • *Department of BioEngineering and BioMedical Sciences, Biological Systems and Engineering Division, Lawrence Berkeley National Laboratory, Berkeley, CA, USA; XiMed Hospitalists, Inc., Scripps Memorial Hospital La Jolla, La Jolla, CA, USA*
AMALA KAJA • *Department of Biochemistry and Molecular Biology, Southern Illinois University School of Medicine, Carbondale, IL, USA*
WANG KE • *Department of Microbiology and Immunology, University of Texas Medical Branch at Galveston, Galveston, TX, USA*
KIMI KONG • *Division of Hematology and Medical Oncology, Department of Medicine and the Mays Cancer Center, University of Texas Health Science Center, San Antonio, TX, USA*
PAN LANG • *Department of Microbiology and Immunology, University of Texas Medical Branch at Galveston, Galveston, TX, USA*
HENRY C. -H. LAW • *Eppley Institute for Research in Cancer and Allied Diseases, Fred and Pamela Buffett Cancer Center, University of Nebraska Medical Center, Omaha, NE, USA*
MATILDE CLARISSA MALFATTI • *Laboratory of Molecular Biology and DNA Repair, Department of Medicine, University of Udine, Udine, Italy*
DREW T. MCDONALD • *Department of BioEngineering and BioMedical Sciences, Biological Systems and Engineering Division, Lawrence Berkeley National Laboratory, Berkeley, CA, USA*
JOY MITRA • *Division of DNA Repair Research, Center for Neuroregeneration, Department of Neurosurgery, Houston Methodist Research Institute, Houston, TX, USA*
SANKAR MITRA • *Division of DNA Repair Research, Center for Neuroregeneration, Department of Neurosurgery, Houston Methodist Research Institute, Houston, TX, USA; Department of BioEngineering and BioMedical Sciences, Biological Systems and Engineering Division, Lawrence Berkeley National Laboratory, Berkeley, CA, USA*
DRAGANA NOE • *Mass Spectrometry and Proteomics Core Facility, University of Nebraska Medical Center, Omaha, NE, USA; Precision Biomarker Laboratories, Cedars-Sinai Medical Center, Los Angeles, CA, USA*
MEGAN PERRY • *Department of Genetics, Cell Biology and Anatomy, Fred and Pamela Buffett Cancer Center, University of Nebraska Medical Center, Omaha, NE, USA*
SUTAPA RAY • *Department of Pediatrics, Division of Internal Medicine, University of Nebraska Medical Center, Omaha, NE, USA*
RABINDRA ROY • *Department of Oncology, Lombardi Comprehensive Cancer Center, Georgetown University, Washington, DC, USA*
BENJAMIN J. RYAN • *Department of Biochemistry and Molecular Biology, Department of Cancer Biology, University of Kansas Medical Center, Kansas City, KS, USA*
ALTAF H. SARKER • *Department of BioEngineering and BioMedical Sciences, Biological Systems and Engineering Division, Lawrence Berkeley National Laboratory, Berkeley, CA, USA*
IBRAHIM M. SAYED • *Department of Pathology, School of Medicine, University of California San Diego, San Diego, CA, USA*
JONAH J. SPENCER • *Department of Biochemistry and Molecular Biology, Department of Cancer Biology, University of Kansas Medical Center, Kansas City, KS, USA*
GIANLUCA TELL • *Laboratory of Molecular Biology and DNA Repair, Department of Medicine, University of Udine, Udine, Italy*

MASON TARPLEY • *Department of Genetics, Cell Biology and Anatomy, Fred and Pamela Buffett Cancer Center, University of Nebraska Medical Center, Omaha, NE, USA*
MIAW-SHEUE TSAI • *Department of BioEngineering and BioMedical Sciences, Biological Systems and Engineering Division, Lawrence Berkeley National Laboratory, Berkeley, CA, USA*
PAM S. WANG • *Department of BioEngineering and BioMedical Sciences, Biological Systems and Engineering Division, Lawrence Berkeley National Laboratory, Berkeley, CA, USA; Oncorus Inc., Cambridge, MA, USA*
TYLER M. WEAVER • *Department of Biochemistry and Molecular Biology, Department of Cancer Biology, University of Kansas Medical Center, Kansas City, KS, USA*
ELIZABETH A. WILLIAMSON • *Division of Hematology and Medical Oncology, Department of Medicine and the Mays Cancer Center, University of Texas Health Science Center, San Antonio, TX, USA*
NICHOLAS T. WOODS • *Eppley Institute for Research in Cancer and Allied Diseases, Fred & Pamela Buffett Cancer Center, University of Nebraska Medical Center, Omaha, NE, USA*
ZHENG XU • *Department of Microbiology and Immunology, University of Texas Medical Branch at Galveston, Galveston, TX, USA*
XUE YAOYAO • *Department of Microbiology and Immunology, University of Texas Medical Branch at Galveston, Galveston, TX, USA*

Part I

Base Excision Repair Assays In Vitro and in Live Cells

Chapter 1

Simultaneous Short- and Long-Patch Base Excision Repair (BER) Assay in Live Mammalian Cells

Rabindra Roy

Abstract

The base excision repair (BER) pathway repairs small, non-bulky DNA lesions, including oxidized, alkylated, and deaminated bases, and is responsible for the removal of at least 20,000 DNA lesions per cell per day. BER is initiated by DNA damage-specific DNA glycosylases that excise the damaged base and generates an abasic (AP) site or single-strand breaks, which are subsequently repaired in mammalian cells either by single-nucleotide (SN) or multiple-nucleotide incorporation via the SN-BER or long-patch BER (LP-BER) pathway, respectively. This chapter describes a plaque-based host cell reactivation (PL-HCR) assay system for measuring BER mechanisms in live mammalian cells using a plasmid-based assay. After transfection of a phagemid (M13mp18) containing a single modified base (representative BER DNA substrates) within a restriction site into human cells, restriction digestions detect the presence or absence (complete repair) of the adduct by the transformation of the digestion products into *E. coli* and counting the transformants as plaques. To monitor the patch size, different plasmids are constructed containing C:-A mismatches within different restriction sites (inhibiting digestion) at various distances on both sides (5′ or 3′) of the modified base-containing restriction sites. Using this assay, the percentage of repair events that occur via 5′ and 3′ patch formation can be quantified.

Key words Base excision repair sub-pathway, Short patch base excision repair, Long patch base excision repair, Host cell reactivation assay, Plaque-based host cell reactivation assay, Plaque assay, Alkylation damage, Oxidative damage

1 Introduction

Cellular DNA is constantly threatened by various endogenous and environmental agents that can damage DNA through various chemical reactions and physical interactions [1]. The endogenous DNA-damaging agents may arise from erroneous DNA replication, chronic inflammatory conditions, or simply cellular respiration [2]. The environmental or exogenous DNA-damaging agents often come from tobacco smoke, radiation, chemotherapy, or a person's diet [3]. Cells have developed multiple DNA repair pathways to repair damaged bases, and failure in these repair processes

Kishor K. Bhakat and Tapas K. Hazra (eds.), *Base Excision Repair Pathway: Methods and Protocols*, Methods in Molecular Biology, vol. 2701, https://doi.org/10.1007/978-1-0716-3373-1_1,

can lead to hastening of the onset of a variety of human diseases, including neurodegeneration, cancer, and premature aging [4, 5].

The base excision repair (BER) pathway repairs small, non-bulky DNA lesions, including oxidized, alkylated, and deaminated bases, and is responsible for the removal of at least 20,000 DNA lesions per cell per day [4, 6]. BER is initiated by DNA adduct-specific DNA glycosylases that excise the damaged base by hydrolytic cleavage of the N-glycosidic bond between the base and the sugar [7]. DNA glycosylase activity leads to the formation of an abasic (AP) site, which is subsequently cleaved by apurinic/apyrimidinic endonuclease 1 (APE1) [8]. The remaining steps of BER are described as either the single-nucleotide BER (SN-BER) or long-patch BER (LP-BER) sub-pathways. In the SN-BER pathway, DNA polymerase β (Pol β) exerts its dRP lyase activity to cleave the 5′-deoxyribose phosphate (5'-dRP) created after AP-site incision by APE1 [9]. Pol β subsequently incorporates 1 nt, and the nick is sealed by DNA ligase III (LIGIII). In the LP-BER pathway, the replicative DNA polymerases δ and ε (Pol δ, Pol ε), in combination with the other replication accessory factors, proliferating cell nuclear antigen (PCNA), and replication factor C (RFC), incorporate nucleotides at the single-nucleotide gap and continue to incorporate 2–10 nts by strand displacement on the 3′ side of the original lesion site [10, 11]. Pol β is also shown to be involved in LP-BER [12, 13]. The flap created by strand displacement by these polymerases is cleaved by flap endonuclease 1 (FEN1), and the resulting nick is sealed by DNA ligase I (Lig I). A third BER sub-pathway termed 5′ gap LP-BER is recently described [14]. This BER pathway engages proteins of BER (namely, DNA glycosylases and APE1), nucleotide excision repair (NER), and double-strand break repair (DSBR). This new BER process involves a 3′ to 5' DNA helicase, RECQ1 (but not other RecQ family helicases, WRN, BLM, RECQ4, and RECQ5) described to mediate the formation of a 9 nt gap at the 5′ end in cooperation with PARP1, RPA, and XPF/ERCC1 (but not XPA) during the repair response, and replicative polymerase(s) (but not Pol β) and FEN1 function to complete the process through incorporation of multiple nucleotides on both 5′ and 3′ sides of the original lesion site, up to 20 nts total. The long patch includes 8 nucleotides on the 5′ side of the lesion, the lesion itself, and 11 nucleotides on the 3′ side of the lesion (Fig. 1) [14]. It is not clear yet how cells decide to proceed via the SN- or LP-BER mechanisms; ATP concentration and the chemistry of 5′-dRP after AP-site incision by APE1 may dictate the selection of BER mechanisms in the cells [15].

Germline mutations or SNPs in many of these BER genes or variations in the activities of several of these BER enzymes are associated with a variety of diseases including cancer [16–21]. However, the lack of an assay that can reliably measure simultaneously the SN- and LP-BER mechanisms in live cells caused limited

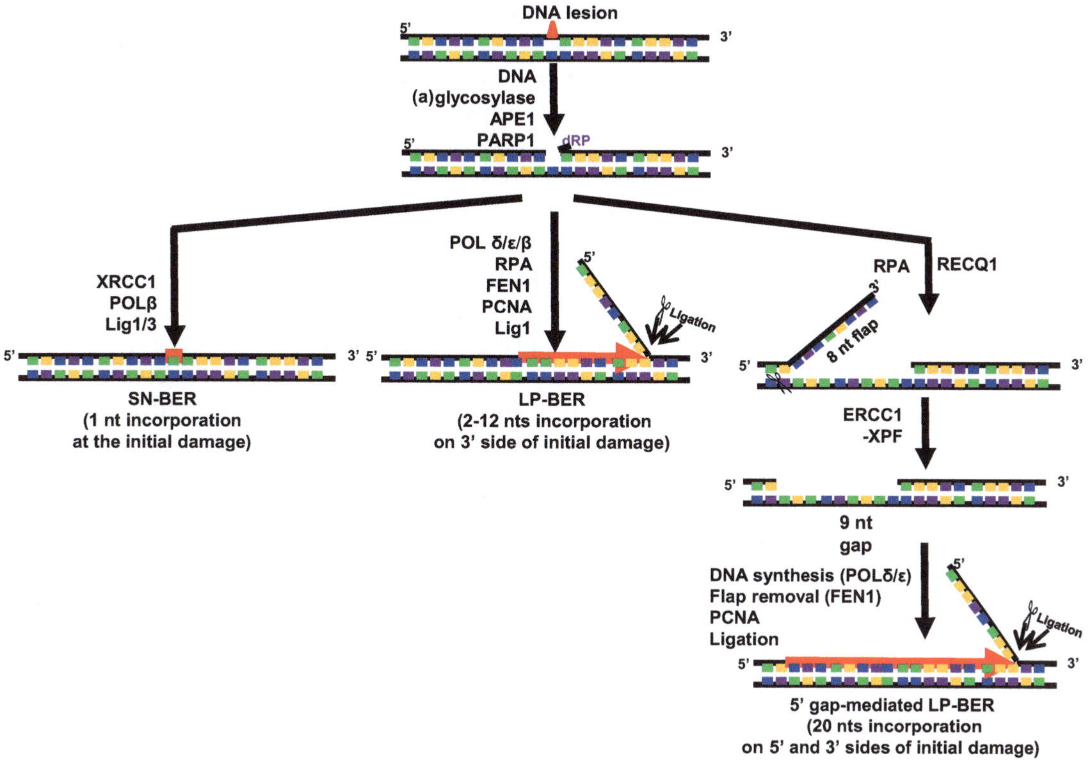

Fig. 1 Base excision repair sub-pathways

epidemiological or clinical studies linking these BER mechanisms to disease.

Historically, most of the BER mechanism studies were carried out using in vitro assays using purified repair proteins and cell lysates but those may not recapitulate physiological DNA repair conditions [22]. To alleviate this problem, host cell reactivation (HCR) assay was developed to study BER in live cells [23, 24]. The HCR assay traditionally transfected damaged plasmid DNA into cultured cells and measured the repair capacity of the cells or repair mechanisms. The major drawback of this system is that the plasmids are treated with DNA damaging agents, which generate a mixture of multiple different types of lesions throughout the plasmid and hinder precise repair mechanism studies for a specific DNA damage type. Recently, a Flow-cytoMetric Host Cell Reactivation (FM-HCR) assay is developed with site-specific single damage-containing plasmids to overcome these challenges [25–28]. However, FM-HCR has a limitation of simultaneous analysis of SN- and LP-BER mechanisms in live cells. Moreover, it requires specialized equipment to perform this analysis. To alleviate these difficulties, we have recently developed a single lesion- and BER sub-pathway-specific PLaque-based HCR (PL-HCR) assay to study BER mechanisms in live cells [14, 29, 30]. Notably, the PL-HCR assay does not require any specialized equipment and computer program/software to perform the assay and analyze data, respectively. Researchers should be able to carry out this assay in laboratories equipped with basic molecular biology and basic cell culture capabilities.

This chapter describes an assay system (PL-HCR) for measuring base excision repair mechanisms in live mammalian cells using plasmid-based representative BER DNA substrates. After transfection of a plasmid (M13mp18) containing a single 1, N^6-ethenoadenine (eA) within a restriction site into human cells, restriction digestions detect the presence or absence (complete repair) of the adduct by the transformation of the digestion products into *E. coli* and counting the transformants. Completely repaired DNA is linearized by restriction digestion and does not form plaques, while unrepaired DNA or products of incomplete BER are resistant to digestion, remain circular, and form plaques. To monitor the patch size, different plasmids are generated containing C:A mismatches within different restriction sites (inhibiting digestion) at various distances and on both sides (5′ or 3′) of the eA-containing restriction sites (representative oligonucleotide sequences used to generate various plasmids shown in Fig. 2b). Similar to eA, other modified bases along with C:A mismatch can be tested for BER mechanisms simply by incorporating them at appropriate restriction sites [14]. Thus, combinations of appropriate restriction enzymes are used for the assay. One enzyme was utilized to probe for repair at the damage site, and another enzyme was utilized to probe for patch formation. Using this assay, the percentage of repair events that occur via 5′ and 3′ patch formation can be quantified. A schematic of the procedure is shown (Fig. 2a).

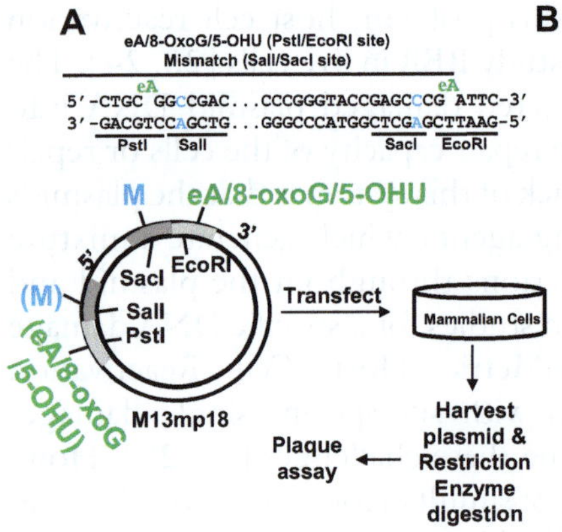

B

Description of the construct	Position of adduct	Position of mismatch	Orientation to adduct
eA (+1 PstI) 5′-GCATGCCTGC(eA)GGTCGACTCTAG-3′	PstI	N/A	N/A
eA (+1 PstI), M (+4 SalI) 5′-GCATGCCTGC(eA)GGCCGACTCTAG-3′	PstI	SalI	3′ (+4)
eA (+1 EcoRI) 5′-CCGAGCTCG(eA)ATTCGTAATC-3′	EcoRI	N/A	N/A
eA (+1 EcoRI), M (-4 SacI) 5′-CCGAGCCCG(eA)ATTCGTAATC-3′	EcoRI	SacI	5′ (-4)

Fig. 2 (**a**) Experimental strategy of PL-HCR for BER sub-pathway analysis. (Modified from Fig. 1a of the article by Woodrick et al. [14]) and (**b**) representative oligonucleotide sequences for construct preparation

2 Materials

2.1 Construction of DNA Carrying a Single Lesion (Undamaged Base/eA/8-oxoG/5-OHU)

1. ATP solutions: 100 mM ATP stock solution is prepared from solid ATP or bought as a pre-made solution. Prepare 1 mM ATP fresh each time from 100 mM ATP solution. Store 100 mM ATP stock solution at −20 °C in aliquots.
2. MicroSpin G-25 columns (Millipore Sigma).
3. 10× annealing buffer (10× AB): 100 mM Tris–HCl (pH 7.5), 500 mM NaCl, distilled water (sterile, Molecular Biology Grade). There is no need to prepare it fresh each time. It can be stored in aliquots at −20 °C.
4. M13mp18 ssDNA: 500 ng/μL M13mp18 ssDNA is commercially available (Bayou Biolabs, Metairie, LA).
5. T7 DNA polymerase (NEB).
6. T4 DNA ligase (NEB).
7. dNTPs (Thermo Fisher).
8. Supercoil-It (Bayou Biolabs).

2.2 In-Cell Repair and Patch Formation Assay

1. Dulbecco's Modified Eagle Medium (DMEM) [+] 4.5 g/L D-glucose [+] L-glutamine [−] sodium pyruvate (Gibco).
2. Fetal bovine serum (FBS; Gibco).
3. 1X Dulbecco's phosphate-buffered saline (DPBS) [−] $CaCl_2$ [−] $MgCl_2$ (Gibco).
4. 1× Opti-MEM (reduced serum medium) [+] L-glutamine [+] HEPES [−] phenol red (Gibco).
5. Penicillin–streptomycin solution (100×; Corning).
6. 1× 0.25% trypsin, 2.21 mM EDTA [−] sodium bicarbonate (Corning).
7. Lipofectamine 2000 (Thermo Fisher).
8. QIAprep Spin Miniprep kit (Qiagen).
9. EcoRI-HF, SacI-HF, PstI-HF, and SalI-HF (NEB).
10. Enzyme Diluent A and C (NEB).
11. Electrocompetent XL1 Blue-MRF' and chemically competent XL1 Blue-MRF' cells (Thermo Fisher).
12. Cuvette 0.2 cm (BioRad).
13. Electroporator (MicroPulser, BioRad).

14. Super optimal broth with catabolite repression (SOC) medium: First prepare super optimal broth (SOB) medium by dissolving thoroughly 20 g tryptone, 5 g yeast extract, and 0.5 g NaCl in 950 mL distilled water. Add 10 mL 250 mM KCl to this solution. Adjust the pH of the medium to 7.0 with 5 N NaOH (~0.2 mL) and autoclave. Cool the medium down to 60 °C or less. Add 5 mL sterile 2 M $MgCl_2$ and 20 mL sterile 1 M glucose.
15. 5 mg/mL tetracycline solution: Add 250 mg tetracycline (stored in −20 °C freezer) to 50 mL ethanol. Vortex as much as possible—some of the solid will go into solution but most of it will dissolve as it sits in the −20 °C (*see* **Note 1**).
16. LB–tetracycline plates: Mix 20 g LB broth and 20 g agar thoroughly in 1 L distilled water in a 2-L flask and autoclave. After autoclaving, let the solution cool at RT for about 20 min. Meanwhile, get the plates ready for pouring. Add 10 mL of tetracycline solution (5 mg/mL) to the 1 L of LB agar (final concentration 50 μg/mL) and swirl. Bring the required volume of tetracycline solution to room temperature before adding to the LB agar solution if it was stored in a −20 °C freezer. Using a 50 mL conical tube, pour LB–Tet agar into the tube and distribute it to 3–4 plates. Keep repeating this until the contents of the flask are finished. One should be able to get about 70–80 plates from 1 L of LB agar.
17. Top agar: Prepare top agar by mixing 5 g LB and 2 g agar thoroughly in 250 mL distilled water in a 500 mL flask and autoclave. After cooling either distribute 3 mL into multiple 15 mL conical tubes for the plaque assay or store the solution at 4 °C. Place 3 mL aliquots in a 50 °C water bath ≈2–3 h before plating (*see* **Note 2**).
18. Plating cells: Add 1 μL frozen stock of chemically competent XL1 Blue-MRF' cells in 5 mL LB–tetracycline liquid media (50 μg/mL tetracycline). Culture in 15 mL round-bottom culture tubes at 37 °C, shaking for 16–18 h. Afterward, the cells can be stored at 4 °C for up to a week.
19. Cool water bath (MultiTemp III, Amersham Biosciences).
20. A 37 °C shaker incubator.
21. A 45 °C water bath near the microbiology workspace and a 37 °C standing incubator for the plates on the day of plating.
22. Tabletop centrifuge (Allegra™ 6R centrifuge, Beckman Coulter, and Eppendorf Centrifuge 5415R).

3 Methods

3.1 Construction of Covalently Closed Circular DNA Carrying a Single Lesion (Undamaged Base/eA/8-oxoG/5-OHU; See Note 3)

The undamaged base/eA/8-oxoG/5-OHU-M13mp18 in vitro construct preparation included three main steps: (A) phosphorylation of the adduct-containing primer oligonucleotide, (B) annealing of the oligonucleotide to the ssDNA, and (C) the primer extension reaction. Each of these steps is described below in detail.

Day 1

3.1.1 Phosphorylation of the Adduct-Containing Oligonucleotides

1. Prepare phosphorylation reactions. For each phosphorylation reaction, add 3 μL 10X PNK buffer, three 3 μL 0.5 M DTT, 13 μL 1 mM ATP, 2 μL oligo (1 μg/μL; for most oligos ≈150–160 μM), 1 μL T4 PNK (10 U/μL), and 8 μL distilled water (sterile, Molecular Biology Grade) in a 1.5 mL microcentrifuge tube.
2. Incubate at 37 °C for 45 min. One reaction will yield ≈36 μL of phosphorylated oligo after purification, and this can be used for six reactions for dsDNA prep. One may expect a yield of a total of ≈15 μg M13mp18 plasmid DNA at the end (*see* **Note 4**).

3.1.2 Separation of Phosphorylated Oligonucleotides from Free ATP Using MicroSpin G-25 Columns (See Notes 5–6)

1. Vortex column and ensure homogenous mixing of the gel matrix. Make sure the majority of the resin is in the tube. Sometimes a relatively good amount of resin is stuck in the inside of the cap at the top of the column.
2. Break off bottom cap and loosen top screw cap.
3. Place column in collection tube (provided) and centrifuge for 1 min at 735 × *g* (2800 rpm). Discard flow-through and collection tube (*see* **Note 7**).
4. Place column in clean 1.5 mL tube. This will be the final tube, so label accordingly.
5. Apply 30 μL phosphorylation reaction from **step 2** to the center of the resin. Be careful not to disturb the resin with the pipette tip.
6. Centrifuge for 2 min at 735 × *g* (2800 rpm). Flow-through contains the purified phosphorylated oligo. The volume will be ≈36 μL; it might be 34 μL, and if it is, add 2 μL water to bring the volume to 36 μL. Discard the column.

3.1.3 Annealing Phosphorylated Oligo to ssDNA Template

1. Prepare annealing reactions (*see* **Note 8**). Assemble each annealing reaction in a 1.5 mL microcentrifuge tube by adding 2 μL 10× annealing buffer (AB), 2 μL 500 ng/μL M13mp18 ssDNA (856 fmol), 6 μL ≈ 10 μM phosphorylated oligo (≈60 pmol), and 10 μL distilled water (sterile, Molecular Biology Grade).

2. Incubate at 80 °C for 5 min.
3. Incubate at 25 °C for 1 h.
4. Incubate on ice for 30 min (*see* **Note 9**).

3.1.4 Extension (See Note 10)

1. Prepare 5× extension buffer (5× EB) on ice (*see* **Note 11**). Add 50 μL 10× T7 DNA polymerase buffer, 10 μL 0.5 M DTT, 7.5 μL 0.1 M ATP, 7.5 μL of each 0.1 M dNTPs, and 2.5 μL BSA (10 mg/mL) in a 1.5 mL microcentrifuge tube.
2. Prepare the extension reaction mix on ice (*see* **Note 12**). Add 6 μL 5× EB, 1 μL T7 DNA polymerase (10 U/μL), 1 μL T4 DNA ligase (400 U/μL), and 2 μL distilled water (sterile, Molecular Biology Grade) in a 1.5 mL microcentrifuge tube.
3. Add 10 μL of the extension reaction mix to each annealing reaction (20 μL) on ice, pipetting up and down to mix. Keep the tube on the ice while mixing.
4. Incubate on ice for 5 min.
5. Incubate at 25 °C for 5 min (*see* **Note 13**).
6. Incubate at 37 °C for 1 h.

3.1.5 Ligation

1. Prepare DNA ligase/ATP mix. For 1 μL mix combine 0.5 μL T4 DNA ligase (400 U/μL) with 0.5 μL of 0.1 M ATP. Add 1 μL of mix to each tube. Mix well (flicking the tubes) and spin down (*see* **Note 14**).
2. Incubate at 14 °C overnight (*see* **Note 15**).

Day 2

3.1.6 Supercoil-It

1. Combine all overnight ligation reactions into one tube.
2. For every 30 μL reaction, add 1.6 μL of 20× Supercoil-It buffer (Bayou Biolabs) to the combined tube and mix well. For example, 12 reactions will result in ≈360 μL in the combined tube. Add (12 × 1.6 μL), which is ≈19 μL of the 20× buffer to that 360 μL tube, bringing the volume to ≈380 μL and the concentration of the buffer to 1×.
3. For every 30 μL reaction, add 0.26 μL of Supercoil-It enzyme (Bayou Biolabs) to the combined tube and mix well. Again, for 12 reactions, add (12 × 0.26 μL), which is ≈3 μL of the enzyme to the combined tube.
4. Incubate at 37 °C for 3 h.

3.1.7 Purification

For purification of the plasmid, use the Qiagen PCR Purification Kit. Process three reactions-worth of a mix per column to maximize the column's binding capacity and not lose too much DNA. Let us take 12 reactions as an example.

1. Aliquot the ≈380 μL Supercoil-It reaction mix into four tubes of ≈95 μL each.
2. Add 475 μL of buffer PB (from the kit) to each tube and mix well.
3. Apply the total 570 μL volume to the supplied purification column after doing the empty (dry) spin to remove the residual ethanol from the wash step and proceed with Qiagen's instructions.
4. Elute the DNA with 50 μL of Molecular Biology Grade water.
5. Combine all four 50 μL elutions into one tube (total volume of 200 μL) and measure the concentration using the Nanodrop spectrophotometer (*see* **Note 16**).

3.2 In-Cell Repair and Patch Formation Assay

The measurement of in-cell repair capacity for SN- and LP-BER mechanisms using the plasmid constructs carrying specific lesions (described above) included three main steps: (A) transfection of mammalian cells with the lesion-containing plasmids, (B) harvesting plasmid DNA from the mammalian cells and restriction enzyme digestion, and (C) transformation of the digested DNA into *E. coli* for plaque assay. Each of these steps is described below in detail.

Day 1–2

3.2.1 Transfection of Mammalian Cells with the Lesion-Containing Plasmids

1. Typically, 1–1.2 × 10^6 HCT 116, HeLa, or mouse embryonic fibroblast cells in 1 mL complete media (*see* **Note 17**) are seeded per well in a 6-well plate overnight at 37 °C.
2. The next day, cells are 70–80% confluent with the monolayer. Wash cells two times with 1XDPBS
3. Add 500 μL Opti-MEM and incubate for 30 min at 37 °C in the cell culture incubator.
4. Prepare transfection reagents. Add 137.5 μL Opti-MEM and 3 μL Lipofectamine 2000 in a 1.5 mL microcentrifuge tube for preparation of Lipo-Mix (140.5 μL) for one well of a 6-well plate. Add 137.5 μL Opti-MEM and 4–20 μL undamaged/damaged plasmid DNA (1.0 ug) in a separate 1.5 mL microcentrifuge tube for preparation of DNA mix (141.5–157.5 μL) for one well of a 6-well plate.
5. Incubate both tubes at room temperature for 5 min.
6. Combine Lipo-mix and DNA mix at a 1:1 ratio: Lipo-mix 140 μL and DNA-mix 140 μL (280 μL total volume).
7. Incubate at room temperature for 20 min.
8. Add 250 μL combined transfection mix to cells in 500 μL Opti-MEM (750 μL total volume).
9. Incubate for 3–4 h at 37 °C in the CO_2 incubator.

10. Add 1.5 mL DMEM supplemented with 15% FBS (final FBS concentration ≈10%) and 1% penicillin/streptomycin and incubate for 21–22 h.

Day 3

3.2.2 Harvesting Plasmid DNA from Mammalian Cells

1. Aspirate media off the wells and add 1 mL DMEM supplemented with 10% FBS and 1% penicillin/streptomycin and incubate for another 3 h.
2. Aspirate media and wash cells with 1X DPBS.
3. Add 300 μL trypsin/EDTA to the 6-well plate.
4. Incubate at 37 °C for 3 min, or until the cells are detached.
5. Add 700 μL 10% FBS and 1% penicillin/streptomycin supplemented with DMEM. Pipette up and down to collect all the cells into a microcentrifuge tube.
6. Centrifuge the cells in a tabletop centrifuge (Allegra™ 6R centrifuge, Beckman Coulter) at 1600 rpm for 5 min (or Eppendorf Centrifuge 5415R for 3 min at 1200 rpm) at 4 °C.
7. Remove media; resuspend the pellet with 500 μL 1× DPBS.
8. Centrifuge the cells as in **step 6**.
9. Remove DPBS (*see* **Note 18**).
10. Isolate the DNA from the pelleted cells, following the Qiagen protocol for Miniprep of plasmid DNA from bacteria. We use Qiagen's QIAprep Spin Miniprep kit (*see* **Note 19**). In the final step, elute the DNA from the spin column with 50 μL Molecular Biology Grade distilled water.

3.2.3 Restriction Enzyme Digestions

1. Prepare master mixes for the number of samples. For each harvested sample there will be five reactions—undigested, EcoRI, SacI, PstI, and SalI digested. For each undigested sample, mix 2 μL 10× rCutSmart™ buffer and 10 μL distilled water (sterile, Molecular Biology Grade). For each of EcoRI-HF, SacI-HF, PstI-HF, and SalI-HF digested samples, mix 2 μL 10× rCutSmart™ buffer, 1 μL enzyme (5 U/μL) and 9 μL distilled water (sterile, Molecular Biology Grade). Dilute EcoRI-HF and PstI-HF with Diluent C and SacI-HF and SalI-HF with Diluent A to working concentrations just before use. Transfer 12 μL of the appropriate mix from the master mix to tubes containing 8 μL miniprep DNA prepared in Subheading 3.2.2. Therefore, for each harvested sample, there are now five tubes. For example, eA 24 h **X** (undigested), eA 24 h **E** (EcoRI), eA 24 h **S** (SacI), eA 24 h **P** (PstI), and eA 24 h **Sal** (SalI).

2. Incubate at 37 °C for 2 h. Cool to room temperature for 10 min (*see* **Note 20**).
3. Dilute each undigested/digested DNA sample 8× (e.g., mix 5 μL undigested/digested DNA with 35 μL water)

3.2.4 Transformation

Transformation Using Electrocompetent Cells

1. Place aliquots of XL1 Blue cells (~1×10^9 cfu/μg), electroporation cuvettes (0.2 cm), and diluted digested DNA on ice. One 40 μL aliquot of XL1 Blue cells is needed per digested sample. The frozen XL1 Blue aliquots take approximately 10–12 min on ice to thaw. Also, label 15 mL round-bottom culture tubes according to the appropriate sample names.
2. Label the tops of the XL1 Blue aliquots with the appropriate names for each sample.
3. Add 2 μL of diluted undigested or digested DNA to the 40 μL aliquot of XL1 blue cells (*see* **Note 21**).
4. Transfer the mixture of DNA and XL1 Blue cells to a 0.2-cm cuvette (BioRad) and tap on the bench to bring the contents down to the bottom of the cuvette. Tap harder if bubbles are seen—this will usually get rid of them. Wipe the outside with a Kimwipe to dry it.
5. Place the cuvette in the electroporator chamber and slide the chamber into electrodes.
6. Switch the electroporator to setting **Ec2** and the measurement units to “ms” (milliseconds).
7. Press “Pulse.”
8. After the number of ms appears, the pulsing is complete. Immediately remove the cuvette and add 1 mL SOC (brought down to room temperature if the SOC stock was stored at 4 °C), pipetting up and down 4–5 times to mix thoroughly (*see* **Note 22**).

Transformation Using Chemically Competent Cells (Alternative to electroporation)

1. Place aliquots of XL1 Blue cells (~1×10^8 cfu/μg) and diluted digested DNA on ice. One 40 μL aliquot of XL1 Blue cells per digested sample. The frozen XL1 Blue aliquots take approximately 10–12 min on ice to thaw. Also, label 15 mL round-bottom culture tubes according to the appropriate sample names.
2. Label the tops of the XL1 Blue aliquots with the appropriate names for each sample.
3. Add 2 μL of diluted undigested or digested DNA to the 40 μL aliquot of XL1 blue cells, and tap the tubes gently for a few seconds (*see* **Note 21**).

4. Keep the tubes on ice for 30 min.
5. Heat shock at 42 °C for 90 s.
6. Return the tubes on ice.
7. Add 450 μL SOC (brought down to room temperature if the SOC stock was stored at 4 °C), and pipette up and down 4–5 times to mix thoroughly (*see* **Note 22**).

3.2.5 Plating (See Notes 23–24)

1. Bring two 200 μL pipettes and 200 μL tips to the microbiology workspace.
2. Warm up the LB–tetracycline plates. Label LB–tetracycline plates with sample names. Warm up and keep the plates at 37 °C to prevent premature solidification of top agar, which will be added later in **step** 7.
3. Bring the first set (X, E, S, P, and Sal) of plates in front.
4. Mix the plating cells by gently shaking and pipette 200 μL plating cells. Set the pipette down on the bench.
5. Pipette 50 μL (or the desired amount; *see* **Note 25**) of transformed cell sample "X" and transfer to 3 mL of warmed top agar (in the 50 °C water bath). Pipette up and down a couple of times.
6. Quickly add the 200 μL plating cells (kept at room temperature) to the mixture and pipette up and down a couple of times.
7. Immediately pour the mixture on the appropriate plate and swirl gently (but quickly) until agar is smooth. Leave the top half on while completing the same **steps 4–6** on the other four plates for the sample set (E, S, P, and Sal).
8. Move all five plates (with tops half off) over to the side. When the next set is complete, put the tops back on fully and stack gently to the side to make room for the next set to dry.
9. Bring the next set (X, E, S, P, and Sal) of plates in front and repeat **steps 4–7**.
10. After all the plates have been completed, wait about 10 min before putting the plates upside down in the 37 °C standing incubator.

3.2.6 Counting and Calculations

After about 12–20 h, take the plates out of the standing incubator and count the plaques.

Calculate the percent repair as follows:

% repair at EcoRI site (total repair) = [(#X – #E)/ #X]*100;
% repair at SacI site (5′ patch formation) = [(#X – #S)/ #X]*100;
% repair at PstI site (total repair) = [(#X – #P)/ #X]*100;
% repair at SalI site (3′ patch formation) = [(#X – #Sal)/ #X]*100

4 Notes

1. Prepare a day before its use since it does not go into solution easily.
2. Top agar can be stored and reheated to liquid in the microwave. Heat 1 min at a time, swirling in between. Only heat as much as is necessary for the agar to be liquid.
3. The protocol can be stopped after the following steps:

   Step B: store phosphorylated oligo at −20 °C for at least a few weeks.

   Step C: store annealed oligo and template buried in ice at 4 °C overnight.
4. Usually, prepare two phosphorylation reactions for a given oligo. Prepare a master mix containing the following components, except the oligo. Distribute the master mix evenly amongst the appropriate number of tubes and add the appropriate oligo to each tube.
5. RPM values written in the protocol are specific to the Eppendorf 5415R microcentrifuge. If one changes centrifuges, check the conversion of "g" to "rpm" for that centrifuge.
6. While performing the purification, adjust the heat block to 80 °C or turn on the dedicated 80 °C heat block if planning to anneal on the same day to avoid a fire hazard.
7. Use column immediately after preparing it. Do not let the resin dry.
8. Prepare reaction mix according to the number of reactions desired for the preparation, and then distribute 20 μL (or ≈19.9 μL) into the appropriate number of tubes. Generally speaking, yields of the covalently closed circular DNA carrying a single lesion will be as follows: 3 reactions = 7.5 μg, 6 reactions = 15 μg.
9. During this final 30-min incubation on ice, thaw dNTPs, BSA, 10× T7 DNA polymerase buffer to get ready for the extension step. DTT and ATP will still be out and thawed (but on ice) from the beginning of the step (Subheading 3.1.1).
10. The extension step is very sensitive because of the exonuclease activity of T7 DNA polymerase. If not followed exactly, one may lose the desired adduct during this step, so take extra caution. Prepare absolutely everything on ice.
11. The given recipe will yield 100 μL of 5× EB, which is enough for 16 extension reactions. If more than 16 annealing reactions are prepared, double the volumes for each component of the 5× EB.

12. Prepare an extension reaction mix for each construct type. For example, if one is making 12 reactions of eA and 6 reactions of 8oxodG, then one would prepare an extension reaction mix for eA (enough for 12) and an extension reaction mix for 8oxodG (enough for 6). Components of the mix will be the same, but one needs to pipette up and down to mix thoroughly after adding the extension reaction mix to each tube. Preparing separate mixes allows one to keep the same pipette tip.
13. Use a cool water bath (MultiTemp III, Amersham Biosciences) for this step. Do not just assume that the room is 25 °C.
14. Since the volume is low and the solution is viscous, prepare about 25% more than the need. For example, prepare the ligation mix for about 15 reactions, if the ligation mix needs to be added to 12 extension reactions.
15. Use a cool water bath (MultiTemp III, Amersham Biosciences) for this step.
16. Ideally, if one has a good yield and has maximized the column's binding capacity, one should get approximately 150–200 ng/μ L. The A260/A280 ratio should be between 1.8 and 2.1.
17. HCT116, HeLa, and MEF cells were cultured in DMEM supplemented with 10% fetal bovine serum (FBS) and 1% penicillin/streptomycin.
18. At this point, one can either proceed with harvesting the DNA or store the pellet at −20 °C and harvest plasmid DNA later.
19. It is shown that both non-replicated and replicated plasmid DNA can be isolated from mammalian cells using the fast and convenient QiAprep Spin Miniprep kit and the conventional time-consuming and labor-intensive Hirt method with comparable efficiency [31, 32].
20. During the 10-min cooldown, get out the desired number of aliquots of electrocompetent or chemically competent XL1 Blue cells on ice to thaw. Also, get the SOC media to come to room temperature.
21. Follow this order—X, E, S, P, and Sal. After adding 2 μL of the Sal digestion to a set, proceed with electroporating each one, X, E, S, P, and Sal, in the order in which the DNA was added. Therefore, each set is done at the same time.
22. After this step, proceed to the next step, plating. Or the tubes can be kept at 4 °C for later use (up to a week.)
23. Approximately 15 min before plating, get one tube of plating cells (after culturing can be stored at 4 °C for up to a week) and the desired number of LB–Tet plates from the 4 °C. The plating cells should come to room temperature. The LB–Tet plates should be placed in the 37 °C standing incubator with

the tops open and facing up. During 15 min, the water that accumulated on the tops should evaporate.

24. If the transformed cells have been stored at 4 °C, bring those out to come to room temperature as well.
25. For plating an experiment for the first time, one needs to do a few of the undigested samples ('X') at various dilutions of the transformed cells (1:100–50 μL, 1:10–50 μL, straight-50 μL, straight-200 μL, etc.) first to figure out the general amount one will need to plate for the rest of the samples, which is done the next day after counting the plaque number from the dilution test. The dilution test helps to get well-separated plaques for the ease of accurate counting.

Acknowledgments

The author thanks all the colleagues who helped to develop and improve this protocol in our laboratory. The work is supported by funding from the National Institute of Health grant R01 CA92306 to R.R. and Cancer Center Support Grant P30 CA051008 for use of Shared Resources.

References

1. Lindahl T, Barnes DE (2000) Repair of endogenous DNA damage. Cold Spring Harb Symp Quant Biol 65:127–133
2. Hussain SP, Harris CC (2007) Inflammation and cancer: an ancient link with novel potentials. Int J Cancer 121(11):2373–2380
3. Poirier MC (2004) Chemical-induced DNA damage and human cancer risk. Nat Rev Cancer 4(8):630–637
4. Lindahl T (1993) Instability and decay of the primary structure of DNA. Nature 362(6422): 709–715
5. de Souza-Pinto NC, Wilson DM 3rd, Stevnsner TV, Bohr VA (2008) Mitochondrial DNA, base excision repair and neurodegeneration. DNA Repair 7(7):1098–1109
6. Sung JS, Demple B (2006) Roles of base excision repair sub-pathways in correcting oxidized abasic sites in DNA. FEBS J 273(8): 1620–1629
7. Zharkov DO, Grollman AP (2005) The DNA trackwalkers: principles of lesion search and recognition by DNA glycosylases. Mutat Res 577(1–2):24–54
8. Wilson DM 3rd, Barsky D (2001) The major human abasic endonuclease: formation, consequences, and repair of abasic lesions in DNA. Mutat Res 485(4):283–307
9. Matsumoto Y, Kim K (1995) Excision of deoxyribose phosphate residues by DNA polymerase beta during DNA repair. Science 269(5224):699–702
10. Klungland A, Lindahl T (1997) Second pathway for completion of human DNA base excision-repair: reconstitution with purified proteins and requirement for DNase IV (FEN1). EMBO J 16(11):3341–3348
11. Pascucci B, Stucki M, Jónsson ZO, Dogliotti E, Hübscher U (1999) Long patch base excision repair with purified human proteins. DNA ligase I as patch size mediator for DNA polymerases delta and epsilon. J Biol Chem 274(47):33696–33702
12. Asagoshi K, Liu Y, Masaoka A, Lan L, Prasad R, Horton JK, Brown AR, Wang XH, Bdour HM, Sobol RW, Taylor JS, Yasui A, Wilson SH (2010) DNA polymerase beta-dependent long patch base excision repair in living cells. DNA Repair (Amst) 9(2):109–119
13. Liu Y, Beard WA, Shock DD, Prasad R, Hou EW, Wilson SH (2005) DNA polymerase beta and flap endonuclease 1 enzymatic specificities sustain DNA synthesis for long patch base excision repair. J Biol Chem 280(5):3665–3674

14. Woodrick J, Gupta S, Camacho S, Parvathaneni S, Choudhury S, Cheema A, Bai Y, Khatkar P, Erkizan HV, Sami F, Su Y, Scharer OD, Sharma S, Roy R (2017) A new sub-pathway of long-patch base excision repair involving 5′ gap formation. EMBO J 36(11): 1605–1622
15. Robertson AB, Klungland A, Rognes T, Leiros I (2009) DNA repair in mammalian cells: base excision repair: the long and short of it. Cell Mol Life Sci 66(6):981–993
16. Osorio A, Milne RL, Kuchenbaecker K, Vaclova T, Pita G, Alonso R, Peterlongo P, Blanco I, de la Hoya M, Duran M, Diez O, Ramon YCT, Konstantopoulou I, Martinez-Bouzas C, Andres Conejero R, Soucy P, McGuffog L, Barrowdale D, Lee A, Swe B, Arver B, Rantala J, Loman N, Ehrencrona H, Olopade OI, Beattie MS, Domchek SM, Nathanson K, Rebbeck TR, Arun BK, Karlan BY, Walsh C, Lester J, John EM, Whittemore AS, Daly MB, Southey M, Hopper J, Terry MB, Buys SS, Janavicius R, Dorfling CM, van Rensburg EJ, Steele L, Neuhausen SL, Ding YC, Hansen TV, Jonson L, Ejlertsen B, Gerdes AM, Infante M, Herraez B, Moreno LT, Weitzel JN, Herzog J, Weeman K, Manoukian S, Peissel B, Zaffaroni D, Scuvera G, Bonanni B, Mariette F, Volorio S, Viel A, Varesco L, Papi L, Ottini L, Tibiletti MG, Radice P, Yannoukakos D, Garber J, Ellis S, Frost D, Platte R, Fineberg E, Evans G, Lalloo F, Izatt L, Eeles R, Adlard J, Davidson R, Cole T, Eccles D, Cook J, Hodgson S, Brewer C, Tischkowitz M, Douglas F, Porteous M, Side L, Walker L, Morrison P, Donaldson A, Kennedy J, Foo C, Godwin AK, Schmutzler RK, Wappenschmidt B, Rhiem K, Engel C, Meindl A, Ditsch N, Arnold N, Plendl HJ, Niederacher D, Sutter C, Wang-Gohrke S, Steinemann D, Preisler-Adams S, Kast K, Varon-Mateeva R, Gehrig A, Stoppa-Lyonnet D, Sinilnikova OM, Mazoyer S, Damiola F, Poppe B, Claes K, Piedmonte M, Tucker K, Backes F, Rodriguez G, Brewster W, Wakeley K, Rutherford T, Caldes T, Nevanlinna H, Aittomaki K, Rookus MA, van Os TA, van der Kolk L, de Lange JL, Meijers-Heijboer HE, van der Hout AH, van Asperen CJ, Gomez Garcia EB, Hoogerbrugge N, Collee JM, van Deurzen CH, van der Luijt RB, Devilee P, Hebon OE, Lazaro C, Teule A, Menendez M, Jakubowska A, Cybulski C, Gronwald J, Lubinski J, Durda K, Jaworska-Bieniek K, Johannsson OT, Maugard C, Montagna M, Tognazzo S, Teixeira MR, Healey S, Investigators K, Olswold C, Guidugli L, Lindor N, Slager S, Szabo CI, Vijai J, Robson M, Kauff N, Zhang L, Rau-Murthy R, Fink-Retter A, Singer CF, Rappaport C, Geschwantler Kaulich D, Pfeiler G, Tea MK, Berger A, Phelan CM, Greene MH, Mai PL, Lejbkowicz F, Andrulis I, Mulligan AM, Glendon G, Toland AE, Bojesen A, Pedersen IS, Sunde L, Thomassen M, Kruse TA, Jensen UB, Friedman E, Laitman Y, Shimon SP, Simard J, Easton DF, Offit K, Couch FJ, Chenevix-Trench G, Antoniou AC, Benitez J (2014) DNA glycosylases involved in base excision repair may be associated with cancer risk in BRCA1 and BRCA2 mutation carriers. PLoS Genet 10(4):e1004256
17. Arora M, Lindgren B, Basu S, Nagaraj S, Gross M, Weisdorf D, Thyagarajan B (2010) Polymorphisms in the base excision repair pathway and graft-versus-host disease. Leukemia 24(8):1470–1475
18. Ladiges W, Wiley J, MacAuley A (2003) Polymorphisms in the DNA repair gene XRCC1 and age-related disease. Mech Aging Dev 124(1):27–32
19. Shao C, Xiong S, Li GM, Gu L, Mao G, Markesbery WR, Lovell MA (2008) Altered 8-oxoguanine glycosylase in mild cognitive impairment and late-stage Alzheimer's disease brain. Free Radic Biol Med 45(6):813–819
20. Obtułowicz T, Winczura A, Speina E, Swoboda M, Janik J, Janowska B, Cieśla JM, Kowalczyk P, Jawien A, Gackowski D, Banaszkiewicz Z, Krasnodebski I, Chaber A, Olinski R, Nair J, Bartsch H, Douki T, Cadet J, Tudek B (2010) Aberrant repair of etheno-DNA adducts in leukocytes and colon tissue of colon cancer patients. Free Radic Biol Med 49(6):1064–1071
21. Sevilya Z, Leitner-Dagan Y, Pinchev M, Kremer R, Elinger D, Rennert HS, Schechtman E, Freedman LS, Rennert G, Paz-Elizur T, Livneh Z (2014) Low integrated DNA repair score and lung cancer risk. Cancer Prev Res (Phila) 7(4):398–406
22. Beard WA, Horton JK, Prasad R, Wilson SH (2019) Eukaryotic base excision repair: new approaches Shine light on mechanism. Annu Rev Biochem 88:137–162
23. Kassam SN, Rainbow AJ (2009) UV-inducible base excision repair of oxidative damaged DNA in human cells. Mutagenesis 24(1):75–83
24. Rainbow AJ, Zacal NJ, Leach DM (2013) Reduced host cell reactivation of oxidatively damaged DNA in aging human fibroblasts. Oncol Rep 29(6):2493–2497
25. Chaim IA, Nagel ZD, Jordan JJ, Mazzucato P, Ngo LP, Samson LD (2017) In vivo measurements of interindividual differences in DNA

glycosylases and APE1 activities. Proc Natl Acad Sci U S A 114(48):E10379–e10388

26. Lee KJ, Piett CG, Andrews JF, Mann E, Nagel ZD, Gassman NR (2019) Defective base excision repair in the response to DNA damaging agents in triple-negative breast cancer. PLoS One 14(10):e0223725
27. Nagel ZD, Engelward BP, Brenner DJ, Begley TJ, Sobol RW, Bielas JH, Stambrook PJ, Wei Q, Hu JJ, Terry MB, Dilworth C, McAllister KA, Reinlib L, Worth L, Shaughnessy DT (2017) Towards precision prevention: technologies for identifying healthy individuals with a high risk of disease. Mutat Res 800-802:14–28
28. Owiti NA, Nagel ZD, Engelward BP (2020) Fluorescence sheds light on DNA damage, DNA repair, and mutations. Trends Cancer 7(3):240–248
29. Dixon M, Woodrick J, Gupta S, Karmahapatra SK, Devito S, Vasudevan S, Dakshanamurthy S, Adhikari S, Yenugonda VM, Roy R (2015) Naturally occurring polyphenol, morin hydrate, inhibits the enzymatic activity of N-methylpurine DNA glycosylase, a DNA repair enzyme with various roles in human disease. Bioorg Med Chem 23(5):1102–1111
30. Choudhury S, Adhikari S, Cheema A, Roy R (2008) Evidence of complete cellular repair of 1, N^6-ethenoadenine, a mutagenic and potential damage for human cancer, revealed by a novel method. Mol Cell Biochem 313(1–2):19–28
31. Ziegler K, Bui T, Frisque RJ, Grandinetti A, Nerurkar VR (2004) A rapid in vitro polyomavirus DNA replication assay. J Virol Methods 122(1):123–127
32. Siebenkotten G, Leyendeckers H, Christine R, Radbruch A (1995) Isolation of plasmid DNA from mammalian cells with QIAprep. QIAGEN News 2:11–12

Chapter 2

In Vitro Assay to Measure APE1 Enzymatic Activity on Ribose Monophosphate Abasic Site

Matilde Clarissa Malfatti, Giulia Antoniali, and Gianluca Tell

Abstract

APE1 (apurinic/apyrimidinic endodeoxyribonuclease 1) is a central enzyme of the base excision repair (BER) pathway playing a pivotal role in protecting mammalian cells against genotoxins and in safeguarding genome stability. Recently, we demonstrated the APE1 ability to process abasic ribonucleotides embedded in DNA. Here, we provide a pipeline of protocols to quantify endodeoxyribonuclease activity by APE1 on these substrates, by using recombinant protein and whole-cell extracts. The repair capacity is measured by using fluorescent oligonucleotide substrates, which are then separated by polyacrylamide gel electrophoresis and detected by imaging scanning. The specificity of APE1 action is demonstrated using specific APE1 enzymatic inhibitors.

Key words Apurinic/apyrimidinic endodeoxyribonuclease 1, Base excision repair, Enzymatic activity, Inhibitor III, Ribose monophosphate abasic site, RNA

1 Introduction

Base excision repair (BER) is a major repair pathway, which corrects a broad range of non-distorting DNA lesions induced by reactive oxygen species and alkylating agents, generated by endogenous and physiological sources, or by environmental genotoxins like oxidative reagents, ionizing radiations, and several chemotherapy agents. Its correct functioning is crucial for genome stability and cell viability and defects associated to BER pathway have been linked to the development of hereditary diseases, age-related degenerative diseases, and human cancers. Studying BER is therefore important for our understanding of how cells may respond to DNA damage and for the development of drugs disrupting the repair mechanism in tumor cells to design novel anticancer strategies based on combination therapy [1].

BER is a multistep process, consisting of two sub-pathways and the orchestration of the different activities of several enzymes with

Kishor K. Bhakat and Tapas K. Hazra (eds.), *Base Excision Repair Pathway: Methods and Protocols*, Methods in Molecular Biology, vol. 2701, https://doi.org/10.1007/978-1-0716-3373-1_2,

partially redundant functions. Briefly, the BER pathway is initiated by a specific DNA glycosylase, which recognizes and removes the modified base cleaving the *N*-glycosidic bond and creating an abasic site (AP). The resulting AP site is then processed by an AP endonuclease that cleaves hydrolytically the phosphodiester bond on the 3′ side of the AP site. The apurinic/apyrimidinic endodeoxyribonuclease 1 (APE1) is the major mammalian enzyme able to excise, at the 5′ position, the generated abasic deoxyribose phosphate leaving a single-strand break, which bears a free hydroxyl (OH) group at the 3′ end and a deoxyribose phosphate (dRP) at the 5′ end. This gap is then filled by different DNA polymerases and the nick is sealed by DNA ligases [1, 2].

During the last years, APE1 has gained great attention as a potential target for chemotherapeutic agents for several types of cancers, and as a consequence, several specific inhibitors (i.e., Inhibitor III [3, 4]) have been developed. Novel unpredicted perspectives in translational medicine are emerging for this protein, which has been endowed with several non-canonical roles besides DNA repair function, particularly in gene expression regulation and RNA metabolism. Recently, we clarified some aspects concerning APE1 activity on RNA, by discovering that APE1 binds structured RNA molecules, cleaves abasic single-stranded RNA (ssRNA) and ribose monophosphate abasic sites embedded in DNA [5], and is involved in RNA-decay having 3′-RNA phosphatase and 3′-exoribonuclease activities [6, 7]. Ribonucleotides embedded in DNA are very frequent, counting of 1 rNMP every 700 dNMPs [8]. The amount of RNA abasic sites in yeast and human cells has been quantified and counted of 3 rAP sites per 1,000,000 ribonucleotides [9]. Although ribose-seq and other techniques have allowed to detect the position and quantify the number of ribonucleotides embedded in genomic DNA [10], it is not still possible to estimate the amount of ribose monophosphate abasic sites embedded in DNA. Up to now, in vitro data obtained from our laboratory have demonstrated that RNase H2 is not able to cleave and promote the repair of ribose monophosphate abasic site embedded in DNA [5]. On the contrary, APE1 is active on this type of damage promoting its repair by BER [5]. In this chapter, we described the technique and basic methods used for (i) measuring the APE1 endodeoxyribonuclease activity on ribose monophosphate abasic site-containing DNA molecules; (ii) preparation of recombinant APE1 protein; (iii) preparation of whole-cell extracts, for in vitro APE1 endodeoxyribonuclease cleavage assay; and (iv) resolving cleaved products by denaturing urea gel (Fig. 1). After incubation with APE1-purified enzyme or cellular extracts, the products of the reaction are separated onto a denaturing polyacrylamide gel, and based on the migration of the cleaved product(s) and their densitometric quantification, APE1 activity can be measured accurately (Fig. 1). Furthermore, potential APE1 small molecule inhibitors

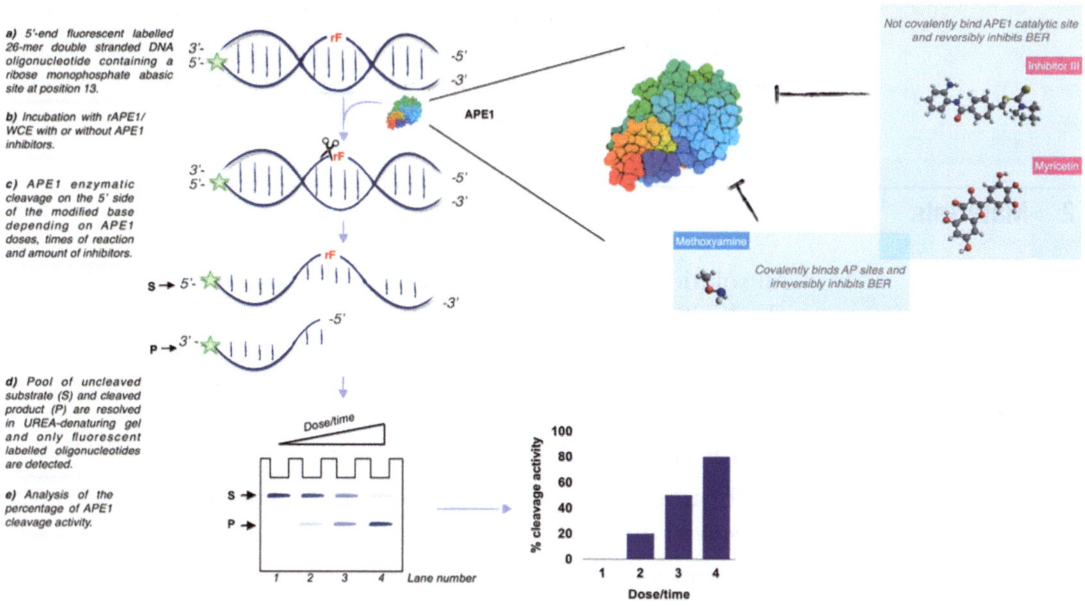

Fig. 1 Schematic representation of each step of endodeoxyribonuclease APE1 assay on abasic monophosphate ribose site embedded in DNA. First, APE1 enzymatic activity is performed *previa* incubation of rAPE1 or APE1 from whole-cell extracts with a 5′-end fluorescent-labelled double-stranded DNA oligonucleotide containing a ribose monophosphate abasic site. Upon APE1 cleavage, the pool containing unprocessed (S) and processed (P) oligonucleotides is resolved onto a urea-denaturing gel. Finally, after gel acquisition, the percentage of APE1 cleavage activity is calculated, as indicated in the main text. To test APE1 cleavage specificity, APE1 inhibitors can be used including myricetin, Inhibitor III, and methoxyamine whose chemical structure and inhibiting function is indicated

can be simultaneously tested either with purified protein or with cellular extracts. Among the most studied APE1 bioactive inhibitors for their potential to enhance cytotoxicity and genotoxicity of alkylating agents, we highlight Inhibitor III [3, 4], myricetin [12], and methoxyamine [12], being the most specific ones (Fig. 1). In this chapter, as an example, we used Inhibitor III. While for Inhibitor III and myricetin, a direct inhibitory effect over the protein catalytic activity was demonstrated, methoxyamine covalently binds to AP sites, thus indirectly blocking the BER process.

Over time, several methods have been described to monitor APE1 enzymatic activity on deoxyribose abasic sites located in synthetic DNA substrates. Moreover, previous protocols were carried out using radioisotopes for the detection of cleavage products [11]. Here, we adopted a specific protocol for the usage of a 5′-end-fluorescent-labelled duplex DNA 25-bp oligonucleotide containing a single ribose monophosphate abasic site embedded in DNA at position 13. Alternatively, the methods can be adapted to study other APE1 substrates with modified ribose monophosphate nucleotides within both DNA or RNA molecules.

The technique can be easily applied to both any kind of cell extracts and highly purified enzymes and the introduction of a fluorescent-labelled substrate gets this assay even more sensitive and reproducible than radioactive substrates.

2 Materials

All solutions should be prepared using ultrapure water and molecular biology-grade reagents.

2.1 Purification of Recombinant APE1 Protein (rAPE1)

1. *E. coli* BL21 (DE3) cells transformed with pGEX-3X/APE1 vector.
2. LB broth media supplemented with 50 μg/mL ampicillin.
3. 1 mM isopropyl β-D-1-thiogalactopyranoside (IPTG) dissolved in water.
4. Protease inhibitor cocktail for bacterial cell lysates.
5. Factor X_a.
6. GSTrap HP (GST affinity chromatography) column.
7. HiTrap Benzamidine FF (HS) (serine protease affinity chromatography) column.
8. HiTrap SP FF (cation exchange chromatography) column.
9. 1× PBS: 8 g NaCl, 0.2 g KCl, 1.44 g Na_2HPO_4, and 0.24 g KH_2PO_4. Bring volume to 1 L with sterilized and deionized water and adjust the pH to 7.4.
10. GSTrap binding buffer: 1× PBS and 1 mM DTT (*see* **Note 1**).
11. GSTrap elution buffer: 1× PBS and 10 mM GSH (*see* **Note 2**).
12. HiTrap Benzamidine binding buffer: 20 mM Na_2HPO_4 pH 7.5, 150 mM NaCl.
13. HiTrap Benzamidine high-salt buffer: 20 mM Na_2HPO_4 pH 7.5, 1 M NaCl.
14. HiTrap Benzamidine low pH elution buffer: 0.05 M glycine pH 3.0.
15. HiTrap binding buffer: 20 mM Na_2HPO_4 pH 8.0, 100 mM NaCl, 1 mM EDTA, 1 mM DTT, 10% v/v glycerol.
16. HiTrap elution buffer: 20 mM Na_2HPO_4 pH 8.0, 1 M NaCl, 1 mM EDTA, 1 mM DTT, 10% v/v glycerol.
17. Sterile disposable.
18. Microbial incubator.
19. Sonicator.
20. Shaker incubator.
21. Spectrophotometer.

22. Fast protein liquid chromatography (FPLC) (ÄKTA™ purifier or equivalent).
23. Ultracentrifuge capable of achieving speeds of up to 20,000 × *g*.

2.2 Whole-Cell Extract (WCE) Preparation

2.2.1 Cell Growth

1. HeLa cells (*see* **Note 3**).
2. Tissue-culture treated petri dish (100 mm diameter).
3. Growth media: 1× DMEM high-glucose supplemented with 10% v/v fetal bovine serum, 2 mM *l*-glutamine and 100 μM penicillin/streptomycin.
4. 1× PBS.
5. 0.05% w/v trypsin supplemented with 0.02% w/v EDTA pH 8.0.
6. Sterile disposable.
7. Centrifuge.

2.2.2 WCE Protein Extraction

1. WCE lysis buffer: 50 mM Tris–HCl pH 7.5, 150 mM NaCl, 1 mM EDTA pH 8.0, 1% v/v Triton X-100 supplemented with protease inhibitors including protease inhibitor cocktail, 1 mM sodium fluoride, 1 mM sodium orthovanadate, 1 mM phenylmethylsulfonyl fluoride (PMSF), 1 mM DTT (*see* **Note 4**).
2. Centrifuge capable of achieving speeds of up to 20,000 × *g*.

2.3 SDS-PAGE

1. Gel casting apparatus inclusive of glass plates, spacers, and a 10-wells comb and the tank for the gel-running apparatus.
2. For the 10% resolving section of the gel: 40% w/v acrylamide/bis acrylamide 37.5:1, 1.5 M Tris–HCl pH 8.8, 10% w/v SDS, 10% w/v APS, and TEMED (*see* **Note 5**).
3. For the 4% stacking section of the gel: 40% w/v acrylamide/bis acrylamide 37.5:1, 0.5 M Tris–HCl pH 6.8, 10% w/v SDS, 10% w/v APS, and TEMED.
4. 10× SDS-PAGE running buffer for protein gel: 30.2 g Trizma, 144 g glycine, 5 g SDS, H_2O to 1 L. Thoroughly mix.
5. Blue Coomassie staining solution: 0.5 g Coomassie Brilliant Blue, 50 mL acetic acid, 500 mL methanol, H_2O to 1 L. Thoroughly mix.
6. Destaining solution: 250 mL methanol, 70 mL acetic acid, H_2O to 1 L. Thoroughly mix.
7. 4× SDS protein loading buffer dye: 0.8 g SDS, 4.6 mL glycerol, 2.4 mL 0.5 M Tris–HCl pH 6.8, 1.4 mL beta-mercaptoethanol, 0.1% w/v bromophenol blue, H_2O to 10 mL.
8. ddH_2O.

9. Heat block.
10. 100% v/v ethanol.
11. 3 MM paper.
12. Isobutanol.
13. Gel tray.

2.4 Oligonucleotide Substrate Preparation

1. Custom-made labeled oligonucleotide stock resuspended at 100 μM in sterilized and DNase- and RNase-free H_2O: 5′-Dye -GGA TCC GGT AGT (rF)TT AGG CCT GAA C-3′ and 5′-GTT CAG GCC TAA CAC TAC CGG ATC C-3′ (where Dye indicates IRDye700 fluorescent dye and rF indicates a ribose sugar [r] and [F] a tetrahydrofuran as abasic site analog).
2. Custom-made complementary oligonucleotide stock resuspended at 100 μM in sterilized and DNase- and RNase-free H_2O: 5′-GTT CAG GCC TAA CAC TAC CGG ATC C-3′.
3. Annealing buffer: 10 mM Tris–HCl pH 7.4 and 10 mM $MgCl_2$ in sterilized and DNase- and RNase-free water.
4. Water bath and tube rack.
5. Sterilized and DNase- and RNase-free water.
6. Plastic DNase- and RNase-free and filtered tips.

2.5 APE1 Endodeoxyribonuclease Cleavage Reaction

1. 10× rAP buffer: 200 mM Tris–HCl pH 7.4, 500 mM KCl, 1% w/v BSA, 0.1% v/v Tween-20, and 10 mM $MgCl_2$.
2. Inhibitor III (stock concentration of 10 mM dissolved in DMSO).
3. DMSO.
4. Ultrapure DNase- and RNase-free water.
5. Plastic DNase- and RNase-free and filtered tips.
6. Heat block.
7. Stop solution containing formamide and 10× orange loading dye (*see* **Note 6**).

2.6 Casting and Running of Denaturing Urea Gel for Determining rAP Cleavage

1. Gel casting apparatus inclusive of glass plates, spacers, and a 10-wells comb and the tank for the gel-running apparatus.
2. 10× TBE: 108 g of Tris base, 55 g of boric acid, 40 mL of 0.5 M EDTA pH 8.0, H_2O to 1 L.
3. 7 M urea denaturing 20% polyacrylamide gel master mix: in a large bottle with a cap, add and mix 42.04 g of urea, 5 mL 10× TBE, 2.5 mL of 2% g/v bis acrylamide solution, 47.5 mL of 40% g/v acrylamide, H_2O to 100 mL (*see* **Note 7**).

4. For the gel: 6 mL of 7 M urea denaturing 20% polyacrylamide gel master mix, 10% g/v APS, and TEMED.
5. 100% v/v ethanol.
6. 3 MM paper.
7. Fluorescent scanner (LI-COR Odyssey® DLxImaging System or equivalent).

3 Methods

APE1 endodeoxyribonuclease activity on a ribose monophosphate abasic site embedded in DNA can be tested both using purified recombinant proteins or whole-cell extracts. Hereafter, methods for APE1 purification and protein cell extracts are described, as follows. The oligonucleotide is employed as a substrate for the enzymatic reactions. As negative control, we suggest the use of commercially available APE1 inhibitors, including Inhibitor III and/or cell extracts depleted for APE1 expression by using sh-RNA.

3.1 Purification and Absorbance Measurement of rAPE1 Concentration

3.1.1 Preparation of E. Coli Lysate Overexpressing APE1

1. Transform BL21(DE3) according to the manufacturer's instructions with APE1 vector to express rAPE1-GST recombinant protein and plate on an LB-dish supplemented with the corresponding antibiotics (e.g., ampicillin) overnight at 37 °C.
2. Inoculate 50 mL liquid LB medium, supplemented with ampicillin, with a single colony obtained from the selection plate and incubate overnight at 37 °C under shaking at 230 rpm.
3. Inoculate 500 mL of LB supplemented with ampicillin with 50 mL of the overnight culture. Grow BL21(DE3) bacteria at 37 °C at 230 rpm measuring the 600-nm absorbance with a spectrophotometer.
4. When OD_{600} is around 0.6–0.8, induce the expression of the recombinant APE1 by adding IPTG and incubate for an additional 4 h at 37 °C at 230 rpm.
5. Harvest cells by centrifugation at 6,000 × *g* for 20 min at 4 °C.
6. Wash the pellet with 1× PBS and centrifuge as in 5.
7. Resuspend the pellet in GSTrap binding buffer added with bacteria protease inhibitor and sonicate for 45″ for six times and 45″ cooling down on ice after each cycle of sonication.
8. Clarify the lysate by centrifugation at 16,000 × *g* at 4 °C for 30 min.
9. Transfer the supernatant to a fresh tube and filter it with a 0.22-μm filter (*see* **Note 8**).

After sample preparation, several steps of purification should be performed and are briefly described below. Every purification step should be performed by using a FPLC instrument following the manufacturer instructions of each column (fluxes and volumes).

3.1.2 GSTrap Purification of rAPE1

1. Wash and equilibrate the GSTrap column with the GSTrap binding buffer.
2. Load the sample.
3. Wash the GSTrap column with the GSTrap binding buffer.
4. Elute rAPE1-GST with GSTrap elution buffer by using a 0–10 mM GSH gradient (*see* **Note 9**).
5. Quantify rAPE1 by using a Nanodrop (*see* **Note 10**) and by SDS-PAGE using a standard curve obtained with BSA.

3.1.3 GST Tag Cleavage and HiTrap Benzamidine Purification

1. Digest eluted rAPE1-GST protein fractions using Factor X_a (*see* **Note 11**).
2. Wash the HiTrap benzamidine column with the HiTrap benzamidine binding buffer.
3. Load the sample and collect flow through containing untagged rAPE1.
4. Wash the column with the high-salt binding buffer and collect flow through containing rAPE1-GST and GST (*see* **Note 12**).
5. Elute Factor X_a with low pH elution buffer.

3.1.4 HiTrap SP FF Purification of rAPE1

1. Wash the HiTrap column with the HiTrap SP binding buffer.
2. Load the sample.
3. Elute rAPE1 with HiTrap SP elution buffer by using a 100–1,000 mM NaCl gradient (*see* **Note 13**).
4. Quantify rAPE1 by using a Nanodrop (Fig. 2a) (*see* **Note 14**) and by SDS-PAGE using a standard curve obtained with BSA (Fig. 2b, c).

3.2 WCE Preparation and Quantification

3.2.1 Harvesting of Cell Culture

It is recommendable to work under sterile conditions and carry out all steps in a laminar-flow hood. Warm growth media and trypsin solution to 37 °C prior to use. For negative control, we recommend preparing cell extracts from APE1 silenced cells (siAPE1) in parallel to transfection control (siSCR).

1. Culture adherent cells to approximately 80% confluence on 100 mm tissue culture petri dishes. Cells should be in log-phase growth and healthy. As an example, 3 × 10^6 HeLa cells are plated *per* dish and cells will be ready 24 h after plating (*see* **Note 15**).
2. Aspirate media, wash the cell monolayer twice with 5 mL PBS, and remove PBS using a vacuum pump.

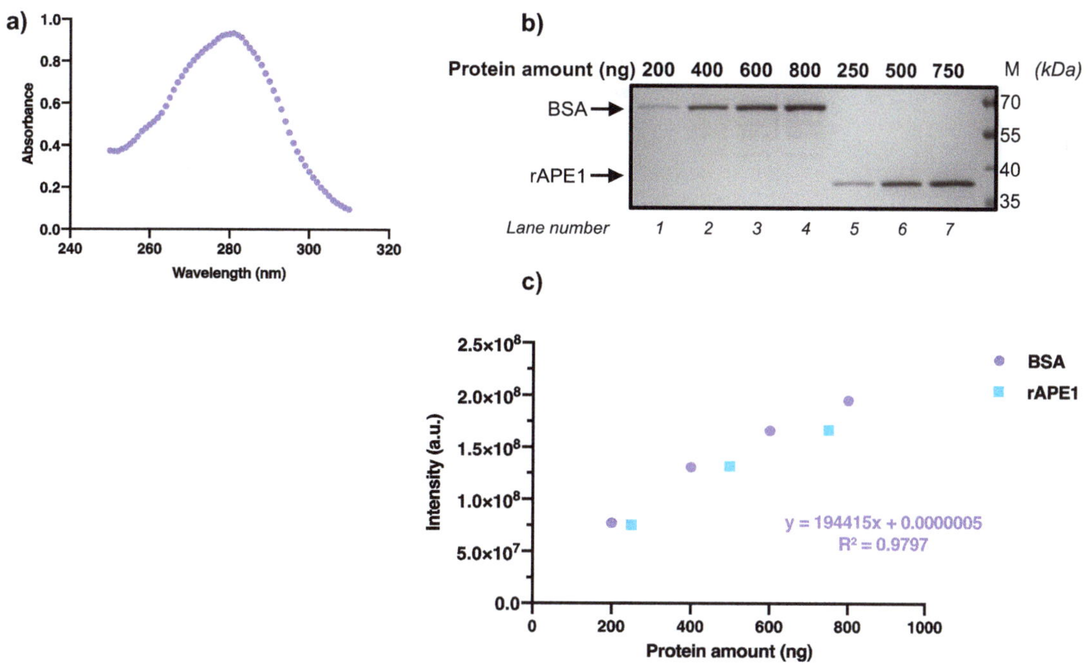

Fig. 2 Recombinant APE1 is accurately quantified before performing enzymatic activity. (**a**) Recombinant APE1 (rAPE1) absorbance spectrum is graphed reporting each value of absorbance detected as a function of light wavelength by using a Nanodrop spectrophotometer. (**b**) Increasing amounts of rAPE1 were loaded onto a 10% SDS-PAGE gel followed by Coomassie staining. A BSA–titration curve was used for the quantification by loading 200, 400, 600, and 800 nanograms. The molecular weight (Mw) expressed in kilodaltons (kDa) is shown on the right of each panel. (**c**) The BSA signal was plotted on a graph and the fitting curve was calculated. The equation and R square were also indicated

3. Add 1 mL of trypsin/EDTA solution to completely cover the monolayer of cells and place the dish in a 37 °C incubator for 2 min (*see* **Note 16**).
4. When trypsinization process is complete, add 5 volumes of medium containing serum to inhibit further tryptic activity. Resuspend cells by gently pipetting and collect cell suspension in a 15 mL tube.
5. Centrifuge the cells at 350 × *g* at RT for 5 min.
6. Remove supernatant using a vacuum pump and resuspend cell pellet in 1 mL PBS and transfer to a 1.5 mL microcentrifuge tube.
7. Centrifuge the microcentrifuge tube at 350 × g at RT for 5 min and discard liquid.
8. Freeze cell pellet at −80 °C or proceed with the preparation of cell extracts (*see* **Note 17**).

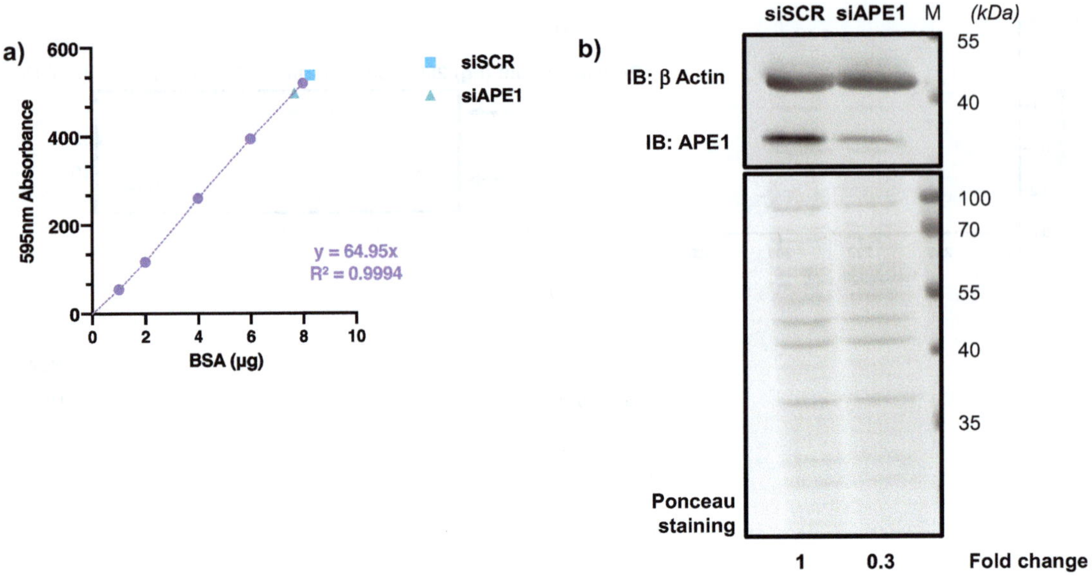

Fig. 3 Quantification of whole-cell extract protein content. (**a**) Each value of BSA signal, detected at 595 nm of wavelength, was plotted on a graph and the fitting curve was calculated. The equation and R square were indicated. Absorbance for both siSCR and siAPE1 samples was indicated and concentrations were calculated by using the BSA–titration curve equation. (**b**) Western blot analysis of APE1 in siSCR and siAPE1 samples. β-Actin was used as a loading control. Ponceau-S staining of the membrane is shown. The molecular weight (Mw) expressed in kilodaltons (kDa) is shown on the right of each panel. Below the panel, the normalized fold change, expressed as the ratio among APE1 levels vs β-actin levels, is indicated

3.2.2 Preparation and Quantification of WCE

All solution should be filtered and kept on ice during the experimental procedure.

1. Add WCE lysis buffer and mix the cell suspension thoroughly avoiding bubble formation (*see* **Note 18**).
2. Keep cell suspension on ice for 30 min.
3. Clarify the lysate by centrifugation at 20,000 × *g* for 20 min to remove insoluble particles and cell debris.
4. Collect supernatant (avoiding the pellet) and transfer it into a new 1.5 mL microcentrifuge tube.
5. Determine protein concentration of the WCE according to Bradford assay using bovine serum albumin as standard (Fig. 3a) and by running an SDS-PAGE (Fig. 3b).

3.3 Casting and Running of SDS-PAGE Gel

3.3.1 Casting of SDS-PAGE Gel

1. Clean gel-running equipment with ethanol and assemble it.
2. In a 15 mL tube, mix the components for the resolving gel and pour into the gel casting apparatus (*see* **Note 19**). Pour an additional 200 µL of isobutanol to flatten and level the resolving section of the gel. Allow the needed time for the gel polymerization. Thereafter, remove the isobutanol and wash

with ddH_2O. Carefully remove any trace of water by absorbing with 3 MM paper.

3. In a 15 mL tube, mix the components for the stacking gel, pour onto the remaining capacity of the gel casting apparatus, and insert the lane comb. Allow the needed time for the gel polymerization.

3.3.2 Running of SDS-PAGE Gel for Determining rAPE1 Concentration

1. Prepare each sample in a small tube mix with a different amount (ng) of BSA/rAPE1 kept at volume with ddH_2O and 4× SDS protein loading buffer dye. Heat the samples at 95 °C for 5 min.
2. After the SDS-PAGE gel has polymerized and has been loaded onto the gel-running apparatus and filled with 1× Tris–glycine SDS-PAGE running buffer, load each sample into the gel wells (*see* **Note 20**).
3. Run the gel at 15 mA for the stacking run and 30 mA for the running run.
4. Once the run is completed, remove the gel from the apparatus and submerge it in a Blue Coomassie staining solution. Incubate the gel under basculation for 10 min.
5. Remove the solution and submerge it in destaining solution. Wash the gel and remove.
6. Submerge it in destaining solution. Incubate the gel under shaking for 2 h up when protein bands will become visible.
7. Scan stained gel by using an imaging system for gel acquisition and band signal quantification (Fig. 2b, c).

3.3.3 Running of SDS-PAGE Gel for WCE

1. Prepare each sample in a small tube by mixing at least 10 μg WCE kept at volume with ddH_2O and 4× SDS protein loading buffer dye. Heat the samples at 95 °C for 5 min.
2. After the SDS-PAGE gel has polymerized and has been loaded onto the gel-running apparatus and filled with 1× Tris–glycine SDS-PAGE running buffer, load each sample into the gel wells.
3. Run the gel at 15 mA for the stacking gel and 30 mA for the running gel.
4. Once the run is completed, remove the gel from the apparatus and set up the membrane-gel sandwich for western blotting analysis.
5. At the end of protein transfer, submerge the nitrocellulose membrane in Ponceau solution for 2 min, and then wash with ddH_2O.
6. Scan membrane by using a white light gel reader (FireReader or equivalent) and incubate the membrane with both antibodies *versus* APE1 and actin as loading control (Fig. 3b).

3.4 Annealing of DNA Oligonucleotides

1. Mix 100 pmol of the rF-containing oligonucleotide with 150 pmol of the complementary oligonucleotide in 40 μL of annealing buffer.
2. Heat the mixture in a water bath set on 95 °C and then allow to cool slowly to room temperature.
3. Final concentration of this stock is 2.50 pmol/μL (*see* **Note 21**).

3.4.1 Set Up the rAP Endoribonuclease Assay by Using rAPE1 or WCE

Here, three experimental procedures are described: (a) dose–response APE1 cleavage activity (Subheading 3.4.2), (b) time course APE1 cleavage activity (Subheading 3.4.3), and (c) time course cleavage activity in the presence of an APE1 inhibitor (Subheading 3.4.4). For all procedures, two alternative approaches are reported using either rAPE1 protein or WCE.

All solutions should be prepared using sterilized DNase- and RNase-free water. It is recommendable to use DNase- and RNase-free plastic and filter tips. Preheat a heat block at 37 °C and perform all the steps on ice.

3.4.2 Dose–Response Analysis

1. Prepare microcentrifuge tubes each corresponding to different doses of rAPE1 or WCE (*see* **Note 22**).
2. Prepare the reaction samples in 10 μL of final volume aliquoting H_2O, 10× rAP buffer and the desired amount of protein or WCE (*see* **Note 23**).
3. Add 1 μL of 250 fmol/μL diluted oligonucleotide.
4. Briefly mix and centrifuge the tubes.
5. Incubate in a heat block at 37 °C for 10 min.
6. Stop the reaction by adding 10 μL of stop solution.
7. Heat the samples at 95 °C for 5 min.
8. Spin and transfer the denatured samples on ice to prevent reannealing (*see* **Note 24**).
9. Analyze products of the cleavage reactions by denaturing acrylamide gel electrophoresis and imaging (typical result is shown in Figs. 4a, b and 5a, b).

3.4.3 Time Course Analysis

1. Prepare microcentrifuge tubes each corresponding to different time points.
2. Premix a unique mix by adding appropriate amounts of 10× rAP buffer, oligonucleotide, rAPE1/WCE, and water and considering all different time points to be measured (*see* **Note 25**). Use the amount of rAPE1 for which 50% of cleavage activity has been detected in a separate dose–response experiment.
3. Briefly mix and centrifuge the tube.
4. Incubate in a 37 °C heat block.

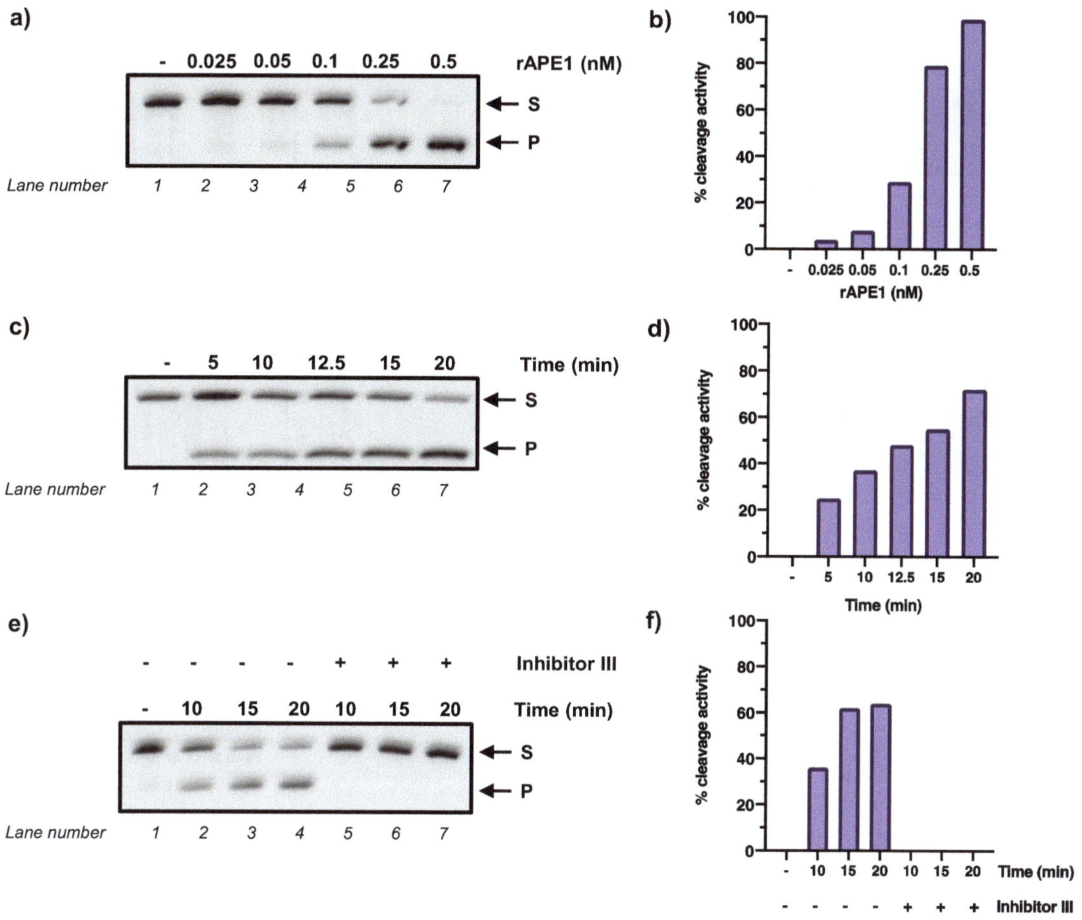

Fig. 4 rAPE1 cleavage activity on rAP site in a dose- and time-dependent manner and inhibition by Inhibitor III. Representative denaturing polyacrylamide gels of rAP site incision by rAPE1 incubated for 10 min with the indicated amount of protein (**a**) or incubated with 0.1 nM of rAPE1 at different time points in the absence (**c**) or in the presence (**e**) of Inhibitor III. Graphs describe the percentage of cleavage activity of rAPE1 (**b**, **d**, **f**), as calculated in Subheading 3.6. S and P are used to indicate substrate and product, respectively

5. Stop the cleavage reaction at chosen timing points by taking 10 μL of reaction from the mix to add in a new microcentrifuge tube in which 10 μL of stop solution has been previously aliquoted.
6. Heat the samples at 95 °C for 5 min.
7. Spin and transfer the denatured samples on ice to prevent reannealing.
8. Analyze products of the cleavage reactions by denaturing acrylamide gel electrophoresis and imaging (typical result is shown in Figs. 4c,d and 5c, d).

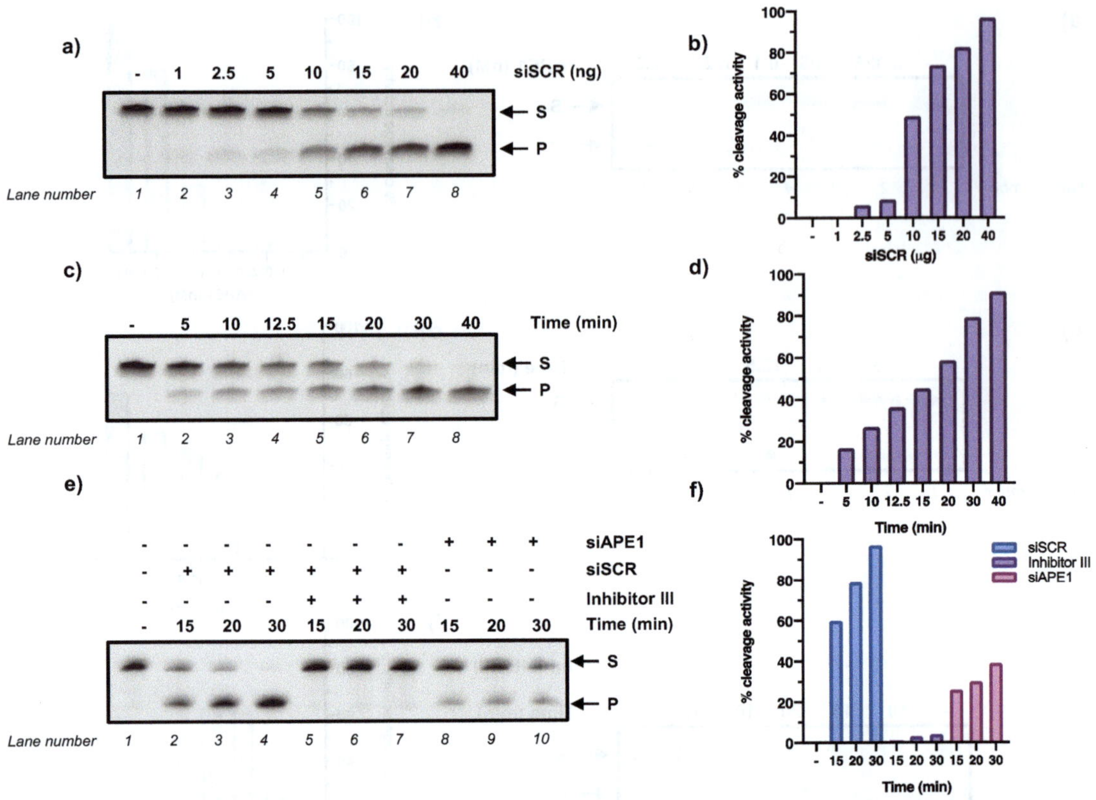

Fig. 5 APE1 cleavage activity from whole-cell extracts on the rAP site in a dose- and time-dependent manner and inhibition by Inhibitor III. Representative denaturing polyacrylamide gels of rAP site incision by APE1 contained in siSCR sample for 10 min with the indicated amount of whole-cell extract (**a**) or incubated with 10 ng of siSCR at different time points (**c**). (**e**) Representative denaturing polyacrylamide gels of rAP site incision by comparing 10 ng of siSCR with siAPE1 samples at different time points in the presence of Inhibitor III. Graphs represent the percentage of cleavage activity by rAPE1, as calculated in Subheading 3.6 (**b**, **d**, **f**). S and P are used to indicate substrate and product, respectively

3.4.4 Specific Inhibition of APE1 Enzymatic Activity by Using Inhibitor III

1. Perform time course experiment by following the instructions stated in Subheading 3.4.3.
2. To block APE1 cleavage activity, preincubate rAPE1 or WCE with an optimal amount of Inhibitor III, APE1 inhibitor, for an optimized timing (*see* **Note 26**).
3. Analyze products of the cleavage reactions by denaturing acrylamide gel electrophoresis and imaging (typical result is shown in Figs. 4e, f and 5e, f).

3.5 Casting and Running of Urea Gel for Determining rAP Cleavage

1. Clean gel-running equipment with ethanol and assemble it.
2. In a 15 mL tube, mix the components and pour into the gel casting apparatus and insert the lane comb. Allow the needed time for the gel polymerization (*see* **Note 27**).

3. After the gel has polymerized, carefully remove the comb and clamps.
4. Place the gel assembly in the electrophoresis apparatus and fill the upper and lower buffer compartments with 0.5× TBE buffer.
5. Rinse the wells thoroughly with a syringe filled with 0.5× TBE to wash away urea (*see* **Note 28**).
6. Pre-run the gel for 10 min to equilibrate the ions in the 0.5× TBE electrophoresis buffer.
7. Load the samples into the gel wells. Four to six μL of the total amount of the reaction can be loaded into gel wells (*see* **Note 29**). The remaining amount can be stored at −20 °C.
8. Run the gel at constant 300 V (*see* **Note 30**).
9. Once the run is completed, remove the gel from the apparatus.
10. Directly scan the gel by using a fluorescence scan reader.

3.6 Analysis of Results

Cleaved products are detected by a fluorescent scanner and quantified by digital imaging. The percentage of APE1 activity is determined using the formula:

$$\%\text{incision} = \left[\frac{\text{Product}}{(\text{Product} + \text{Substrate})}\right] \times 100$$

4 Notes

1. DTT should be added in each buffer immediately prior to use.
2. GSH should be added immediately prior to use.
3. Cells should be healthy, exponentially growing, and mycoplasma-free to obtain active extracts.
4. Prepare protease inhibitors for storage at −20 °C and use a fresh aliquot every month depending on the number of freeze-thaw cycles. WCE lysis buffer should be complemented with protease inhibitor immediately prior to use. Buffers containing protease inhibitors should only be used freshly once.
5. Acrylamide is a powerful neurotoxin and should be handled with extreme caution. TEMED should be added to the gel, mix immediately prior to pouring it into the apparatus, preferentially under a chemical hood.
6. An aliquot of stop solution should be prepared freshly immediately prior to use.
7. Mix thoroughly by using a magnetic stirrer until urea is dissolved in solution. If necessary, heat the solution at a maximal temperature of 50 °C.

8. Quantify bacterial protein extracts (e.g., using a Bradford method) and eventually dilute them to have an approximative concentration of 1 mg/mL.
9. Analyze the fractions by SDS-PAGE and Coomassie staining. Quantify eluted fractions on an SDS-PAGE gel using a BSA standard curve and pool relevant fractions.
10. rAPE1 concentration can be measured to the Nanodrop by inserting $\varepsilon/1000 = 56.76$ and molecular weight = 35′555 Da, obtained by Expasy.
11. Follow the manufacturer' instructions for dose and timing.
12. Analyze the fractions by SDS-PAGE and Coomassie staining. Quantify eluted fractions on an SDS-PAGE gel using a BSA standard curve and pool relevant fractions.
13. Analyze the fractions by SDS-PAGE and Coomassie staining. Quantify eluted fractions on an SDS-PAGE gel using a BSA standard curve and pool relevant fractions.
14. rAPE1 concentration can be measured to the Nanodrop by inserting $\varepsilon/1000 = 56.76$ and molecular weight = 35′555 Da, obtained by Expasy.
15. Although this protocol describes cell growth in a 100 mm petri dish, the size of cell culture can be changed multiplying the volume of the solutions by the relative size of the culture.
16. The time required to detach cells from the culture surface depends on the cell type, population density, serum concentration in the growth medium, and trypsin concentration. In general, the time of exposure should be kept to a minimum since trypsin can cause cellular damage.
17. Cell pellet can be stored indefinitely at −80 °C but should be left for at least 10 min prior to proceed with cell protein extract preparation to promote cell lysis.
18. The amount of buffer depends on the cell number harvested. Ensure that the cell pellet is thoroughly resuspended in buffer and increase a larger volume if required.
19. Pour the resolving gel to approximately 75% the capacity of the gel casting apparatus and leave the remaining 25% of the top for the stacking gel.
20. A protein ladder marker lane should be added to monitor the rAPE1 position on the gel. The molecular weight of rAPE1 is around 36,000 Da.
21. A 250 fmol/μL intermediate solution is prepared freshly in ultrapure DNase- and RNase-free water.
22. Always consider including at least one negative control sample with no recombinant APE1 protein or WCE to detect a spontaneous degradation of the oligonucleotide.

23. The amount of recombinant protein and WCE to be used should be optimized. It is advisable to test the concentration range in which a 10% till to 90% of cleavage is observed.
24. Samples may be stored at −20 °C and run on the gel later.
25. Use the amount of APE1 for which 50% of cleavage activity has been detected in dose–response experiments.
26. As for the time course experiment, start with an amount of protein for which you calculated an amount of 50% of cleavage activity. In the case of Inhibitor III, a final concentration of 230 μM was used to inhibit 0.1 nM of rAPE1 and 10 ng of WCE, for 15 min at 37 °C.
27. Pour urea gel several hours before running to allow a complete polymerization of the gel.
28. If the wells are not well-washed from urea, detected lanes will appear crocked.
29. The first and last wells of the gel will become distorted when electrophoresis is underway; thus, it is better to not load any sample.
30. To avoid overheating of the gel during the run, it is suggested to run it at 4 °C.

Acknowledgements

The authors thank all the members of the GT lab for constructive feedback during the development of this work. GT is supported by a grant from the Associazione Italiana per la Ricerca sul Cancro (**IG19862**).

References

1. Malfatti MC, Antoniali G, Codrich M et al (2020) New perspectives in cancer biology from a study of canonical and non-canonical functions of base excision repair proteins with a focus on early steps. Mutagenesis 35:129–149
2. Antoniali G, Malfatti MC, Tell G (2021) Chapter 17. Emerging concepts on the early steps of base excision repair pathway with a focus on gene expression. In: DNA damage, DNA repair and disease, vol 2, pp 24–47
3. Rai G, Vyjayanti VN, Dorjsuren D et al (2012) Synthesis, biological evaluation, and structure-activity relationships of a novel class of apurinic/apyrimidinic endonuclease 1 inhibitors. J Med Chem 55:3101–3112
4. Poletto M, Malfatti MC, Dorjsuren D et al (2016) Inhibitors of the apurinic/apyrimidinic endonuclease 1 (APE1)/nucleophosmin (NPM1) interaction that display anti-tumor properties. Mol Carcinog 55:688–704
5. Malfatti MC, Balachander S, Antoniali G et al (2017) Abasic and oxidized ribonucleotides embedded in DNA are processed by human APE1 and not by RNase H2. Nucleic Acids Res 45:11193–11212
6. Tell G, Wilson DM, Lee CH (2010) Intrusion of a DNA repair protein in the RNome world: is this the beginning of a new era? Mol Cell Biol 30:366–371
7. Malfatti MC, Antoniali G, Codrich M, Tell G (2021) Coping with RNA damage with a focus on APE1, a BER enzyme at the crossroad

between DNA damage repair and RNA processing/decay. DNA Repair (Amst) 104:103133

8. Sparks JL, Chon H, Cerritelli SM et al (2012) RNase H2-initiated ribonucleotide excision repair. Mol Cell 47:980–986
9. Liu Y, Rodriguez Y, Ross RL et al (2020) RNA abasic sites in yeast and human cells. Proc Natl Acad Sci U S A 117:20689–20695
10. Cile Mingard C, Wu J, Mckeague M, Sturla SJ (2020) Next-generation DNA damage sequencing. Chem Soc Rev 49:7354–7377
11. Esqueda A, Mohammed MZ, Madhusudan S, Neamati N (2012) Purification and specific assays for measuring APE-1 endonuclease activity. Methods Mol Biol 928:161–174
12. Simeonov A, Kulkarni A, Dorjsuren D et al (2009) Identification and characterization of inhibitors of human apurinic/apyrimidinic endonuclease APE1. PLoS One 4:e5740

Chapter 3

Highly Sensitive Radioactivity-Based DNA 3′-Phosphatase Activity Assay for Polynucleotide Kinase 3′-Phosphatase

Anirban Chakraborty and Tapas K. Hazra

Abstract

Endogenous and exogenous genotoxic agents can generate various types of non-ligatable DNA ends at the site of strand break in the mammalian genome. If not repaired, such lesions will impede transcription and replication and can lead to various cellular pathologies. Among various "dirty" DNA ends, 3′-phosphate is one of the most abundant lesions generated in the mammalian cells. Polynucleotide kinase 3′-phosphatase (PNKP) is the major DNA end-processing enzyme for resolving 3′-phosphate termini in the mammalian cells, and thus, it is involved in DNA base excision repair (BER), single-strand break repair, and classical nonhomologous end joining (C-NHEJ)-mediated DNA double-strand break (DSB) repair. The 3′-OH ends generated following PNKP-mediated processing of 3′-P are utilized by a DNA polymerase to fill in the gap, and subsequently, the nick is sealed by a DNA ligase to complete the repair process. Here we describe two novel assay systems to detect phosphate release by PNKP's 3′-phosphatase activity and PNKP-mediated in vitro single-strand break repair with minimal repair components (PNKP, DNA polymerase, and DNA ligase) using either purified proteins or cell-free nuclear extracts from mammalian cells/tissues. These assays are highly reproducible and sensitive, and the researchers would be able to detect any significant difference in PNKP's 3′-phosphatase activity as well as PNKP-mediated single-strand break repair activity in diseased mammalian cells/tissues vs normal healthy controls.

Key words Reactive oxygen species, Base excision repair, Single strand break repair, PNKP, 3′--phosphate, Double strand break repair, DNA polymerase, DNA ligase

1 Introduction

Various endogenous and exogenous agents pose a continuous threat to the integrity of cellular DNA [1–3]. Reactive oxygen species (ROS), generated as by-products of cellular respiration, inflammatory response, and from exposure to ionizing radiation and various environmental toxicants, are the major sources of genome damage in living organisms [4]. The DNA damage induced by ROS includes a number of oxidized DNA base lesions and DNA single- and double-strand breaks [5, 6]. Base excision repair (BER) is the major pathway for repairing the vast majority of

Kishor K. Bhakat and Tapas K. Hazra (eds.), *Base Excision Repair Pathway: Methods and Protocols*, Methods in Molecular Biology, vol. 2701, https://doi.org/10.1007/978-1-0716-3373-1_3,

endogenously induced small DNA base lesions as well as single-strand breaks [7]. The bifunctional DNA glycosylases recognize oxidized DNA base lesions, excise the damaged base, and also, cleave the DNA backbone by their intrinsic AP lyase activities, leaving either a 3′-phospho-α, β-unsaturated aldehyde (PUA) or 3′-phosphate [1, 8]. In *E. coli*, these 3′-blocking groups are efficiently removed by the AP endonucleases (APEs), Xth and Nfo [5]. The mammalian APE, APE1, can efficiently remove the PUA; however, it has very weak 3′-phosphatase activity [5]. Since oxidized base-specific mammalian DNA glycosylases, NEIL1 and NEIL2, generate 3′-phosphate groups following AP lyase activity, these cannot be processed by APE1 [5]. These 3′-phosphate groups are processed by polynucleotide kinase 3′-phosphatase (PNKP) [9, 10] which is a bifunctional end-processing enzyme in mammalian cells with 3′-phosphatase and 5′-kinase activities and is a primary contributor of DNA 3′-phosphatase activity associated with mammalian chromatin [11]. We have further established that it is the major 3′-phosphatase in the mammalian cells [12, 13]. Furthermore, PNKP has been shown to function in the repair of DNA single-strand breaks generated by ionizing radiation [14] and we have recently established the critical role of PNKP in classical nonhomologous end joining (C-NHEJ)-mediated DNA double-strand break (DSB) repair [15]. The unprocessed 3′-phosphates pose a strong block to elongating RNA polymerase II (RNAPII) [16] and prolonged stall of RNAPII may lead to its degradation via polyubiquitination [17]. Thus, PNKP-mediated DNA end processing is critical for cellular survival. The 3′-OH terminus generated following PNKP-mediated end processing is used by a repair polymerase to fill in the gap. A DNA ligase eventually seals the nick by forming a phosphodiester bond.

Unrepaired DNA lesions may lead to various pathologies, including premature aging, neurodevelopmental disorders/neurodegeneration, and cancer [18–21]. Using a novel in vitro assay system to specifically detect the phosphate release by purified PNKP and cell-free nuclear extract (mammalian cell lineages, mouse brain tissues, and human postmortem patients' brains), we have demonstrated that PNKP's 3′-phosphatase activity is potentiated by wild-type ataxin-3 (ATXN3), a polyglutamine (CAG) repeat-containing protein mutated in spinocerebellar ataxia type 3 (SCA3) [13]. The pathogenic form of ATXN3 with polyQ expansion blocks PNKP's activity, resulting in DNA single- and double-strand break accumulation and chronic activation of the DNA damage–response pathways [13, 17, 22]. Similarly, we reported that the mutant form of huntingtin (HTT), which undergoes trinucleotide repeat expansion in Huntington's disease (HD), abrogates PNKP's 3′-phosphatase activity in cell and HD mouse models [23] leading to strand break accumulation. Also, we have developed a novel in vitro assay system to reconstitute single-strand break repair using minimal repair constituents like PNKP, a DNA polymerase, and a DNA ligase, either in purified form or in cell-free nuclear extracts. We have used this assay system to establish total in vitro BER activity of purified NEIL2 immunocomplex (IC) from mammalian cells on AP site containing substrates [24] and further reported compromised BER activity in vitro on a plasmid substrate due to its lack of association with downstream repair proteins, viz., PNKP, Lig IIIα, and particularly DNA polymerase β (Pol β) [24]. Also, using this assay system, we have provided evidence for compromised repair efficiency in SCA3 patients and HD mouse brains compared to their control counterparts [17, 23].

In this chapter, we focus on providing details of the molecular methods used in our studies to measure 3′-phosphatase activity of PNKP and single-strand break repair activity involving recombinant PNKP, DNA polymerase, and DNA ligase on 3′-phosphate containing single-strand break that mimic DNA substrate.

2 Materials

2.1 In Vitro Assay for PNKP's 3′-Phosphatase Activity

1. Oligo sets used for preparing radiolabeled and non-radiolabeled (cold) substrate for PNKP's 3′-phosphatase activity (*see* **Note 1**):
   (a) U-26 (uracil containing 26-nucleotide oligo): 5′-U A CTC GTG TGC CGT GTA GAC CGT GCC-3′.
   (b) F-25 (25-nucleotide oligo): 5′-GCT TAG CTT GGA ATC GTA TCA TGT A-3′.
   (c) G-51 (51-nucleotide complementary oligo): 5′-GGC ACG GTC TAC ACG GCA CAC GAG TGT ACA TGA TAC GAT TCC AAG CTA AGC-3′.
   (d) 3′P-25 (3′-phosphate containing 25-nucleotide oligo): 5′-GCT TAG CTT GGA ATC GTA TCA TGT A-(3′ phosphate)
   (e) 5′P-25 (5′-phosphate containing 25-nucleotide oligo): (5′ phosphate)-A CTC GTG TGC CGT GTA GAC CGT GCC-3′
   (f) G-51 (51-nucleotide complementary oligo): 5′-GGC ACG GTC TAC ACG GCA CAC GAG TGT ACA TGA TAC GAT TCC AAG CTA AGC-3′.
2. dH2O (nuclease-free).
3. 10× T4 PNK buffer.
4. ATP, [γ-^{32}P].
5. T4 PNK.
6. 5 M NaCl.

7. 1 M Tris–HCl pH 8.0.
8. 10× T4 DNA ligase buffer.
9. T4 DNA ligase.
10. 10 mM ATP.
11. 10× uracil DNA glycosylase (Udg) buffer.
12. Udg and Fpg (formamidopyrimidine DNA glycosylase).
13. Dry heat block with thermostat control.
14. Micro Bio Spin ™ P-6 columns (Bio-Rad).
15. Tabletop microcentrifuge.
16. Urea.
17. 40% acrylamide/bis solution (29:1).
18. 10× Tris–borate–EDTA (TBE) buffer: 1 M Tris–HCl, 1 M boric acid, 0.02 M EDTA–disodium salt. Dilute 1:10 with milliQ water to make working 1× solution.
19. 10% ammonium persulfate (APS).
20. Tetramethylethylenediamine (TEMED) (electrophoresis grade).
21. Gel casting empty cassettes (1.0 mm lane thickness).
22. 2× loading dye: 800 μL formamide (70% final), 0.02 N NaOH (stock 0.1 N), pinch of bromophenol blue (BPB).
23. 10× PNKP assay buffer: 400 mM HEPES pH 7.5, 500 mM KCl, 100 mM $MgCl_2$, 1 mM EDTA, 50 mM DTT, 1 mM spermidine, and 50% glycerol. Dilute the 10× buffer with nuclease-free dH20 to make 1× working buffer.
24. 10 mg/mL bovine serum albumin, acetylated.
25. Gel running tank and power supply.
26. Incubator oven with thermostat control for running gel at 65 °C constant temperature.
27. Whatman paper.
28. Nylon bag.
29. Heat sealer.
30. Typhoon phospho-imager.
31. Exposure cassettes and phospho-imager screens.

2.2 Assay of In Vitro Single-Strand Break Repair Activity

1. Oligo sets for preparing substrate for PNKP-mediated single-strand break repair activity (*see* **Notes 1** and **2**):
   (a) 3′P-25 (25-mer 3′-phosphate containing oligo): 5′-GCT TAG CTT GGA ATC GTA TCA TGT A-(3′ phosphate)

   (b) 5′P-25 (25-mer 5′-phosphate containing oligo): (5′ phosphate)-A CTC GTG TGC CGT GTA GAC CGT GCC-3′
   (c) G-51 (51-mer complementary oligo): 5′-GGC ACG GTC TAC ACG GCA CAC GAG TGT ACA TGA TAC GAT TCC AAG CTA AGC-3′.
2. dH2O (nuclease-free).
3. dCTP, [α-^{32}P].
4. 5 M NaCl.
5. 1 M Tris–HCl pH 8.0.
6. 10 mM ATP.
7. 10 mM of dCTP, dATP, dGTP, dTTP (all individual stocks).
8. Dry heat block with thermostat control.
9. Tabletop microcentrifuge.
10. Urea.
11. 40% acrylamide/bis solution (29:1).
12. 10× Tris–borate–EDTA (TBE) buffer: 1 M Tris–HCl, 1 M boric acid, 0.02 M EDTA–disodium salt. Dilute 1:10 with milliQ water to make working 1× solution.
13. 10% ammonium persulfate (APS).
14. Tetramethylethylenediamine (TEMED) (electrophoresis grade).
15. Gel casting empty cassettes (Invitrogen, Carlsbad, CA [1.0 mm]).
16. 2× loading dye: 800 μL formamide (70% final), 0.02 N NaOH (stock 0.1 N), pinch of bromophenol blue (BPB).
17. 10× total repair assay buffer: 400 mM HEPES pH 7.5–7.9, 500 mM KCl, 100 mM $MgCl_2$, 1 mM EDTA, 50 mM DTT, 1 mM spermidine, and 50% glycerol. Dilute the 10× buffer with nuclease-free dH20 to make 1× working buffer.
18. 10 mg/mL bovine serum albumin, acetylated.
19. Gel running tank (Invitrogen) and power supply.
20. Incubator oven with thermostat control for running gel at 65 °C constant temperature.
21. Whatman paper.
22. Nylon bag.
23. Heat sealer.
24. Typhoon phospho-imager.
25. Exposure cassettes and phospho-imager screens.

3 Methods

3.1 In Vitro Assay of 3′-Phosphatase Activity of PNKP

3.1.1 Labeling the 5′ End of U-26 Uracil-Containing Oligo with ATP γ-^{32}P

1. Mix in 50 μL volume in the following order of adding:

   38.5 μL dH2O

   5.0 μL 10× T4 PNK buffer

   2.5 μL U-26 (working stock: 10 pmoles/μL; total = 25 pmoles)

   3.0 μL γ-ATP ^{32}P (total = 30 μCi) (*see* **Note 3**)

   μl T4 PNK (10 units/μL; 10 units used per reaction),
2. Incubate at 37 °C dry heat block for 25 min (*see* **Note 4**).
3. Stop the reaction by heating the reaction mixture at 85 °C for 5 min to inactivate T4 PNK, and then either put in ice for short-term storage or the reaction can be continued.

3.1.2 Annealing with F-25 and G-51 Complementary Oligos

1. Add to the previous reaction mixture in the following order (volume = 60 μL):

   50.0 μL from previous reaction

   1.25 μL 5 M NaCl

   1.50 μL 1 M Tris–HCl pH 8.0

   3.0 μL F-25 (10 pmoles/μL; total = 30 pmoles)

   3.0 μL complementary G-51 (10 pmoles/μL; total = 30 pmoles)

   1.25 μL dH_2O
2. Put the boiling water in a beaker and leave to cool down (*see* **Note 5**).

3.1.3 Ligation (See Note 6)

1. Mix the components in the following order (volume = 70 μL):

   55.0 μL previously annealed mixture

   4.0 μL dH2O

   7.0 μL 10× T4 DNA ligase buffer

   μL 10 mM ATP,

   μL T4 ligase (100 units/μL, total = 300 units),
2. Incubate overnight in a water bath at room temperature.

3.1.4 Udg and Fpg Digestion of the Ligated Product (See Note 7)

1. Mix in the following order to set up the digestion reaction (volume = 80 μL):

   67.0 μL ligated DNA mixture

   8.0 μL 10× Udg buffer

   1.5 μL Udg (5 units/μL, total = 7.5 units)

   1.5 μL Fpg (8 units/μL, total = 12 units)

   2.0 μL dH2O

2. Incubate at 37 °C for 1 h (*See* **Note 8**).
3. Inactivate Udg and Fpg at 65 °C for 15 min in dry heat block.

The reaction product can be stored at −20 °C or it can be processed immediately for better results.

3.1.5 Removal of the Free, Unincorporated ATP-γ-^{32}P from the Udg/Fpg-Digested Reaction Mixture

Use of the Bio-Rad Micro Bio Spin ™ P-6 columns is recommended for this purification step. Manufacturer's protocol can be followed for this step. Briefly, the following steps can be carried out:

1. Invert the column multiple times to mix the resin well before use. Remove the trapped air bubbles inside the resin by gentle tapping. Snap open the lower tip and the top cap and place the spin column in a supplied collection tube for Tris buffer to drain by gravity.
2. Add 500 μL of dH_2O to wash the residual buffer and drain the liquid by spinning at 1000 *g* for 2 min in a tabletop centrifuge at room temperature. Repeat this step four times.
3. Place the column in a fresh 1.5 mL microcentrifuge tube. Apply the radiolabeled substrate on the resin without touching the resin material (*see* **Note 9**).
4. Spin at 1000 *g* for 4 min and collect the purified radiolabeled substrate. This substrate may be kept frozen in aliquots at − 20 °C (the process of substrate preparation is shown schematically in Fig. 1a) (*see* **Note 10**).

3.1.6 Checking the Purity of the Radiolabeled Substrate

1. Mix the following reagents to prepare 20% urea polyacrylamide gel (recipe for 16 mL) (*see* **Note 11**):

   6.8 g urea (7 M final)

   8.0 mL 40% acrylamide/bis solution (20% final)

   1.6 mL 10× TBE
2. Keep at 65 °C overnight to melt or briefly microwave (50% power: consecutive 30″-20″-10″ heating cycles).
3. Keep on ice before pouring the solution, add the compounds below in the order, and cast the gel in the empty cassette:

   150 μL 10% APS

   10 μL TEMED
4. Prepare the running buffer (1× TBE) and pre-warm at 65 °C in an oven.
5. Mix the sample (1 μL of the purified substrate, 19 μL dH_2O, and 20 μL 2× loading dye to make the volume of 40 μL) and heat for 5 min at 95 °C (in the thermal block) and then load half the volume and save the other half (for rerun, if the gel run is not good) (*see* **Note 12**).
6. Run the gel at a 65 °C oven at constant 250 volts (*see* **Note 13**).

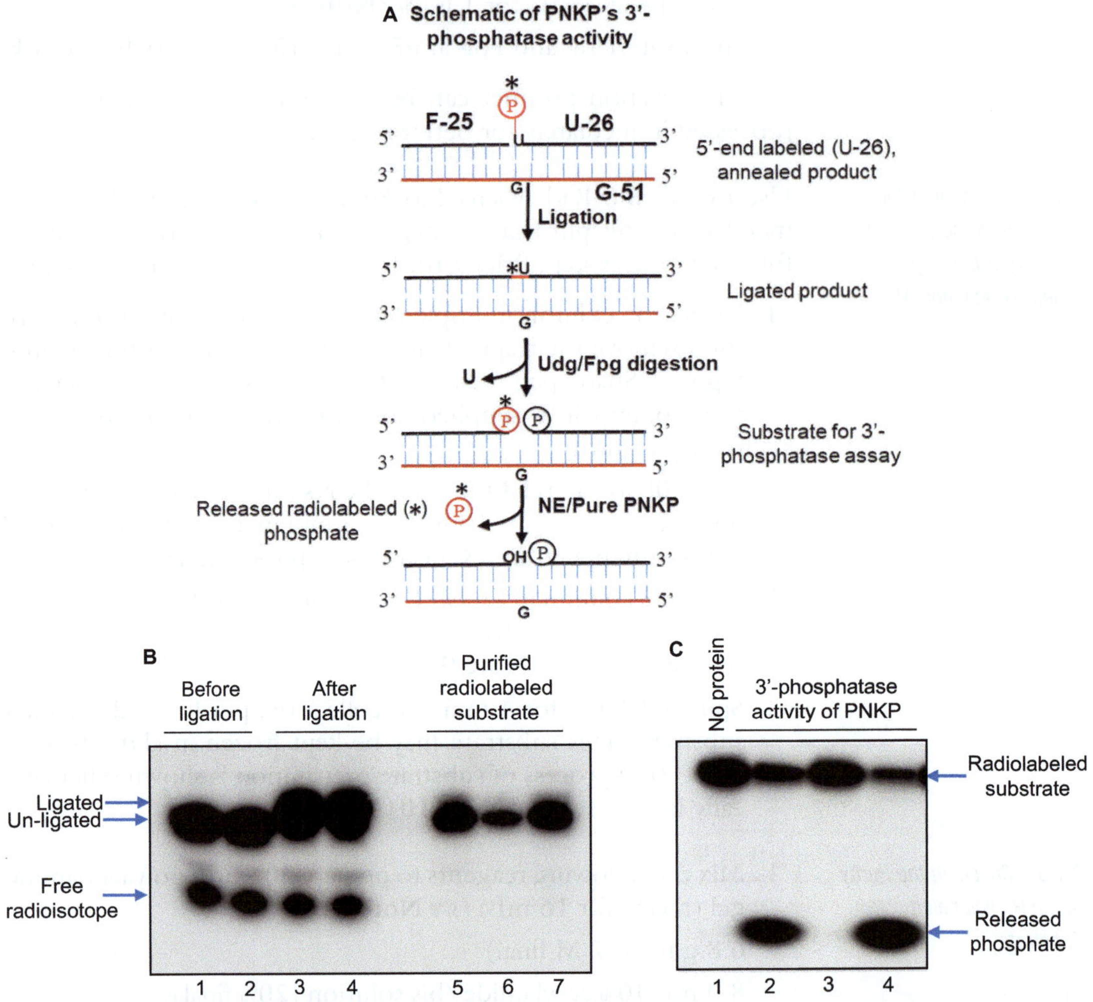

Fig. 1 In vitro measurement of PNKP's 3′-phosphatase activity. (**a**) Schematic representation of the various steps involved in the preparation of substrate for PNKP's 3′-phosphatase activity. (**b**) Lns 1–4: Assessment of the DNA ligation efficiency in 20% urea-PAGE. The upper ligated bands are shown in lns 3–4 compared to the 5′-end radiolabeled annealed unligated product in lns 1–2. Lns 5–7: Purified radiolabeled substrate for 3′-phosphatase assay. (**c**) Lns 1–4: Representative denaturing gel showing PNKP's 3′-phosphatase activity. Ln 1: No protein control, PNKP activity in the nuclear extract prepared from control (ln 2) vs PNKP-depleted (ln 3) HEK293 cell. Ln 4: Purified PNKP (25 fmol) as positive control

7. Transfer the gel to a piece of Whatman paper and place inside a nylon bag and seal with a heat sealer.
8. Place the gel in the cassette, fix with a tape, put the phospho-imager screen on top of the gel, expose at least 30–60 min, and scan with the Typhoon phospho-imager. One representative figure is shown (Fig. 1b).

3.1.7 Preparation of the Cold Substrate (for Evaluation of the Radiolabeled Product Formation in the Linear Range)

1. Add the components in the following order:

   1.25 μL 3′ P-25 oligo (100 pmole/μL working stock prepared in dH_2O just before use)

   1.50 μL 5′ P-25 oligo (100 pmole/μL working stock prepared in dH_2O just before use)

   1.25 μL G-51 oligo (100 pmole/μL working stock prepared in dH_2O just before use)

   0.25 μL 5 M NaCl

   0.625 μL 1 M Tris–HCl pH 8.0

   20.125 μL dH_2O (up to 25 μL)

2. Keep for 5 min in boiling water in a beaker.
3. Leave on the bench to cool down slowly at room temperature overnight.
4. Keep frozen at −20 °C.

3.1.8 Measuring the Concentration of Pure Proteins

Measure the concentration of the pure proteins or nuclear extracts by Bradford method, using BSA as a standard (the protein measured should be in the linear range of the standard curve).

3.1.9 Assay of 3′-Phosphatase Activity of PNKP

1. Prepare diluted stocks of 0.5 ng/μL of purified PNKP in 1× PNKP assay buffer for reaction. Similarly, if using nuclear extract, dilute to 250 ng/μL (*see* **Note 14**).

   Prepare reaction mixture by mixing the following reagents on ice:

   2 μL of 10× buffer (final working concentration 1×)

   2 μL of acetylated BSA (10 mg/mL stock, final 20 μg/reaction)

   2 μL of ATP (10 mM stock, final 1 mM/reaction)

   1 μL of cold substrate preparation (final amount = 5 picomoles) (*see* **Note 15**).

   1–2 μL of labeled substrate

   2–4 μL of purified PNKP (1–2 ng)/1–2 μL nuclear extract (250–500 ng) (*see* **Note 16**).

   dH_2O to adjust the volume to 20 μL in total,

2. Start the reaction by mixing the protein and master mix (all the reagents mixed together except the protein/nuclear extract) (*see* **Note 17**). Incubate the reaction mixture at 37 °C for 10 min (for pure PNKP) or 15 min (for nuclear extract). However, the incubation time should be adjusted according to the activity of the pure protein/nuclear extract.

3. Stop the reaction by adding an equal volume of 2× loading dye and heating at 95 °C for 5 min. Load a half volume of samples on 20% urea-PAGE (pre-run the gel for 15 minutes at a 65 °C incubator oven) (*see* **Note 18**).
4. Run the gel at constant 250 volts at a 65 °C incubator oven.
5. Place the gel on Whatman paper, put in a nylon bag, seal, and expose in a cassette with a phospho-imager screen for at least 1 h followed by image scan with a Typhoon phospho-imager. One representative figure is shown (Fig. 1c).

3.1.10 Quantitation of the 3′-Phosphatase Activity

The intensity of the substrate and the released phosphate bands are quantitated using the ImageJ software (NIH) and the activity is expressed as "% product released" using the formula: *(Intensity of released phosphate band/Sum of intensity of substrate and released phosphate bands)*100.*

3.2 PNKP-Mediated Single-Strand Break Repair Activity Assay In Vitro

3.2.1 Preparation of Substrate by Annealing the Oligos

1. Add the reaction components in the following order (*see* **Note 19**):

   1.25 μL 3′P-25 oligo (100 pmole/μL working stock, total = 125 pmoles)

   1.50 μL 5′P-25 oligo (100 pmole/μL working stock, total = 150 pmoles)

   1.25 μL G-51 Oligo (100 pmole/μL working stock, total = 125 pmoles)

   0.25 μL 5 M NaCl

   0.625 μL 1 M Tris–HCl pH 8.0

   20.125 μL dH2O (up to 25 μL)
2. Keep for 5 min in boiling water in a beaker.
3. Leave on the bench to cool down slowly at room temperature overnight.
4. Keep frozen at −20 °C.

3.2.2 Measuring the Concentration of Pure Proteins (See Note 20)

Measure the concentration of the pure proteins or nuclear extracts by the Bradford method, using BSA as a standard (the protein measured should be in the linear range of the standard curve).

3.2.3 Assay of Single-Strand Break Repair Activity In Vitro (a Schematic Representation Is Shown in Fig. 2a)

1. Prepare diluted stocks of PNKP (25 fmol/μL), DNA polymerase β (25 fmol/μL), and ligase IIIα (100–200 fmol/μL) in 1× total repair assay buffer for reaction. Similarly, if using a nuclear extract, dilute to 500 ng/μL (*see* **Note 21**).
2. Prepare reaction mixture by mixing the following reagents on ice:

   1 μL of double-stranded oligo substrate (stock = 5 pmole/μL; total = 5 pmoles/reaction)

2 μL 10× basic buffer (use +/− KCl-containing buffer, final salt concentration ≤ 100 mM after adding the nuclear extract or purified proteins)

2 μL acetylated BSA (10 mg/ml stock, final = 20 μg/reaction)

2 μL ATP (10 mM stock, final = 1 mM/reaction)

0.2 μL cold dCTP (freshly prepared 0.5 mM stock in 1× Tris–EDTA buffer, final = 5 μM) (*see* **Note 22**)

0.2 μL cold dATP (freshly prepared 5 mM stock in 1× Tris–EDTA buffer, final = 50 μM)

0.2 μL cold dTTP (freshly prepared 5 mM stock in 1× Tris–EDTA buffer, final = 50 μM)

0.2 μL cold dGTP (freshly prepared 5 mM stock in 1× Tris–EDTA buffer, final 50 μM)

μL dCTP [α-^{32}P] (stock is 3.3 pmole/μL, final = 0.33 pmole/reaction) (*see* **Note 23**),

H_2O—up to 20 μL (after taking into account the pure proteins/nuclear extract volume as detailed below).

3. Distribute the master mix (all reagents except the protein) in the reaction tubes.
4. Start the reaction by mixing the protein or nuclear extract on the walls of the reaction tube and spinning briefly. (*see* **Note 24**).
5. Incubate the reaction mixture at 30 °C (*see* **Note 25**) for 45 min. However, the incubation time should be adjusted according to the activity of the pure protein/nuclear extract. The time can be increased up to 1 h if feeble activity is observed or can be reduced to 30 min if robust activity is seen.
6. Stop the reaction by adding an equal volume of 2× loading dye and heating at 95 °C for 5 min. Load a half volume of samples on 20% urea-PAGE (pre-run the gel for 15 min at a 65 °C incubator oven) (*see* **Note 18**).
7. Run the gels (*see* Subheading 6 of 3.1 for gel preparation) at constant 250 volts until the dye front leaves the gel (at a 65 °C incubator oven) (*see* **Note 26**).
8. Place the gel on Whatman paper, put in a nylon bag, seal, and expose in a cassette with a phospho-imager screen for at least 1 h followed by an image scan with a Typhoon phospho-imager. One representative figure is shown (Fig. 2b).

3.2.4 Quantitation of the Total Repair Activity

The intensity of the lower unligated and the upper ligated bands were quantitated using the ImageJ software (NIH) and the total repair activity is expressed as % of the ligated product using the formula: *(Intensity of ligated band/Sum of intensity of un-ligated and ligated bands) *100.*

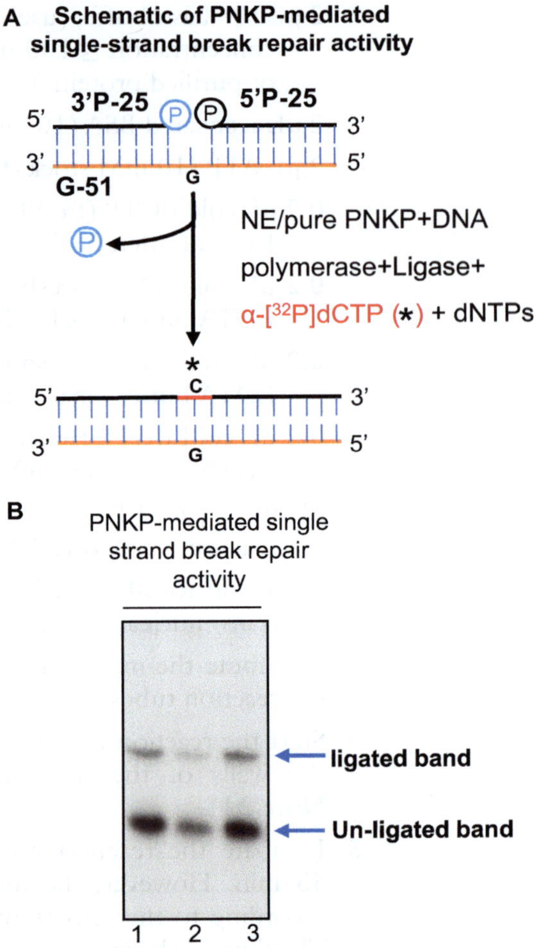

Fig. 2 PNKP-mediated single-strand break repair (SSBR) activity. (**a**) Schematic illustration of the steps involved in the PNKP-mediated SSBR activity. (**b**) Ln 1: SSBR using purified proteins (25 fmol PNKP, 25 fmol DNA polymerase β, 100 fmol DNA ligase IIIα), nuclear extract (500 ng) from WT mouse brain tissue (ln 2), and HEK293 cell (ln 3). In each case, the lower 25-nt band indicates DNA polymerase-mediated radiolabeled dCTP-incorporated unligated band following PNKP-mediated 3′-phosphate release and the upper 51-nt band indicates the full-length ligated repaired product

4 Notes

1. HPLC purification is required for all these oligos to avoid contamination with undesired sequences.
2. If a researcher is interested to investigate the PNKP-independent repair where 3′-phosphate removal is not necessary, he/she may use the following oligo instead of 3′P-25.

3′OH-25 (3′-OH containing 25-nucleotide oligo): 5′-GCT TAG CTT GGA ATC GTA TCA TGT A-3′

Furthermore, if only ligase activity needs to be assayed, the following oligo should be used instead of 3′P-25 or 3′ OH-25.

3′OH-26 (3′-OH containing 26-nucleotide oligo without AP site): 5′-GCT TAG CTT GGA ATC GTA TCA TGT AC-3′

3. ATP, [γ-^{32}P]- stock: 250 μCi in 25 μL, so 10 μCi/μL. Recommended use in each reaction is 3 μL = 30 μCi; 30 μCi/25 pmole oligos. This recipe is followed if the radioactive is within its first half-life period. The amount used should be doubled once the radioactive enters its second half-life.
4. This is the optimum recommended reaction time. If the labeling reaction does not produce good yield, then the incubation time may be increased up to 45 min.
5. The temperature of the water should be at least between 85 and 95 °C. Slow annealing overnight at room temperature works best. Alternatively, the reaction should be carried out for a minimum of 6 h for effective annealing.
6. Before starting the ligation reaction, keep 2.5 μL from the annealed mixture in a 1.5 mL microcentrifuge tube to load on a 20% urea-PAGE later to check efficient labeling-annealing. This can be stored frozen at this stage.
7. Before starting the Udg–Fpg digestion, keep 2.5 μL of the ligation mixture in a 1.5 mL microcentrifuge tube to load on a gel later to check ligation efficiency. This can be frozen at this stage.
8. Udg and Fpg are very robust *E. coli* enzymes, and thus, they may lead to extensive digestion and degradation of the product. In that case, the incubation time could be reduced to 45 min.
9. We recommend distributing the reaction mixture into two P-6 spin columns for efficient purification. So 40 μL of the reaction mixture can be dispensed into each P-6 column after buffer exchange. Also, we recommend nuclease-free dH_2O for elution. Hence, buffer should be exchanged from Tris to dH_2O in the spin columns.
10. Before freezing the substrates, 1 μL of the purified substrate should be saved in a 1.5 mL microcentrifuge tube for checking the purity of the substrate.
11. Avoid using additional dH_2O to make up the volume to 16 mL because volume increases in the presence of urea.
12. Urea deposits into the lanes making it difficult to load the gel lanes with samples. It is thus recommended to flush the lanes of

the gel with a syringe filled with 1× TBE. Also, the earlier saved unligated and ligated products can also be run in the same gel to troubleshoot which reaction of this multistep reaction was inefficient, if any.

13. Bromophenol blue runs between 5 and 25 bp on 20% urea-PAGE (this should be taken into account while running the gel).
14. PNKP has robust 3′-phosphatase activity. Thus, 25–50 fmol (1–2 ng) of the purified protein shows significant 3′--phosphatase activity. Using a higher dose of PNKP makes the reaction out of the linear range, and thus, it becomes difficult to compare the activities since almost all the substrate molecules are cleaved. Also, depending on the salt concentration in the protein preparation or the nuclear extract, −/+ KCl containing 1× buffer can be used for dilution. It is essential to maintain a uniform salt concentration in all the reactions for reliable comparison among the samples.
15. Five picomole is the starting amount of the cold substrate. If the reaction goes out of linear range due to robust PNKP activity, the amount of the cold substrate could be increased up to 7.5 picomoles. On the other hand, if the phosphate release is modest due to weak PNKP activity, the cold substrate should be reduced to 2.5 picomoles.
16. Purified protein/nuclear extract and the radiolabeled substrate amounts should be adjusted depending on the initial result to get optimum PNKP activity in the linear reaction range. This is not a strict recipe.
17. A recommended practice would be to initially distribute the protein/nuclear extract at the bottom of the tube and then add the master mix on the wall of the reaction tube and briefly spin to mix the components together. In this way, it will be possible to maintain a uniform reaction start time.
18. Excess free polyacrylamide goes out of the gel in this pre-run period and, thus, helps an efficient urea-PAGE run.
19. Mix the oligos in a smaller volume, for efficient annealing.
20. Nuclear extract should be freshly prepared from cells or from frozen nuclear pellet (saved in dry ice first for rapid freezing and then frozen in −80 °C). Otherwise, ligase activity will be lost and the ligation reaction does not work efficiently.
21. If the nuclear extract needs to be diluted for the assay, it should be diluted in minimal nuclear extraction buffer (20 mM HEPES pH 7.6, 0.15 M NaCl, 2 mM DTT, 1 mM EDTA, 0.1 mM EGTA, 10% glycerol) without protease inhibitor cocktail tablets to maintain the ligase activity. In that case, the total repair reaction buffer needs to be adjusted adequately to have a

final salt concentration ≤ 100 mM and 5% glycerol in the reaction. Otherwise, the repair reactions are not efficient.

22. For ligation product formation in the linear range, nonradioactive (cold) dCTP needs to be added to the reaction mix but in 1/tenth concentration of the other dNTPs so that there is possibility of enough incorporation of the radioactive dCTP.
23. If radioactive dCTP enters the second half-life, use 0.2 μL (= 0.66 pmol)/reaction. Also, the radioactivity must be added last to the master mix.
24. Moreover, 1 to 4 μL (0.5–2 μg) of nuclear extract can be added depending on the activity. The starting amount can be 1 μg.
25. The total repair reaction is carried out at 30 °C for maintaining efficient ligase activity throughout the incubation period. PNKP and DNA polymerases are optimally active at 37 °C. However, to take into consideration the ligase activity, the optimal incubation temperature is fixed at 30 °C.
26. Two radiolabeled bands are expected: 25 nt lower band (incorporated dCTP-α-^{32}P, but not ligated fragment) and 51 nt upper band (the ligated total repair product). Those two radioactive bands must be well separated to be clearly visible and can be quantitated for PNKP-mediated repair assay.

Acknowledgements

This work was supported by the National Institute of Health Grant 2R01 NS073976 to TH.

References

1. Hazra TK, Das A, Das S, Choudhury S, Kow YW, Roy R (2007) Oxidative DNA damage repair in mammalian cells: a new perspective. DNA Repair (Amst) 6(4):470–480. https://doi.org/10.1016/j.dnarep.2006.10.011
2. Vermeij WP, Hoeijmakers JHJ, Pothof J (2014) Aging: not all DNA damage is equal. Curr Opin Genet Dev 26:124–130. https://doi.org/10.1016/j.gde.2014.06.006
3. Tubbs A, Nussenzweig A (2017) Endogenous DNA damage as a source of genomic instability in cancer. Cell 168(4):644–656. https://doi.org/10.1016/j.cell.2017.01.002
4. Gates KS (2009) An overview of chemical processes that damage cellular DNA: spontaneous hydrolysis, alkylation, and reactions with radicals. Chem Res Toxicol 22(11):1747–1760. https://doi.org/10.1021/tx900242k
5. Wiederhold L, Leppard JB, Kedar P, Karimi-Busheri F et al (2004) AP endonuclease-independent DNA base excision repair in human cells. Mol Cell 15(2):209–220. https://doi.org/10.1016/j.molcel.2004.06.003
6. Chang HYH, Pannunzio NR, Adachi N, Lieber MR (2020) Non-homologous DNA end joining and alternative pathways to double-strand break repair. Nat Rev Mol Cell Biol 18:495–506
7. Mullins EA, Rodriguez AA, Bradley NP, Eichman BF (2019) Emerging roles of DNA glycosylases and the base excision repair pathway. Trends Biochem Sci 44(9):765–781. https://doi.org/10.1016/j.tibs.2019.04.006
8. Wallace SS (2014) Base excision repair: a critical player in many games. DNA Repair (Amst)

19:14–26. https://doi.org/10.1016/j.dnarep.2014.03.030

9. Jilani A, Ramotar D, Slack C, Ong C et al (1999) Molecular cloning of the human gene, PNKP, encoding a polynucleotide kinase 3′--phosphatase and evidence for its role in repair of DNA strand breaks caused by oxidative damage. J Biol Chem 274(34):24176–24186. https://doi.org/10.1074/jbc.274.34.24176
10. Karimi-Busheri F, Daly G, Robins P, Canas B et al (1999) Molecular characterization of a human DNA kinase. J Biol Chem 274(34): 24187–24194. https://doi.org/10.1074/jbc.274.34.24187
11. Habraken Y, Verly WG (1988) Further purification and characterization of the DNA 3′--phosphatase from rat-liver chromatin which is also a polynucleotide 5′-hydroxyl kinase. Eur J Biochem 171(1–2):59–66. https://doi.org/10.1111/j.1432-1033.1988.tb13758.x
12. Mandal SM, Hegde ML, Chatterjee A, Hegde PM et al (2012) Role of human DNA glycosylase Nei-like 2 (NEIL2) and single strand break repair protein polynucleotide kinase 3′--phosphatase in maintenance of mitochondrial genome. J Biol Chem 287(4):2819–2829. https://doi.org/10.1074/jbc.M111.272179
13. Chatterjee A, Saha S, Chakraborty A, Silva-Fernandes A et al (2015) The role of the mammalian DNA end-processing enzyme polynucleotide kinase 3′-phosphatase in spinocerebellar ataxia type 3 pathogenesis. PLoS Genet 11(1):e1004749. https://doi.org/10.1371/journal.pgen.1004749
14. Whitehouse CJ, Taylor RM, Thistlethwaite A, Zhang H et al (2001) XRCC1 stimulates human polynucleotide kinase activity at damaged DNA termini and accelerates DNA single-strand break repair. Cell 104(1):107–117. https://doi.org/10.1016/s0092-8674(01)00195-7
15. Chakraborty A, Tapryal N, Venkova T, Horikoshi N et al (2016) Classical non-homologous end-joining pathway utilizes nascent RNA for error-free double-strand break repair of transcribed genes. Nat Commun 7:13049. https://doi.org/10.1038/ncomms13049
16. Neil AJ, Belotserkovskii BP, Hanawalt PC (2012) Transcription blockage by bulky end termini at single-strand breaks in the DNA template: differential effects of 5′ and 3′ adducts. Biochemistry 51(44):8964–8970. https://doi.org/10.1021/bi301240y
17. Chakraborty A, Tapryal N, Venkova T, Mitra J et al (2020) Deficiency in classical nonhomologous end-joining-mediated repair of transcribed genes is linked to SCA3 pathogenesis. Proc Natl Acad Sci U S A 117(14):8154–8165. https://doi.org/10.1073/pnas.1917280117
18. Wallace SS, Murphy DL, Sweasy JB (2012) Base excision repair and cancer. Cancer Lett 327(1–2):73–89. https://doi.org/10.1016/j.canlet.2011.12.038
19. Hegde ML, Mantha AK, Hazra TK, Bhakat KK et al (2012) Oxidative genome damage and its repair: implications in aging and neurodegenerative diseases. Mech Ageing Dev 133(4): 157–168. https://doi.org/10.1016/j.mad.2012.01.005
20. Vijg J, Suh Y (2013) Genome instability and aging. Annu Rev Physiol 75:645–668. https://doi.org/10.1146/annurev-physiol-030212-183715
21. Rulten SL, Caldecott KW (2013) DNA strand break repair and neurodegeneration. DNA Repair (Amst) 12(8):558–567. https://doi.org/10.1016/j.dnarep.2013.04.008
22. Gao R, Liu Y, Silva-Fernandes A, Fang X et al (2015) Inactivation of PNKP by mutant ATXN3 triggers apoptosis by activating the DNA damage-response pathway in SCA3. PLoS Genet 11(1):e1004834. https://doi.org/10.1371/journal.pgen.1004834
23. Gao R, Chakraborty A, Geater C, Pradhan S et al (2019) Mutant huntingtin impairs PNKP and ATXN3, disrupting DNA repair and transcription. elife 8:e42988. https://doi.org/10.7554/eLife.42988
24. Dey S, Maiti AK, Hegde ML, Hegde PM et al (2012) Increased risk of lung cancer associated with a functionally impaired polymorphic variant of the human DNA glycosylase NEIL2. DNA Repair (Amst) 11(6):570–578. https://doi.org/10.1016/j.dnarep.2012.03.005

Chapter 4

Generation of Recombinant Nucleosomes Containing Site-Specific DNA Damage

Benjamin J. Ryan, Tyler M. Weaver, Jonah J. Spencer, and Bret D. Freudenthal

Abstract

Eukaryotic DNA exists in chromatin, where the genomic DNA is packaged into a fundamental repeating unit known as the nucleosome. In this chromatin environment, our genomic DNA is constantly under attack by exogenous and endogenous stressors that can lead to DNA damage. Importantly, this DNA damage must be repaired to prevent the accumulation of mutations and ensure normal cellular function. To date, most in-depth biochemical studies of DNA repair proteins have been performed in the context of free duplex DNA. However, chromatin can serve as a barrier that DNA repair enzymes must navigate in order find, access, and process DNA damage in the cell. To facilitate future studies of DNA repair in chromatin, we describe a protocol for generating nucleosome containing site-specific DNA damage that can be utilized for a variety of in vitro applications. This protocol describes several key steps including how to generate damaged DNA oligonucleotides, the expression and purification of recombinant histones, the refolding of histone complexes, and the reconstitution of nucleosomes containing site-specific DNA damage. These methods will enable researchers to generate nucleosomes containing site-specific DNA damage for extensive biochemical and structural studies of DNA repair in the nucleosome.

Key words DNA damage, Base excision repair, Nucleosomes, Chromatin

1 Introduction

DNA is constantly exposed to endogenous and exogenous stressors that lead to the accumulation of DNA damage. Oxidative stress is a primary source of DNA damage that contributes approximately 10^4 lesions per cell, per day [1]. Well over 100 different forms of oxidative DNA base lesions exist including the prevalent examples of 8-oxo-7,8-dihydro-2′-deoxyguanosine (8oxoG), 2,6-diamino-4-hydroxy-5-formamidopyrimidine (FapyG), 8-oxox-7,8-dihydro-2′-deoxyadenosine (8oxoA), thymine glycol, and 5-hydroxy-

Benjamin J. Ryan and Tyler M. Weaver contributed equally to this work

Kishor K. Bhakat and Tapas K. Hazra (eds.), *Base Excision Repair Pathway: Methods and Protocols*, Methods in Molecular Biology, vol. 2701, https://doi.org/10.1007/978-1-0716-3373-1_4,

2′-deoxycytidine (OH5C). In general, oxidative damage results in changes to the structure of the DNA base. These alterations increase the likelihood that the damaged base will form noncanonical base pairing that can result in the accumulation of genomic transversions and/or transitions [2, 3]. Ultimately, these oxidative DNA lesions must be repaired to prevent the accumulation of mutations that can lead to genomic instability and several different disease states [4, 5].

The base excision repair (BER) pathway is responsible for the repair of most oxidative DNA base damage, requiring efficient recognition of oxidative DNA damage [6, 7]. We direct readers to other excellent reviews of BER for more details about the pathway [8–10]. Briefly, BER is initiated by one of several different DNA glycosylases, which recognize damaged DNA bases [7, 11]. The DNA glycosylase then excises the damaged DNA base resulting in the formation of a baseless sugar moiety known as an apurinic/apyrimidinic site (AP site). The resulting AP site is then processed by AP endonuclease 1 (APE1), resulting in a 5′-nick in the DNA backbone. The 5′-nicked substrate is then processed by two distinct enzymatic activities of polymerase (Pol) β. The lyase activity of DNA Pol β removes the 5′-nick generating a one-nucleotide gapped DNA substrate. Pol β then utilizes its DNA synthesis activity to fill the gap with a non-damaged nucleotide, generating a 3′-nick. After incorporation of the non-damaged nucleotide, the XRCC1-DNA ligase III complex seals the 3′-nick, ultimately restoring the coding potential of the DNA.

Mechanistic insight into the function of BER enzymes has largely come from in vitro studies using free duplex DNA (see reference [12] and associated references). However, cellular DNA often exists as chromatin, where genomic DNA is packaged into a fundamental repeating unit known as the nucleosome [13, 14]. The nucleosome is composed of an octameric core of histone proteins, H2A, H2B, H3, and H4, that wraps ~147 bp of DNA [15]. The robust interaction between histone proteins and the nucleosomal DNA can act as a significant barrier to BER proteins, which must be overcome to effectively complete repair of the damaged DNA. We direct the readers to several excellent reviews for further information on how nucleosome structure regulates BER enzymes [16–22]. Over the past 20 years, initial progress has been made to define how BER proteins repair DNA damage in the context of the nucleosome [23–66]. However, rapid progress has been hampered by the lack of comprehensive step-by-step protocols for generating nucleosomes containing site-specific DNA damage.

To overcome this challenge, we present a step-by-step protocol for the reconstitution of recombinant NCPs containing site-specific DNA damage. This protocol was adapted from the landmark method from Dyer and Luger et al. described over 15 years ago, as well as the many labs who have made significant progress on methods for generating damaged DNA substrates for the in vitro reconstitution of nucleosomes [39, 43, 64, 65, 67–70]. Initially, we describe a ligation-based method for generating a 147-bp Widom 601 strong positioning DNA substrate containing site-specific DNA damage [71]. We then describe the expression and purification of human histone proteins from *E. coli* and the salt dialysis method for refolding recombinant histone proteins into histone H2A/H2B dimers and histone H3/H4 tetramers. Finally, we describe a salt dialysis method for reconstituting NCPs using the Widom 601 strong positioning DNA containing site-specific DNA damage and how to purify these NCP substrates for downstream in vitro biochemical or structural biology experiments. Importantly, the utility of this step-by-step protocol is highlighted by recent work from our lab using this method to generate NCPs containing THF, enabling us to determine the structural basis for APE1 processing DNA damage in the NCP [72].

2 Materials

2.1 Instruments and Equipment

1. PCR thermal cycler.
2. Thermo Fisher Owl A-series horizontal gel system or equivalent.
3. Bioreactor or shaking incubator.
4. QSonica Q500 Sonic dismembrator or equivalent.
5. Stir plate.
6. Gravity-flow column with filter unit.
7. 3,000-Da MWCO Dialysis Membrane.
8. 10,000-Da MWCO Centrifugal Filter Unit.
9. AKTA pure protein purification system or equivalent.
10. Superdex 200 Increase 10/300 GL or equivalent.
11. Biocomp Gradient Master 108 or equivalent.
12. Beckman Coulter 38.5 mL Open-Top Ultra-Clear Centrifuge tube or equivalent.
13. SW 32 Ti Swinging-Bucket Rotor or equivalent.
14. Optima XE-100 ultracentrifuge or equivalent.

2.2 Reagents

1. Project-specific oligonucleotides (design described in Subheading 3.1.1).
2. Agarose.
3. New England Biolabs T4 DNA ligase or equivalent.
4. New England Biolabs 10× T4 DNA ligase buffer or equivalent.

5. New England Biolabs 6× gel loading dye or equivalent.
6. Ethidium bromide or equivalent DNA visualizing reagent.
7. pET3a-histone H2A plasmid.
8. pET3a-histone H2B plasmid.
9. pET3a-histone H3 plasmid.
10. pET3a-histone H4 plasmid.
11. New England Biolabs BL21 (DE3) pLysS competent cells or equivalent.
12. Novagen Rosetta™ 2(DE3) pLysS competent cells or equivalent.
13. Luria broth (LB): Casein Digest Peptone 10 g/L, sodium chloride 10 g/L, yeast extract 5 g/L.
14. LB agar plates containing 100 μg/mL ampicillin.
15. LB agar plates containing 100 μg/mL ampicillin and 25 μg/mL chloramphenicol.
16. Centrum multivitamin or equivalent, 1 tablet dissolved in 50 mL ddH_2O.
17. Triton X-100.
18. Dimethyl sulfoxide (DMSO).
19. Cytiva Q-Sepharose Fast Flow anion-exchange chromatography resin or equivalent.
20. Cytiva SP-Sepharose Fast Flow cation-exchange chromatography resin or equivalent.

2.3 Buffers, Solutions, and Media

1. 1× TS: 10 mM Tris–HCl (pH 7.5) and 10 mM NaCl.
2. 1× TBE: 90 mM Tris base, 90 mM boric acid, and 2 mM EDTA (pH 8.3).
3. 10% denaturing urea-PAGE gel solution: 10% 29:1 acrylamide/bis-acrylamide, 8 M urea, and 1× TBE.
4. 1× TE: 10 mM Tris–HCl (pH 7.5) and 1 mM EDTA.
5. Gel extraction buffer: 200 mM NaCl, 1 mM EDTA.
6. 10× M9 minimal media (pH 7.2): 335 mM $Na_2HPO_4{\bullet}7H_2O$, 220 mM KH_2PO_4, 10 mM NaCl, 20 mM NH_4Cl, supplemented with 0.5% glucose, 2 mM $MgSO_4$, 0.2 mM $CaCl_2$, and 1.0% vitamin solution (*see* **Note 1**).
7. 20% glucose solution (w/v).
8. 1 M $MgSO_4$.
9. 1 M $CaCl_2$.
10. 100 mg/mL ampicillin.
11. 25 mg/mL chloramphenicol.

12. 1 M isopropyl β- d-1-thiogalactopyranoside (IPTG).
13. Histone lysis buffer: 50 mM Tris–HCl pH 7.5, 100 mM NaCl, 1 mM benzamidine, 5 mM beta mercaptoethanol, 1 mM EDTA.
14. Guanidinium buffer: 20 mM Tris–HCl pH 7.5, 6 M guanidinium hydrochloride, 10 mM DTT.
15. 8 M urea.
16. 8 M urea, 100 mM NaCl.
17. 8 M urea, 200 mM NaCl.
18. 8 M urea, 300 mM NaCl.
19. 8 M urea, 400 mM NaCl.
20. 8 M urea, 500 mM NaCl.
21. 8 M urea, 600 mM NaCl.
22. 8 M urea, 700 mM NaCl.
23. High-salt reconstitution buffer: 20 mM Tris–HCl pH 7.5, 2 M NaCl, 1 mM EDTA, 0.5 mM benzamidine, 1 mM DTT (added fresh).
24. No-salt reconstitution buffer: 20 mM Tris–HCl pH 7.5, 1 mM EDTA, 0.5 mM benzamidine, 1 mM DTT (added fresh).
25. 10% sucrose in 1× TE.
26. 40% sucrose in 1× TE.
27. 5% native PAGE gel: 5% 59:1 acrylamide/bis-acrylamide and 0.2× TBE.

3 Methods

3.1 Generation and Purification of Damaged DNA Substrates

Several methods exist for incorporating site-specific DNA damage into the 601 strong positioning DNA sequence. These methods include oligonucleotide synthesis strategies [24, 43, 57, 64], PCR amplification [37, 65], nickase-based methods [39, 40, 73], and ligation-based methods [66, 70, 74, 75]. We point the readers to an excellent and comprehensive review on the different methods for generating site-specific DNA damage for nucleosome reconstitution [76]. Here, we describe a T4-DNA ligase-based method for site-specific incorporation of the apurinic/apyrimidinic site analogue tetrahydrofuran (THF) into the Widom 601 strong positioning DNA sequence containing a 6-carboxyfluorescein (6-FAM) label [71]. This modular methodology can be performed at large scales in a cost-effective manner, ultimately providing an optimal way of generating damaged DNA for nucleosome reconstitutions (for additional benefits of the T4 DNA ligase-based system, see Subheading 3.1.1).

3.1.1 Design of Damaged DNA Substrates

The T4 DNA ligase-based method described here allows for the site-specific incorporation of DNA damage within a 147-bp Widom 601 strong positioning DNA sequence. The first step is to design oligonucleotide (oligo) fragments that correspond to each strand of the 601 DNA sequence. A diagram of the oligo design can be found in Fig. 1a. For simplicity, we refer to these two DNA strands as the J-strand (non-damaged strand) and the I-strand (damaged strand; *see* **Note 2**), which is modeled after the nomenclature used from the first crystal structure of a nucleosome containing the Widom 601 sequence [77]. The J-strand is split into two oligonucleotides that consist of 74 bp (J-strand oligo 1) and 73 bp (J-strand oligo 2) of DNA. The I-strand is then split into three oligonucleotides that consist of 50 bp (I-strand oligo 1), 50 bp (I-strand oligo 2), and 47 bp (I-strand oligo 3) of DNA. The length of these component oligos are kept under 100 bps to enable easy access to commonly used oligo synthesis companies, such as Integrated DNA Technologies. The sequences for each of the five oligos are as follows:

- J-strand oligo 1:

  5′ ATCGGATGTATATATCTGACACGTGCCTGGAGACTA GGGAGTAATCCCCTTGGCGGTTAAAACGCGGGGGACAG**3′**

- J-strand oligo 2:

  5′/Phos/ CGCGTACGTGCGTTTAAGCGGTGCTAGAGC TGTCTACGACCAATTGAGCGGCCTCGGCACCGGGATTCT CGAT**3′**

- I-strand oligo 1:

  5′[FAM] ATCGAGAATCCCGGTGCCGAGGCCGCTCAA TTGGTCGTAGACAGCTC**3′**

- I-strand oligo 2:

  5′/Phos/ TAGCACCGCTTAAACGCACGTACGCGCTGT CCCCCGCGTTTTAACCGCCA**3′**

- I-strand oligo 3:

  5′/Phos/ AGGGGATTACTCCCTAGTCTCCAGGCACGT GTCAGATATAT**/THF/**CATCCGAT**3′**

The oligo design is such that each oligo has at least 25 bp of complimentary DNA between every J- and I-strand DNA oligo. Importantly, J-strand oligo 2, I-strand oligo 2, and I-strand oligo 3 also all contain 5′-phosphate groups at the end of the DNA that is required for ligation (Fig. 1a). When mixed at equimolar ratios and annealed, the resulting 147-bp DNA sequence contains three ligation spots with a 5′-phosphate juxtaposed to a 3′-OH. The three nicks can then be sealed by T4 DNA ligase generating the full-

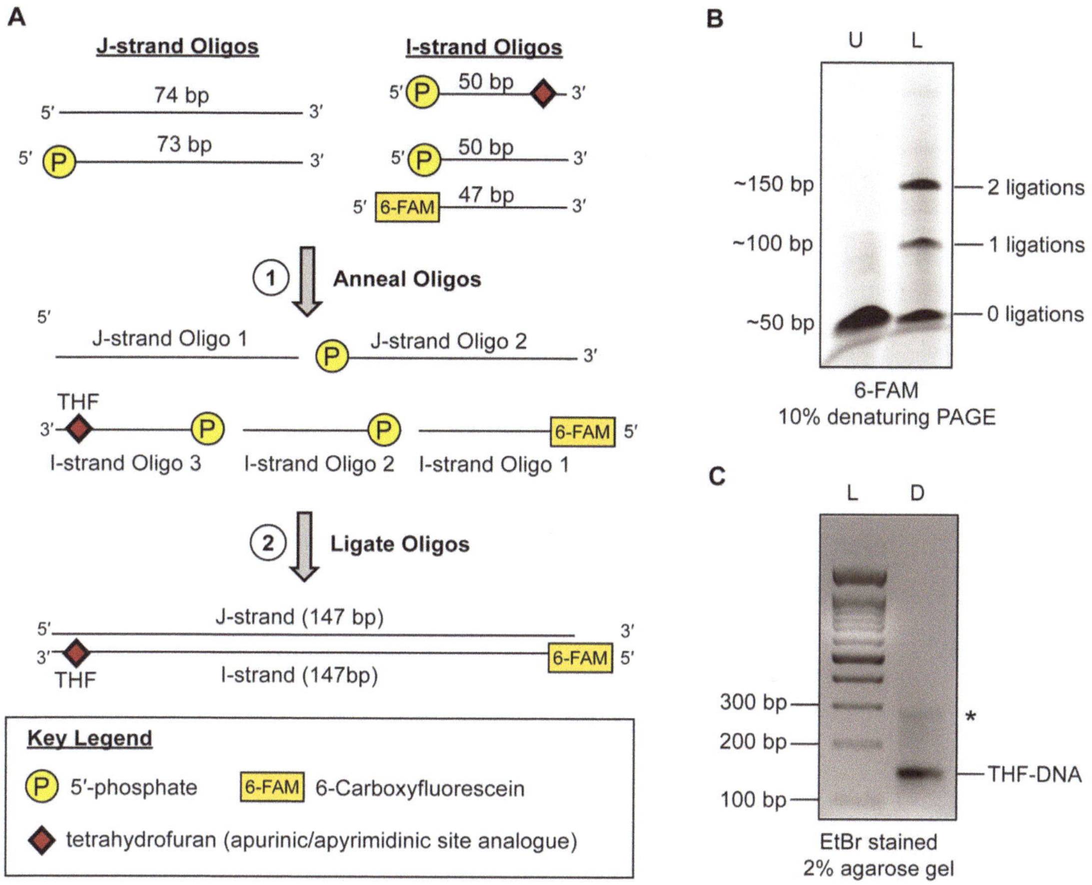

Fig. 1 Generation and purification of damaged DNA substrates using the ligation-based method. (**a**) Diagram for the individual oligo generation of damaged DNA substrates. (**b**) A representative 10% denaturing PAGE gel for the ligation reaction. Lane "U" is the starting unligated reaction and lane "L" is the ligated reaction. The DNA was visualized using the 6-FAM label. (**c**) A representative 2% agarose gel of the purified and annealed damaged DNA substrate. Lane "L" is a 100-bp DNA ladder and lane "D" is the purified and annealed damaged DNA. The DNA was visualized with ethidium bromide. The * denotes a minor contamination that is commonly seen after purification of the ligated DNA

length 147-bp DNA sequence containing DNA damage. In the example shown in Fig. 1a, the THF is the ninth position from the end of the DNA on the I-strand oligo 3. However, this ligation-based method is built to enable movement of the THF to any location on the I-strand or J-strand of the DNA, with only minor exceptions (*see* **Note 3**).

The benefits of using the T4-DNA ligase-based approach are numerous. First, the relatively small oligo size means purchasing numerous oligos containing DNA damage is cost-effective, which is not the case with full-length 147-bp DNA oligos containing DNA damage. Next, the ligation reactions and purification scheme can be performed in large-scale batches enabling the generation of large quantities of damaged DNA. Next, the modularity of the system means that moving the DNA damage to different locations in the 601 DNA sequence requires simply ordering a single additional oligo. Finally, the modularity also allows for the incorporation of various fluorescent tags, protein attachment modifications, and additional DNA damage types. This ultimately means that nucleosome substrates containing DNA damage can be generated for a variety of BER substrates and different in vitro experimental techniques.

3.1.2 Ligation of Damaged DNA Substrates

1. Dissolve oligos with 1× TS to a final concentration of 1 nmol/μL.
2. Mix the oligos in a 1:1 stoichiometry at a concentration of 1 nmol/μL (*see* **Note 4**).
3. Anneal the oligos using a thermal cycler by heating to 95 °C before cooling to 4 °C at a rate of 0.1 °C/s.
4. Dilute the annealed oligos to a final volume of 190 μL with H_2O and 10× T4 DNA ligase buffer (NEB).
5. Ligate the annealed DNA oligos by the addition of 10 μL of T4 DNA ligase (NEB) for a final reaction volume of 200 μL.
6. Incubate the ligation reactions at 25 °C for 24–48 h.
7. Run a 10% denaturing urea-PAGE gel (10% 29:1 acrylamide/bis acrylamide, 8 M urea, and 1× TBE) to monitor the progress of the ligation reaction (Fig. 1b; *see* **Note 5**).

3.1.3 Purification of Damaged DNA Substrates

1. Pour a 10% denaturing urea-PAGE gel (10% 29:1 acrylamide/bis-acrylamide, 8 M urea, and 1× TBE).
2. Load the denaturing urea-PAGE gel with an equal volume of ligated DNA mixed with 6× gel loading dye.
3. Run the denaturing urea-PAGE gel at 125 V for 1 h.
4. Visualize the ligated DNA via the fluorescent tag, or by staining with ethidium bromide. An example gel of the ligation reaction before purification can be found in Fig. 1b.
5. Excise the ligated 147-bp DNA from the denaturing urea-PAGE gel and cut the excised fragments into small pieces.
6. Extract the ligated 147-bp DNA from the gel fragments by soaking in 20 mL of DNA extraction buffer for 24 h. Repeat the extraction at least twice to ensure collection of all the damaged DNA.
7. Buffer exchange the extracted DNA with 1× TE at least five times using a centrifugal filter unit (10,000-Da MWCO).
8. Concentrate the extracted DNA in the centrifugal filter unit (10,000-Da MWCO) to a final concentration of 10 μM (*see* **Note 6**).

3.1.4 Annealing and Storage of Damaged DNA Substrates

1. Anneal the purified 147-bp oligos using a thermal cycler by heating to 95 °C before cooling to 4 °C at a rate of 0.1 °C/s.
2. Determine the concentration of resuspended 147-bp DNA spectroscopically using the theoretical extinction coefficient for the respective 147-bp DNA sequence. Typical yields are ~10–15% of starting material.
3. Aliquot the 147-bp damaged DNA in 5 nmol aliquots and store frozen at −20 °C indefinitely (*see* **Notes 7** and **8**). An example of a purified 147-bp damaged DNA can be found in Fig. 1c.

3.2 Histone Expression and Purification

Generating recombinant nucleosomes requires milligram quantities of histone H2A, H2B, H3, and H4 proteins. To generate these quantities, recombinant histones are expressed in *E. coli* and purified extensively using a protocol modified from Dyer and Luger et al. [67]. The expression, purification, and long-term storage process for each individual histone is similar, with only minor differences highlighted below.

3.2.1 Histone Plasmid Generation

The inserts corresponding to the coding region of human histone H2A (UniProt identifier: P0C0S8), H2B (UniProt identifier: P62807), H3 C110A (UniProt identifier Q71DI3), and H4 (Uniprot identifier: P62805) genes were cloned into a tagless pET3a expression vector.

3.2.2 Histone Expression

1. Perform a fresh transformation using NEB BL21 (DE3) pLysS competent cells for histone H2A, H3, and H4 growths. For histone H2B, perform a fresh transformation using Novagen Rosetta™ 2 (DE3) pLysS competent cells. Perform the transformation following the manufacturer's protocol for the respective cells.
2. Plate the transformation on an LB agar plate with antibiotic specific for that histone:
   - H2A: 100 μg/mL ampicillin
   - H2B: 100 μg/mL ampicillin 25 μg/mL chloramphenicol
   - H3: 100 μg/mL ampicillin
   - H4: 100 μg/mL ampicillin
3. Take a streak of colonies from the transformation and inoculate 35 mL of LB containing antibiotics (50 ug/mL ampicillin and/or 12.5 ug/mL chloramphenicol).
4. Grow the inoculated culture in a shaking incubator at 37 °C until turbid, which generally takes ~3–4 h.
5. Inoculate 100 mL of M9 minimal media culture containing antibiotics (50 ug/mL ampicillin and/or 12.5 ug/mL chloramphenicol) with 5 mL of the turbid culture. We typically

inoculate a single 100 mL M9 minimal media culture per five bioreactor bottles.

6. Grow the inoculated culture in a shaking incubator at 37 °C for 12–16 h.
7. Inoculate each 2-L bioreactor bottle containing 1.5 L of M9 minimal media and antibiotics (50 ug/mL ampicillin and/or 12.5 ug/mL chloramphenicol) with 15 mL of overnight culture that was inoculated and grown in **steps 5** and **6**.
8. Grow the large, inoculated cultures in a bioreactor at 37 °C to an $OD_{600} = 0.4$
9. Induce expression of the histone with the following IPTG (from a 1 M IPTG stock) concentrations and induction duration:
   - H2A: 0.4 mM IPTG for 4 h
   - H2B: 0.4 mM IPTG for 4 h
   - H3: 0.4 mM IPTG for 3 h
   - H4: 0.2 mM IPTG for 3 h
10. Harvest cells by centrifugation 1500 RCF at 25 °C in a swinging bucket rotor.
11. Resuspend each histone pellet with histone lysis buffer (15 mL per 1.5 L culture), and store frozen at −80 °C.

3.2.3 Histone Lysis and Extraction from Inclusion Bodies

1. Completely thaw the resuspended histone pellet at room temperature. This generally takes approximately 1 h.
2. Dilute the resuspended cells with histone lysis buffer to a final volume of 160 mL.
3. Lyse the resuspended cells via sonication (QSonica Q500) on ice for three rounds of 10 s on and 50 s off (amplitude = 90%).
4. Repeat the sonication procedure three times.
5. Transfer the 160 mL of cell lysate to four conical tubes.
6. Clear the cell lysate via high-speed centrifugation using a fixed-angle rotor at 24,000 xG for 20 min at 25 °C.
7. Discard the supernatant and keep the pellet.
8. Resuspend each pellet with 25 mL of histone lysis buffer supplemented with 1% Triton X-100.
9. Centrifuge the resuspended pellet in a fixed angle rotor at 24,000 xG for 20 min at 25 °C.
10. Repeat **steps 8** and **9**.
11. Discard the supernatant and resuspend each pellet in 25 mL of histone lysis buffer without Triton X-100.
12. Centrifuge the resuspended pellet in a fixed-angle rotor at 24,000 xG for 20 min at 25 °C.

13. Discard the supernatant and consolidate each pellet into a beaker.
14. Add 1 mL of DMSO to the pellet and break the pellet up using a spatula.
15. Add a small stir bar to the beaker and stir the pellet for 30 min using a stir plate at 25 °C.
16. Add 30 mL of guanidinium buffer dropwise (in a continuous manner) to the pellet and DMSO mixture using a Pasteur pipette.
17. Extract the histones from the pellet by stirring vigorously for 1 h at 25 °C.
18. Centrifuge the solution containing extracted histones at 24,000 xG for 20 min at 25 °C.
19. Save the supernatant.
20. Perform a second round of histone extraction by repeating **steps 14** through **19**.
21. Combine the supernatant containing the extracted histones and dialyze against 8 M urea overnight (*see* **Notes 9** and **10**).

3.2.4 Histone Purification

1. Equilibrate 10 mL Q-Sepharose Resin with 8 M urea.
2. Mix the supernatant containing the extracted histones with the equilibrated Q-Sepharose resin (*see* **Note 11**).
3. Incubate stirring for 30 minat 25 °C.
4. Add the slurry containing the extracted histones and Q-Sepharose resin to a gravity-flow column.
5. Collect and save the Q-Sepharose column flow-through containing the histone.
6. Wash the Q-Sepharose resin with 40 mL of 8 M urea and combine the wash with the flow-through from **step 5**.
7. Equilibrate 25 mL S-Sepharose resin with 8 M urea.
8. Mix the Q-Sepharose flow-through and wash containing the histone with 25 mL of S-Sepharose resin.
9. Add the slurry containing the extracted histones and S-Sepharose resin to a gravity-flow column.
10. Wash the S-Sepharose resin with 40 mL of 8 M urea.
11. Elute the histone from the S-Sepharose resin by addition of 50 mL of 8 M urea containing increasing concentrations of NaCl. Collect each NaCl elution.
    - 8 M urea, 100 mM NaCl
    - 8 M urea, 200 mM NaCl
    - 8 M urea, 300 mM NaCl

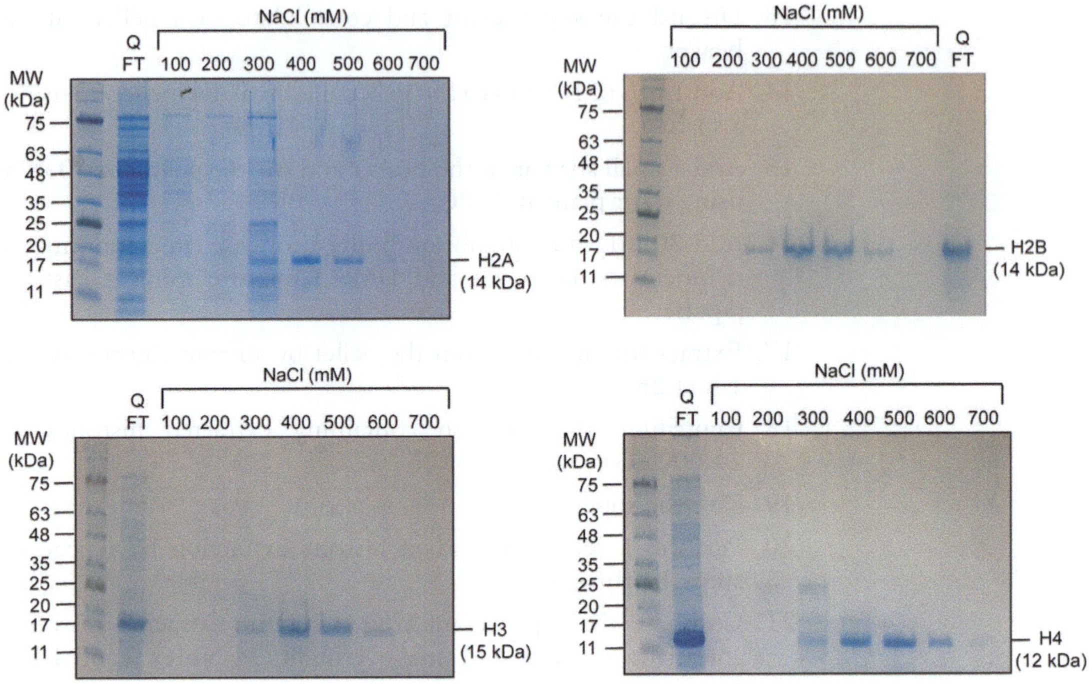

Fig. 2 Representative 12.5% SDS-PAGE gels for the purification of histone H2A (top left), H2B (top right), H3 (bottom left), and H4 (bottom right). The proteins were visualized by Coomassie blue staining

- 8 M urea, 400 mM NaCl
- 8 M urea, 500 mM NaCl
- 8 M urea, 600 mM NaCl
- 8 M urea, 700 mM NaCl

12. Run a sample of each NaCl elution on a 12% SDS-PAGE gel with a protein ladder.
13. Pool fractions containing histone protein based on the molecular weight of each histone. A representative gel for the purification of each individual histone is shown in Fig. 2.
14. Pool the purified histone fractions and dialyze against 4 L of water. Exchange the water for fresh water five times, with at least two of those exchanges lasting overnight.
15. Aliquot into the purified histone fractions into 50 mL conical tubes and flash-freeze using liquid nitrogen.
16. Lyophilize the frozen, purified histone.
17. Aliquot the lyophilized histone and store at −20 °C.

3.3 Generation of H2A/H2B Dimer and H3/H4 Tetramer

After purifying each individual histone, histone octamers or histone H2A/H2B dimers and H3/H4 tetramers can be generated, purified, and stored for rapid reconstitution of NCPs. Below, we outline the salt dialysis method for refolding H2A/H2B dimers and H3/H4 tetramers. We prefer refolding H2A/H2B dimers and H3/H4 tetramers instead of histone octamers due to the increased yield of in-tact complexes.

3.3.1 Refolding of H2A/H2B Dimer

1. Resuspend the lyophilized H2A and H2B in guanidinium buffer to a final concentration of 2 mg/mL.
2. Incubate the resuspended H2A and H2B in guanidinium buffer at room temperature for 2 h.
3. Determine the concentration of resuspended H2A and H2B spectroscopically using the following theoretical molar extinction coefficients:
   - H2A: 4470 M^{-1} cm^{-1}
   - H2B: 7450 M^{-1} cm^{-1}
4. Mix equimolar amounts of H2A and H2B in dialysis tubing (3,000-Da MWCO).
5. Dialyze against 1 L of high-salt refolding buffer (ice cold, 4 °C) three times, for a minimum of 8 h each exchange.

3.3.2 Refolding of H3/H4 Tetramer

1. Resuspend lyophilized H3 and H4 in guanidinium buffer to a final concentration of 2 mg/mL.
2. Incubate the resuspended H3 and H4 in guanidinium buffer at room temperature for 2 h.
3. Determine the concentration of resuspended H3 and H4 spectroscopically using the following theoretical molar extinction coefficients:
   - H3: 4470 M^{-1} cm^{-1}
   - H4: 5960 M^{-1} cm^{-1}
4. Mix equimolar amounts of histone H3 and H4 in dialysis tubing (3,000-Da MWCO).
5. Dialyze against 1 L of high-salt ice-cold (4 °C) refolding buffer three times, with at least two of those exchanges overnight (*see* **Note 12**).

3.3.3 Purification of H2A/H2B Dimer and H3/H4 Tetramer

1. Concentrate the refolded H2A/H2B dimer and H3/H4 tetramer to ~100 μM (*see* **Note 13**).
2. Purify H2A/H2B dimer or H3/H4 tetramer over a Superdex 200 Increase 10/300 GL gel filtration column.
3. Confirm the purity and stoichiometry of the H2A/H2B dimer or H3/H4 tetramer by running an SDS PAGE gel of the S200 gel filtration fractions (Fig. 3; *see* **Note 14**).
4. Combine fractions containing purified H2A/H2B dimer or H3/H4 tetramer and concentrate using a centrifugal filter unit (10,000-Da MWCO) to at least 100 μM.

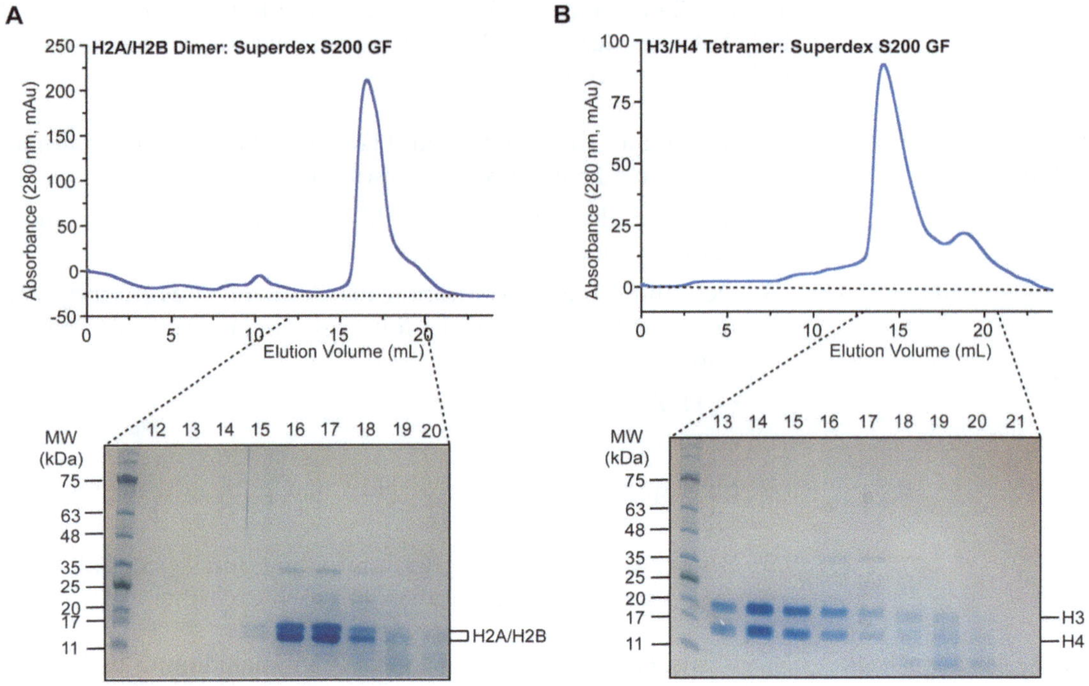

Fig. 3 Purification of H2A/H2B dimer and H3/H4 tetramer. (**a**) Representative H2A/H2B dimer chromatogram from a Superdex S200 gel filtration run. The resulting fractions were resolved on a 12.5% SDS-PAGE gel and visualized by Coomassie blue staining. (**b**) Representative H3/H4 tetramer chromatogram from a Superdex S200 gel filtration run. The resulting fractions were resolved on a 12.5% SDS-PAGE gel and visualized by Coomassie blue staining

5. Mix the concentrated H2A/H2B dimer and H3/H4 tetramer with an equal volume of 100% glycerol and store indefinitely at −20 °C.

3.4 Reconstituting Nucleosomes Containing DNA Damage

NCP reconstitution is done via a salt dialysis method using the damaged DNA, H2A/H2B dimer, and H3/H4 tetramer previously purified. After nucleosome reconstitution, the nucleosomes containing DNA damage are purified by sucrose gradient ultracentrifugation to separate free linear DNA and other subnucleosomal species, which makes them suitable for quantitative in vitro experiments.

3.4.1 Nucleosome Reconstitution

1. Thaw the damaged DNA, H2A/H2B dimer, and H3/H4 tetramer on ice.
2. Determine the concentration of the damaged DNA spectroscopically using the theoretical molar extinction coefficient for the specific DNA substrate.

3. Dilute the damaged DNA to a concentration of 2.5 μM using high-salt buffer.
4. Determine the concentration of H2A/H2B dimer and H3/H4 tetramer spectroscopically using the following theoretical molar extinction coefficients:
   - H2A/H2B dimer: 11,920 M^{-1} cm^{-1}
   - H3/H4 tetramer: 20,860 M^{-1} cm^{-1}
5. Dilute the H2A/H2B dimer to 5 μM and H3/H4 tetramer to 2.5 μM using high-salt buffer.
6. Mix the damaged DNA, H2A/H2B dimer, and H3/H4 tetramer in a 1:2:1 molar ratio and place in dialysis tubing (10,000 Da MWCO).
7. Place the dialysis tubing in a beaker containing 300 mL of high-salt buffer and equilibrate for 30 min.
8. Dilute the high-salt buffer with no-salt buffer stepwise to the following concentrations:
   - 1.5 M NaCl, 150 mL of no-salt buffer, 2 h
   - 1 M NaCl, 150 mL of no-salt buffer, 2 h
   - 0.75 M NaCl, 300 mL of no-salt buffer, 2 h
   - 0.5 M NaCl, 300 mL of no-salt buffer, 2 h
   - 0.25 M NaCl, 1200 mL of no-salt buffer, overnight
   - 0.125 M NaCl, 2400 mL of no-salt buffer, 2 h

3.4.2 Nucleosome Purification

1. Remove the reconstituted nucleosome from the dialysis tubing and place in a conical tube.
2. Concentrate the nucleosome down to ~0.5 mL using a centrifugal filter unit (10,000-Da MWCO).
3. Heat-shock the nucleosome at 55 °C for 30 min (*see* **Note 15**). A representative gel of the nucleosome prior to purification can be seen in Fig. 4a.
4. Make six 10–40% sucrose gradients in 38.5 mL Beckman Coulter Ultra-Clear Centrifuge tubes using a Biocomp Gradient Master 108 (*see* **Note 16**).
5. Layer the 0.5 mL of damaged nucleosome to the top of a sucrose gradient.
6. Ultracentrifuge the sucrose gradients at 125,000 xG for 40–42 h in an SW 32 rotor at 4 °C.
7. Fractionate the sucrose gradient containing nucleosome by pulling 1 mL from the top of the tube and storing in 1.5 mL Eppendorf tubes.
8. Measure the absorbance at 260 nm for each fraction to identify those that contain nucleic acids and run the fractions

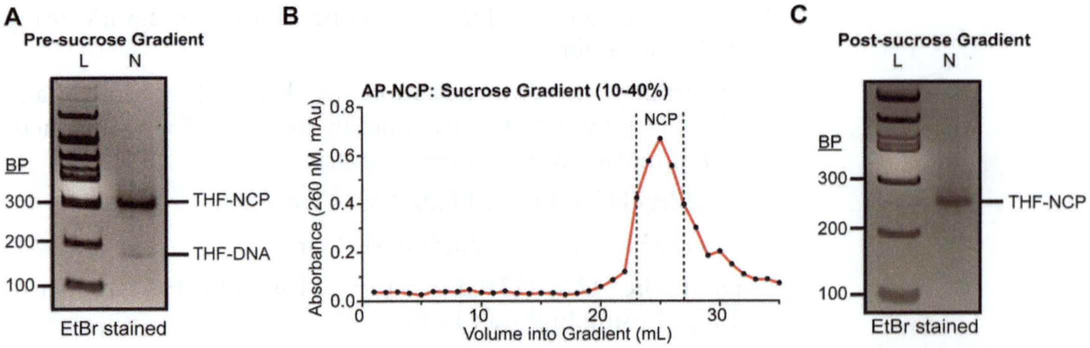

Fig. 4 Generation and purification of nucleosomes containing site-specific DNA damage. (**a**) A representative 6% native PAGE (59:1) gel of the reconstituted nucleosome before purification (**b**) A representative chromatogram of the nucleosome after fractionation from a 10–40% sucrose gradient. (**c**) A representative 6% native PAGE (59:1) gel of the reconstituted nucleosome after purification

containing nucleic acid on a native PAGE gel (5% 59:1 acrylamide/bis-acrylamide and 0.2× TBE). A representative chromatogram of the nucleosome fractions after sucrose gradient ultracentrifugation can be seen in Fig. 4b.

9. Pool the fractions that contain purified nucleosome and buffer exchange five times into 1× TE using a centrifugal filter unit (10,000-Da MWCO). A representative gel of the purified nucleosome containing DNA damage can be seen in Fig. 4c.
10. Concentrate the nucleosome to at least 1 μM and store at 4°C (*see* **Notes 17** and **18**).

4 Notes

1. The glucose, $MgSO_4$, $CaCl_2$, and vitamin solution are added to the M9 media after autoclaving as they can precipitate and/or degrade during autoclaving.
2. The DNA damage can be placed on either I-strand or J-strand depending on the desired location in the nucleosome. If moving the DNA damage to the J-strand, we suggest changing the J-strand to three oligos and the I-strand to two oligos.
3. Avoid placing the THF or other DNA damage within 5 bp of any T4-DNA ligase sites. If unavoidable, the reactions will result in a significantly reduced yield of full-length ligated 147-bp DNA.
4. We perform each individual reaction with 10 nmols of each individual oligo, which should theoretically yield 10 nmols of the final ligated product (assuming 100% yield). An average prep size is ~50–100 nmols, or 5–10 individual reactions.

5. The ligation reaction generally does not reach 100% completion. If desired, an additional 5uL of T4 DNA ligase can be added after the first 24 h to ensure as much of the DNA substrate has been ligated as possible.
6. Concentrating the extracted DNA to 10 μM enables efficient annealing of the I- and J-strand oligos in the subsequent annealing step.
7. The purified damaged DNA should be stored in aliquots that match the average prep size of the downstream nucleosome reconstitution. We typically perform 5 nmol nucleosome reconstitutions.
8. We have stored damaged DNA substrates for up to a year without issues. After long-term storage, we suggest checking the quality of the DNA on an agarose gel before proceeding to nucleosome reconstitution.
9. Urea can spontaneously degrade into isocyanic acid, which reacts with lysine and arginine side chains to form carbamylation [78]. To prevent carbamylation of histones during purification, 8 M urea solutions are made fresh and deionized with a Dowex Amberlite resin.
10. The extracted histones should be dialyzed against enough 8 M urea to bring the concentration of guanidinium–HCl below 250 mM. This step is critical for the subsequent purification of the histone via ion-exchange chromatography as the histones will not bind the S-Sepharose resin above 300 mM NaCl.
11. The Q-Sepharose is used to bind contaminating proteins from the histone extraction step.
12. It is not uncommon to see precipitant forming in the dialysis membrane after the first 24 h spinning in the high-salt buffer. This is often a result of impurities in the purification of the individual histones.
13. The H2A/H2B dimer and H3/H4 tetramer is concentrated to no more than 100 μM to ensure separation from contaminants during gel filtration.
14. It is not uncommon to see minor contaminates in the dimer or tetramer. In our experience, these can be removed at subsequent purification steps and do not significantly affect reconstitution of the nucleosome.
15. Purifying the damaged DNA via denaturing PAGE can lead to an excess of either the I- or J-strand. Excess I- or J-strand ssDNA will readily form histone octamers wrapped by single-stranded DNA [79]. The heat shock at 55 °C will ensure the nucleosome is properly positioned on the 601 DNA, while also denaturing any ssDNA–histone octamer complexes that are difficult to purify via sucrose gradient ultracentrifugation.

16. Another method for purifying nucleosomes is via electrophoresis using a Bio-Rad Model 491 Prep Cell [80].
17. The nucleosome concentration is determined spectroscopically by diluting 2× with 4 M NaCl and measuring the absorbance at 260 nM. Theoretical extinction coefficients should be determined using the damaged DNA sequence.
18. The purified nucleosomes should be stored at 4 °C in 1× TE at a concentration of at least 1 μM. The purified nucleosomes should be stable for at least 1 month. For longer-term storage (>1 month), we suggest storing the purified nucleosomes at 4 °C in 1× TE at a concentration of at least 10 μM.

Acknowledgements

This research was supported by the National Institute of General Medical Science R35-GM128562 (B.J.R., T.M.W., J.J.S., and B.D. F) and the National Institute of General Medical Science F32-GM140718 (T.M.W.). We also thank Drs. Karolin Luger, Catherine Musselman, and Michael Poirier for the histone plasmids used in this study.

References

1. Lindahl T, Nyberg B (1972) Rate of depurination of native deoxyribonucleic acid. Biochemistry 11(19):3610–3618
2. Sekiguchi M, Tsuzuki T (2002) Oxidative nucleotide damage: consequences and prevention. Oncogene 21(58):8895–8904
3. Kreutzer DA, Essigmann JM (1998) Oxidized, deaminated cytosines are a source of C→ T transitions in vivo. Proc Natl Acad Sci 95(7): 3578–3582
4. Cooke MS, Evans MD, Dizdaroglu M, Lunec J (2003) Oxidative DNA damage: mechanisms, mutation, and disease. FASEB J 17(10): 1195–1214
5. Wallace SS, Murphy DL, Sweasy JB (2012) Base excision repair and cancer. Cancer Lett 327(1–2):73–89. https://doi.org/10.1016/j.canlet.2011.12.038
6. Krokan HE, Bjoras M (2013) Base excision repair. Cold Spring Harb Perspect Biol 5(4): a012583. https://doi.org/10.1101/cshperspect.a012583
7. Kim Y-J, Wilson DM III (2012) Overview of base excision repair biochemistry. Curr Mol Pharmacol 5(1):3–13
8. Whitaker AM, Schaich MA, Smith MS, Flynn TS, Freudenthal BD (2017) Base excision repair of oxidative DNA damage: from mechanism to disease. Front Biosci (Landmark edition) 22:1493
9. Wallace SS (2014) Base excision repair: a critical player in many games. DNA Repair 19:14–26
10. Beard WA, Horton JK, Prasad R, Wilson SH (2019) Eukaryotic base excision repair: new approaches shine light on mechanism. Annu Rev Biochem 88:137–162
11. Hegde ML, Hazra TK, Mitra S (2008) Early steps in the DNA base excision/single-strand interruption repair pathway in mammalian cells. Cell Res 18(1):27–47. https://doi.org/10.1038/cr.2008.8
12. Schermerhorn KM, Delaney S (2014) A chemical and kinetic perspective on base excision repair of DNA. Acc Chem Res 47(4): 1238–1246. https://doi.org/10.1021/ar400275a
13. Kornberg RD (1974) Chromatin structure: a repeating unit of histones and DNA. Science 184(4139):868–871
14. Woodcock C, Safer J, Stanchfield J (1976) Structural repeating units in chromatin: I. evidence for their general occurrence. Exp Cell Res 97(1):101–110

15. Luger K, Mäder AW, Richmond RK, Sargent DF, Richmond TJ (1997) Crystal structure of the nucleosome core particle at 2.8 Å resolution. Nature 389(6648):251–260
16. Caffrey PJ, Delaney S (2020) Chromatin and other obstacles to base excision repair: potential roles in carcinogenesis. Mutagenesis 35(1): 39–50
17. Kennedy EE, Caffrey PJ, Delaney S (2018) Initiating base excision repair in chromatin. DNA Repair 71:87–92
18. Li C, Delaney S (2019) Challenges for base excision repair enzymes: acquiring access to damaged DNA in chromatin. Enzymes 45: 27–57
19. Kumar N, Raja S, Van Houten B (2020) The involvement of nucleotide excision repair proteins in the removal of oxidative DNA damage. Nucleic Acids Res 48(20):11227–11243
20. Kutuzov M, Belousova E, Ilina E, Lavrik O (2020) Impact of PARP1, PARP2 & PARP3 on the base excision repair of nucleosomal DNA. Adv Exp Med Biol 1241:47–57
21. Rodriguez Y, Hinz JM, Smerdon MJ (2015) Accessing DNA damage in chromatin: preparing the chromatin landscape for base excision repair. DNA Repair 32:113–119
22. Meas R, Wyrick JJ, Smerdon MJ (2019) Nucleosomes regulate base excision repair in chromatin. Mutat Res Rev Mutat Res 780: 29–36. https://doi.org/10.1016/j.mrrev.2017.10.002
23. Bilotti K, Kennedy EE, Li C, Delaney S (2017) Human OGG1 activity in nucleosomes is facilitated by transient unwrapping of DNA and is influenced by the local histone environment. DNA Repair (Amst) 59:1–8. https://doi.org/10.1016/j.dnarep.2017.08.010
24. Li C, Delaney S (2019) Histone H2A variants enhance the initiation of base excision repair in nucleosomes. ACS Chem Biol 14(5): 1041–1050. https://doi.org/10.1021/acschembio.9b00229
25. Olmon ED, Delaney S (2017) Differential ability of five DNA glycosylases to recognize and repair damage on Nucleosomal DNA. ACS Chem Biol 12(3):692–701. https://doi.org/10.1021/acschembio.6b00921
26. Tarantino ME, Dow BJ, Drohat AC, Delaney S (2018) Nucleosomes and the three glycosylases: high, medium, and low levels of excision by the uracil DNA glycosylase superfamily. DNA Repair (Amst) 72:56–63. https://doi.org/10.1016/j.dnarep.2018.09.008
27. Beard BC, Stevenson JJ, Wilson SH, Smerdon MJ (2005) Base excision repair in nucleosomes lacking histone tails. DNA Repair (Amst) 4(2): 203–209. https://doi.org/10.1016/j.dnarep.2004.09.011
28. Beard BC, Wilson SH, Smerdon MJ (2003) Suppressed catalytic activity of base excision repair enzymes on rotationally positioned uracil in nucleosomes. Proc Natl Acad Sci U S A 100(13):7465–7470. https://doi.org/10.1073/pnas.1330328100
29. Hinz JM, Rodriguez Y, Smerdon MJ (2010) Rotational dynamics of DNA on the nucleosome surface markedly impact accessibility to a DNA repair enzyme. Proc Natl Acad Sci U S A 107(10):4646–4651. https://doi.org/10.1073/pnas.0914443107
30. Meas R, Smerdon MJ (2016) Nucleosomes determine their own patch size in base excision repair. Sci Rep 6:27122. https://doi.org/10.1038/srep27122
31. Meas R, Smerdon MJ, Wyrick JJ (2015) The amino-terminal tails of histones H2A and H3 coordinate efficient base excision repair, DNA damage signaling and postreplication repair in Saccharomyces cerevisiae. Nucleic Acids Res 43(10):4990–5001. https://doi.org/10.1093/nar/gkv372
32. Nakanishi S, Prasad R, Wilson SH, Smerdon M (2007) Different structural states in oligonucleosomes are required for early versus late steps of base excision repair. Nucleic Acids Res 35(13):4313–4321. https://doi.org/10.1093/nar/gkm436
33. Rodriguez Y, Duan M, Wyrick JJ, Smerdon MJ (2018) A cassette of basic amino acids in histone H2B regulates nucleosome dynamics and access to DNA damage. J Biol Chem 293(19): 7376–7386. https://doi.org/10.1074/jbc.RA117.000358
34. Rodriguez Y, Hinz JM, Laughery MF, Wyrick JJ, Smerdon MJ (2016) Site-specific acetylation of histone H3 decreases polymerase beta activity on nucleosome Core particles in vitro. J Biol Chem 291(21):11434–11445. https://doi.org/10.1074/jbc.M116.725788
35. Rodriguez Y, Smerdon MJ (2013) The structural location of DNA lesions in nucleosome core particles determines accessibility by base excision repair enzymes. J Biol Chem 288(19): 13863–13875. https://doi.org/10.1074/jbc.M112.441444
36. Howard MJ, Rodriguez Y, Wilson SH (2017) DNA polymerase beta uses its lyase domain in a processive search for DNA damage. Nucleic Acids Res 45(7):3822–3832. https://doi.org/10.1093/nar/gkx047
37. Rodriguez Y, Horton JK, Wilson SH (2019) Histone H3 lysine 56 acetylation enhances AP endonuclease 1-mediated repair of AP sites in

nucleosome core particles. Biochemistry 58(35):3646–3655. https://doi.org/10.1021/acs.biochem.9b00433

38. Rodriguez Y, Howard MJ, Cuneo MJ, Prasad R, Wilson SH (2017) Unencumbered pol beta lyase activity in nucleosome core particles. Nucleic Acids Res 45(15):8901–8915. https://doi.org/10.1093/nar/gkx593
39. Banerjee DR, Deckard CE 3rd, Elinski MB, Buzbee ML, Wang WW, Batteas JD, Sczepanski JT (2018) Plug-and-play approach for preparing chromatin containing site-specific DNA modifications: the influence of chromatin structure on base excision repair. J Am Chem Soc 140(26):8260–8267. https://doi.org/10.1021/jacs.8b04063
40. Banerjee DR, Deckard CE, Zeng Y, Sczepanski JT (2019) Acetylation of the histone H3 tail domain regulates base excision repair on higher-order chromatin structures. Sci Rep 9(1):1–11
41. Huggins CF, Chafin DR, Aoyagi S, Henricksen LA, Bambara RA, Hayes JJ (2002) Flap endonuclease 1 efficiently cleaves base excision repair and DNA replication intermediates assembled into nucleosomes. Mol Cell 10(5):1201–1211
42. Nilsen H, Lindahl T, Verreault A (2002) DNA base excision repair of uracil residues in reconstituted nucleosome core particles. EMBO J 21(21):5943–5952
43. Menoni H, Gasparutto D, Hamiche A, Cadet J, Dimitrov S, Bouvet P, Angelov D (2007) ATP-dependent chromatin remodeling is required for base excision repair in conventional but not in variant H2A.Bbd nucleosomes. Mol Cell Biol 27(17):5949–5956. https://doi.org/10.1128/MCB.00376-07
44. Menoni H, Shukla MS, Gerson V, Dimitrov S, Angelov D (2012) Base excision repair of 8-oxoG in dinucleosomes. Nucleic Acids Res 40(2):692–700. https://doi.org/10.1093/nar/gkr761
45. Prasad A, Wallace SS, Pederson DS (2007) Initiation of base excision repair of oxidative lesions in nucleosomes by the human, bifunctional DNA glycosylase NTH1. Mol Cell Biol 27(24):8442–8453. https://doi.org/10.1128/MCB.00791-07
46. Odell ID, Barbour JE, Murphy DL, Della-Maria JA, Sweasy JB, Tomkinson AE, Wallace SS, Pederson DS (2011) Nucleosome disruption by DNA ligase III-XRCC1 promotes efficient base excision repair. Mol Cell Biol 31(22):4623–4632. https://doi.org/10.1128/MCB.05715-11
47. Odell ID, Newick K, Heintz NH, Wallace SS, Pederson DS (2010) Non-specific DNA binding interferes with the efficient excision of oxidative lesions from chromatin by the human DNA glycosylase, NEIL1. DNA Repair (Amst) 9(2):134–143. https://doi.org/10.1016/j.dnarep.2009.11.005
48. Jagannathan I, Pepenella S, Hayes JJ (2011) Activity of FEN1 endonuclease on nucleosome substrates is dependent upon DNA sequence but not flap orientation. J Biol Chem 286(20):17521–17529. https://doi.org/10.1074/jbc.M111.229658
49. Yang C, Sengupta S, Hegde PM, Mitra J, Jiang S, Holey B, Sarker AH, Tsai MS, Hegde ML, Mitra S (2017) Regulation of oxidized base damage repair by chromatin assembly factor 1 subunit a. Nucleic Acids Res 45(2):739–748. https://doi.org/10.1093/nar/gkw1024
50. Fu I, Smith DJ, Broyde S (2019) Rotational and translational positions determine the structural and dynamic impact of a single ribonucleotide incorporated in the nucleosome. DNA Repair (Amst) 73:155–163. https://doi.org/10.1016/j.dnarep.2018.11.012
51. Hinz JM (2014) Impact of abasic site orientation within nucleosomes on human APE1 endonuclease activity. Mutat Res Fund Mol Mecha Mutag 766–767:19–24. https://doi.org/10.1016/j.mrfmmm.2014.05.008
52. Hinz JM, Mao P, McNeill DR, Wilson DM 3rd (2015) Reduced nuclease activity of Apurinic/Apyrimidinic endonuclease (APE1) variants on nucleosomes: identification of access residues. J Biol Chem 290(34):21067–21075. https://doi.org/10.1074/jbc.M115.665547
53. Maher RL, Prasad A, Rizvanova O, Wallace SS, Pederson DS (2013) Contribution of DNA unwrapping from histone octamers to the repair of oxidatively damaged DNA in nucleosomes. DNA Repair (Amst) 12(11):964–971. https://doi.org/10.1016/j.dnarep.2013.08.010
54. Maher RL, Wallace SS, Pederson DS (2019) The lyase activity of bifunctional DNA glycosylases and the 3′-diesterase activity of APE1 contribute to the repair of oxidized bases in nucleosomes. Nucleic Acids Res 47(6):2922–2931. https://doi.org/10.1093/nar/gky1315
55. Maher RL, Marsden CG, Averill AM, Wallace SS, Sweasy JB, Pederson DS (2017) Human cells contain a factor that facilitates the DNA glycosylase-mediated excision of oxidized bases from occluded sites in nucleosomes. DNA Repair (Amst) 57:91–97. https://doi.org/10.1016/j.dnarep.2017.06.029

56. Caffrey PJ, Delaney S (2021) Nucleosome Core particles lacking H2B or H3 tails are altered structurally and have Differential Base excision repair fingerprints. Biochemistry 60(3):210–218
57. Kennedy EE, Li C, Delaney S (2019) Global repair profile of human alkyladenine DNA glycosylase on nucleosomes reveals DNA packaging effects. ACS Chem Biol 14(8):1687–1692
58. Caffrey PJ, Kher R, Bian K, Li D, Delaney S (2020) Comparison of the base excision and direct reversal repair pathways for correcting 1, N 6-Ethenoadenine in strongly positioned nucleosome Core particles. Chem Res Toxicol 33(7):1888–1896
59. Cannan WJ, Tsang BP, Wallace SS, Pederson DS (2014) Nucleosomes suppress the formation of double-strand DNA breaks during attempted base excision repair of clustered oxidative damages. J Biol Chem 289(29): 19881–19893. https://doi.org/10.1074/jbc.M114.571588
60. Bennett L, Madders E, Parsons JL (2019) HECTD1 promotes base excision repair in nucleosomes through chromatin remodelling. Nucleic Acids Res 48:1301. https://doi.org/10.1093/nar/gkz1129
61. Cannan WJ, Rashid I, Tomkinson AE, Wallace SS, Pederson DS (2017) The human ligase IIIalpha-XRCC1 protein complex performs DNA Nick repair after transient unwrapping of Nucleosomal DNA. J Biol Chem 292(13): 5227–5238. https://doi.org/10.1074/jbc.M116.736728
62. Chafin DR, Vitolo JM, Henricksen LA, Bambara RA, Hayes JJ (2000) Human DNA ligase I efficiently seals nicks in nucleosomes. EMBO J 19(20):5492–5501
63. Kutuzov M, Belousova E, Kurgina T, Ukraintsev A, Vasileva I, Khodyreva S, Lavrik O (2021) The contribution of PARP1, PARP2 and poly (ADP-ribosyl) ation to base excision repair in the nucleosomal context. Sci Rep 11(1):1–17
64. Bilotti K, Tarantino ME, Delaney S (2018) Human oxoguanine glycosylase 1 removes solution accessible 8-oxo-7, 8-dihydroguanine lesions from globally substituted nucleosomes except in the dyad region. Biochemistry 57(9): 1436–1439
65. Ura K, Araki M, Saeki H, Masutani C, Ito T, Iwai S, Mizukoshi T, Kaneda Y, Hanaoka F (2001) ATP-dependent chromatin remodeling facilitates nucleotide excision repair of UV-induced DNA lesions in synthetic dinucleosomes. EMBO J 20(8):2004–2014
66. Ren M, Shang M, Wang H, Xi Z, Zhou C (2021) Histones participate in base excision repair of 8-oxodGuo by transiently cross-linking with active repair intermediates in nucleosome core particles. Nucleic Acids Res 49(1):257–268
67. Dyer PN, Edayathumangalam RS, White CL, Bao Y, Chakravarthy S, Muthurajan UM, Luger K (2003) Reconstitution of nucleosome core particles from recombinant histones and DNA. Methods Enzymol 375:23–44
68. Kosmoski JV, Smerdon MJ (1999) Synthesis and nucleosome structure of DNA containing a UV photoproduct at a specific site. Biochemistry 38(29):9485–9494
69. Sczepanski JT, Wong RS, McKnight JN, Bowman GD, Greenberg MM (2010) Rapid DNA-protein cross-linking and strand scission by an abasic site in a nucleosome core particle. Proc Natl Acad Sci U S A 107(52): 22475–22480. https://doi.org/10.1073/pnas.1012860108
70. Duan M-R, Smerdon MJ (2010) UV damage in DNA promotes nucleosome unwrapping. J Biol Chem 285(34):26295–26303
71. Lowary P, Widom J (1998) New DNA sequence rules for high affinity binding to histone octamer and sequence-directed nucleosome positioning. J Mol Biol 276(1):19–42
72. Weaver TM, Hoitsma NM, Spencer JJ, Gakhar L, Schnicker NJ, Freudenthal BD (2022) Structural basis for APE1 processing DNA damage in the nucleosome. Nat Commun 13(1):5390. https://doi.org/10.1038/s41467-022-33057-7
73. Deckard CE III, Sczepanski JT (2021) Reversible chromatin condensation by the DNA repair and demethylation factor thymine DNA glycosylase. Nucleic Acids Res 49(5): 2450–2459
74. Bai J, Zhang Y, Xi Z, Greenberg MM, Zhou C (2018) Oxidation of 8-Oxo-7,8-dihydro-2′-deoxyguanosine leads to substantial DNA-histone cross-links within nucleosome Core particles. Chem Res Toxicol 31(12): 1364–1372. https://doi.org/10.1021/acs.chemrestox.8b00244
75. Li F, Zhang Y, Bai J, Greenberg MM, Xi Z, Zhou C (2017) 5-Formylcytosine yields DNA-protein cross-links in nucleosome Core particles. J Am Chem Soc 139(31): 10617–10620. https://doi.org/10.1021/jacs.7b05495
76. Taylor J-S (2015) Design, synthesis, and characterization of nucleosomes containing site-specific DNA damage. DNA Repair 36:59–67

77. Vasudevan D, Chua EY, Davey CA (2010) Crystal structures of nucleosome core particles containing the '601'strong positioning sequence. J Mol Biol 403(1):1–10

78. Stark GR, Stein WH, Moore S (1960) Reactions of the cyanate present in aqueous urea with amino acids and proteins. J Biol Chem 235(11):3177–3181

79. Adkins NL, Swygert SG, Kaur P, Niu H, Grigoryev SA, Sung P, Wang H, Peterson CL (2017) Nucleosome-like, single-stranded DNA (ssDNA)-histone octamer complexes and the implication for DNA double strand break repair. J Biol Chem 292(13):5271–5281

80. Kujirai T, Arimura Y, Fujita R, Horikoshi N, Machida S, Kurumizaka H (2018) Methods for preparing nucleosomes containing histone variants. In: Histone variants. Springer, pp 3–20

Chapter 5

A DNA Cleavage Assay Using Synthetic Oligonucleotide Containing a Single Site-Directed Lesion for In Vitro Base Excision Repair Study

Bo Hang

Abstract

Many chemicals cause mutation or cancer in animals and humans by forming DNA lesions, including base adducts, which play a critical role in mutagenesis and carcinogenesis. A large number of such adducts are repaired by the DNA glycosylase-mediated base excision repair (BER) pathway, and some are processed by nucleotide excision repair (NER) and nucleotide incision repair (NIR). To understand what structural features determine repair enzyme specificity and mechanism in chemically modified DNA in vitro, we developed and optimized a DNA cleavage assay using defined oligonucleotides containing a single, site specifically placed lesion. This assay can be used to investigate novel activities against any newly identified derivatives from chemical compounds, substrate specificity and cleavage efficiency of repair enzymes, and quantitative structure–function relationships. Overall, the methodology is highly sensitive and can also be modified to explore whether a lesion is processed by NER or NIR activity, as well as to study its miscoding properties in translesion DNA synthesis (TLS).

Key words Base excision repair (BER), Glycosylases, AP endonucleases, Nucleotide incision repair (NIR), Nucleotide excision repair (NER), Translesion DNA synthesis (TLS), DNA adducts, Phosphoramidite chemistry, Defined oligonucleotide, DNA cleavage assay, Polyacrylamide gel electrophoresis (PAGE)

1 Introduction

There are several major repair pathways that excise mutagenic or replication blocking modified bases or a modifying group from DNA, including the base excision repair (BER), nucleotide excision repair (NER), and nucleotide incision repair (NIR) [1, 2]. In general, most of base mismatches and chemically modified DNA lesions with minimal helix distorting are processed by BER, which is initiated by damage-specific DNA glycosylases, and the resulting apurinic/apyrimidinic (AP) site is processed in mammalian cells by short-patch or long-patch pathways [1, 2, 3]. NER is responsible

Kishor K. Bhakat and Tapas K. Hazra (eds.), *Base Excision Repair Pathway: Methods and Protocols*, Methods in Molecular Biology, vol. 2701, https://doi.org/10.1007/978-1-0716-3373-1_5,

for removing bulky and helix-distorting DNA lesions including those formed by ultraviolet (UV) light, carcinogens in cigarette smoke, and endogenous processes [1, 2]. NIR is a specialized form of DNA repair, which was first reported by Hang et al. in 1996 describing an unusual mechanism for the major human AP endonuclease in processing benzene metabolite *p*-benzoquinone (*p*BQ)-derived exocyclic adducts [4, 5]; later, several oxidative cytosine bases were also found by Gros et al. to be substrates for AP endonuclease using the same mechanism of action [6, 7].

DNA synthesizers appeared in the 1980s. With this power tool, it was possible to design oligonucleotides containing normal and modified bases for which phosphoramidites could be made. As stated by Drs. Delaney and Essigmann [8], the advancements in single chemical DNA adducts made by synthetic chemists, molecular biologists, and biophysicists have made possible to investigate DNA damage genotoxicity, mutagenicity, and repair. With global chemical modification, the resulting DNA lesion mixture is often complex so that it is difficult to analyze which specific chemical modification in DNA is responsible for mutagenic or other biological events. In the past, considerable effort was made by laboratories to explore repair activities towards newly synthesized derivatives of exogenous and endogenous carcinogens and define substrate specificity and efficiency of enzymes that function in initial recognition and excision in BER. Such knowledge has been greatly improved by the use of cell-free extracts (CFEs) in the initial stage of research when purified proteins were not available [e.g., 4, 5, 9, 10, 11]. However, sometimes it is far more difficult to interpret such results. With the pure enzymes one can answer many research questions in defining substrate specificity and range, analyzing structure–function relationships, reconstituting repair pathways, and performing site-directed mutagenesis [12]. It is noteworthy that there are still identified DNA adducts that have not been studied for their repairability; also, not every repair enzyme/pathway has been examined and characterized for their activity against all the known biologically important adducts.

There were several biochemical assays evaluating BER activity that have been described in the literature, including base release assay using globally modified radiolabeled polynucleotides or DNA substrates (base release is a function of DNA glycosylase), DNA cleavage assay based on radiolabeled oligonucleotide containing a site-directed lesion, approaches using circular DNA (plasmid genome) containing a lesion of interest, immunoassay for lesion detection and level measurement in DNA, and other assays that may be applicable to both in vitro and in vivo studies. Although valuable information has been obtained with these assays, some of these approaches may lead to ambiguous and nonquantitative results, as a significant problem is the mixed DNA damage and

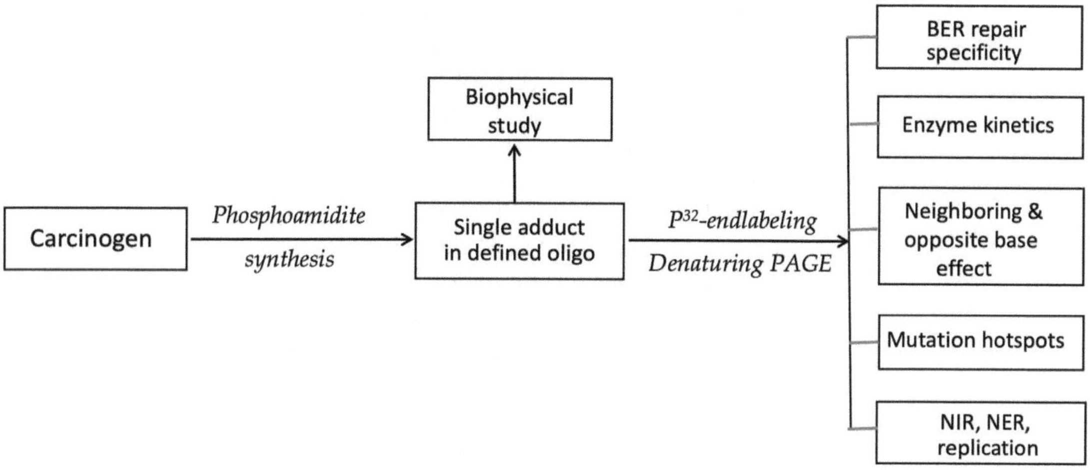

Fig. 1 The general scheme of the DNA cleavage assay and its in vitro applications in BER, NIR, NER, and replication studies

heterogeneity of the substrates for enzymes. Without doubt, a DNA substrate containing a site-directed single lesion is amenable to a fine analysis of the repair activity.

Over the years, our group has focused on the in vitro biochemical studies of many environmental carcinogen-derived DNA adducts or other lesions [4, 13, 14, 15, 16, 17] and identified many new DNA substrates for prokaryotic and eukaryotic DNA glycosylases, damage-specific AP endonucleases, and NER enzymes using the oligomers with an adduct as substrates. Our priority was to identify those repair activities towards the biologically relevant DNA adducts, which is critical for understanding the biological impact of such adducts in the etiological mechanisms of chemical carcinogenesis. Our next goal was to carry out quantitative structure–function studies by comparing repair specificity and efficiency towards structurally analogous adducts, such as exocyclic etheno, hydroxymethyl etheno, ethano, and *p*BQ adducts [14].

The oligonucleotide cleavage assay outlined in this chapter is designed to study novel DNA repair activity, enzyme specificity, and kinetic properties, as well as other factors that may affect repair, such as thermodynamic impact and neighboring and opposite base effect (Fig. 1). This assay is a preferred in vitro method in laboratories studying DNA repair. In this assay, construction of a defined oligonucleotide with a single site-directed DNA lesion is central. For de novo synthesis, we first synthesize a fully protected phosphoramidite of the modified nucleoside of interest and incorporate the latter into DNA oligomers, which can then be utilized for biochemical and biophysical studies. For testing enzymatic activity, the oligomer with a modified base is 5′- labeled with ^{32}P-ATP and annealed to its complementary strand. A purified repair protein or CFE is then mixed with the oligomer substrate and incubated at

37 °C for a varying amount of time. The reaction mixture, along with 5′-^{32}P-ATP-labeled size markers, is loaded onto a denaturing polyacrylamide gel electrophoresis (PAGE), allowing the cleavage product(s), either from direct cleavage or induced chemically from the newly generated AP site, to be separated from the noncleaved intact substrate. The respective gel bands are scanned and quantified to calculate repair efficiency. This assay is highly sensitive and dependable and rapid (1–2 days) if one does not count the time for chemical synthesis. The major limiting factor is the phosphoramidite synthesis, if needed, requiring a skilled synthetic chemist. The use of radioactive labeling can also potentially limit the application of this assay. The same methodology has also been successfully applied to the analysis of NIR and NER processes towards various substrates, as described below. In addition, with the same modified oligomer substrates and denaturing gel system, replication-based in vitro studies can also be performed.

2 Materials

2.1 Chemical Synthesis of Defined Oligonucleotide Containing a Site-Directed Lesion

This step is needed if no commercial phosphoramite for the lesion of interest is available. For many years, a number of syntheses of phosphoramites of various modified deoxynucleosides and incorporation into defined oligomers have successfully been carried out by Dr. B. Singer and colleagues at the Lawrence Berkeley National Laboratory, e.g., the adducts derived from benzene metabolite *p*BQ, 1,*N*^{6}-*p*BQdA, 3,*N*^{4}-*p*BQdC and 1,*N*^{2}-*p*BQdG [18, 19] and *N*^{2}-(4-hydroxyphenyl)-2′-dG (*N*^{2}-4-HOPh-dG) that has been identified in vivo [20], the hydroxymethyl etheno dA (7-hm-εA) and dC (8-hm-εC) adducts from glycidaldehyde [21, 22], and the ethano adduct 1,*N*^{6}-EdA from the therapeutic agent 1,3-bis (2-chloroethyl)nitrosourea (BCNU) [23, 24]. Many other phosphoramites of modified nucleosides have been synthesized and incorporated into defined oligomers by several research groups [e.g., 25, 26, 27, 28, 29, 30, 31, 32]. Nowadays, various phosphoramidites containing biologically interesting base modifications are now available from commercial sources such as Glen Research. The whole synthesis is generally time-consuming with different schemes and requires a skilled synthetic chemist. Therefore, the synthesis is not described in this chapter as a step-by-step process, rather a principle of the technique (Fig. 2a). It should be mentioned that, in addition to the phosphoramidite approach, researchers have also developed a method called postsynthetic modification of DNA bearing a convertible nucleoside to site-specifically replace the exocyclic groups of DNA bases with *N*-, *O*-, and *S*-substitutions (see ref. [8] for more details).

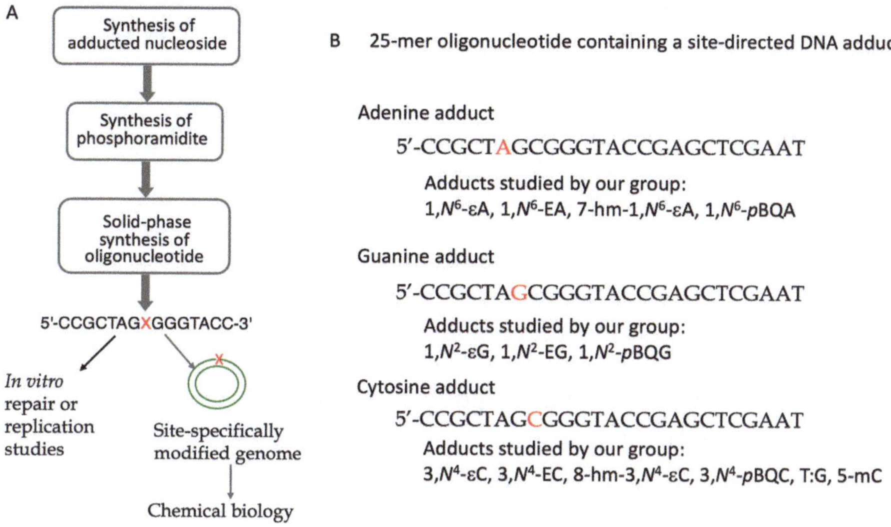

Fig. 2 (**a**) The general steps of chemical synthesis of defined oligonucleotides containing a single site-directed lesion or adduct based on phosphoramidite chemistry. (**b**) The defined 25-mer oligonucleotide containing a single, site-specifically placed base adduct. The examples of adducts studied by Singer, Hang, and colleagues over the years are listed

2.2 The Basic Steps of Synthesis of a Modified Phosphoramidite and Its Insertion into a Defined Oligomer Involve

1. Synthesis of a modified deoxynucleoside: This is achieved by the reaction of a carcinogen or its metabolite of interest with a deoxynucleoside at an appropriate pH and the resulting products are separated and the modified deoxynucleoside is isolated and purified.
2. Synthesis of the modified phosphoramidite: The above modified deoxynucleoside is converted to a suitably protected form of 5′-O-DMT-3′-O-phosphoramidite, which is treated to give the pure phosphoramidite. It is also necessary to protect any other functional groups that may interfere with the general chemistry of oligomer synthesis. This is a key step whereby the modified base in the phosphoramidite has to be kept stable to both the reagents used in synthesis and those involved in the deprotection of the oligomer. Substantial efforts have also been made by chemists to develop alternative protecting groups.
3. Solid-phase synthesis of the oligonucleotide: This is done by standard phosphoramidite chemistry and automated DNA synthesizer. The deprotection of the oligomer is carried out with an appropriate strategy. The oligomer is finally purified with both gel electrophoresis and HPLC. The DMT-off oligonucleotide is analyzed by electrospray mass spectrometry; incorporation of the correct modified base is further verified by enzyme digestion of the oligomer followed by reverse-phase HPLC analysis.

3 Methods

3.1 The Enzymatic Assay: Radiolabeling of Template Oligonucleotide and Annealing to Complementary Strand

This is a standard procedure, and other protocol(s) or reagents can be followed or replaced for 5′-labeling of the oligomer substrate. The latter is then annealed to its complementary strand to make a duplex for biochemical studies. The most widely used sequence in our group was a 25-mer sequence (Fig. 2b), which were normally synthesized with a modified base at its sixth dA position, seventh dG position, and eighth dC position.

All the steps of the following protocol should be carried out in a radioactive control room or laboratory. The radioactive material used requires additional instructions for proper handling and disposal.

1. Take out the frozen [γ-^{32}P] ATP vial to thaw behind the plexiglass screen before starting.
2. Prepare the T4 polynucleotide kinase (PNK) mixture on ice (210 μL in total):

   1 μL T4 PNK (30 U, USB, 70031)

   33 μL 10× kinase buffer

   176 μL ddH_2O
3. Set up kinasing reaction by adding the following items to an Eppendorf tube behind the plexiglass screen (33 μL total) and incubate at 37 °C for 30 min:

   μL PNK mixture from above

   7 μL oligo (0.1 OD/mL for a 25-mer oligomer)

   5 μL [γ-^{32}P] ATP (50 mCi, Amersham, PB10218)
4. Inactivate the PNK activity by heating the sample at 95–100 °C for 3 min. Centrifuge the reaction mixture briefly.
5. Set up the annealing reaction by adding the following items to the above tube to make a 80 μL solution:

   8 μL 10× annealing buffer

   12 μL complementary oligo (0.1 OD/mL for 25-mer; *see* **Note 1**)

   27 μL ddH_2O

   10× anneal buffer (1000 μL): 100 μL 1 M Tris–HCl pH 7.8 (100 mM)

   20 μL 0.5 M EDTA (10 mM)

   200 μL 5 M NaCl (1 M NaCl)

   680 μL ddH_2O

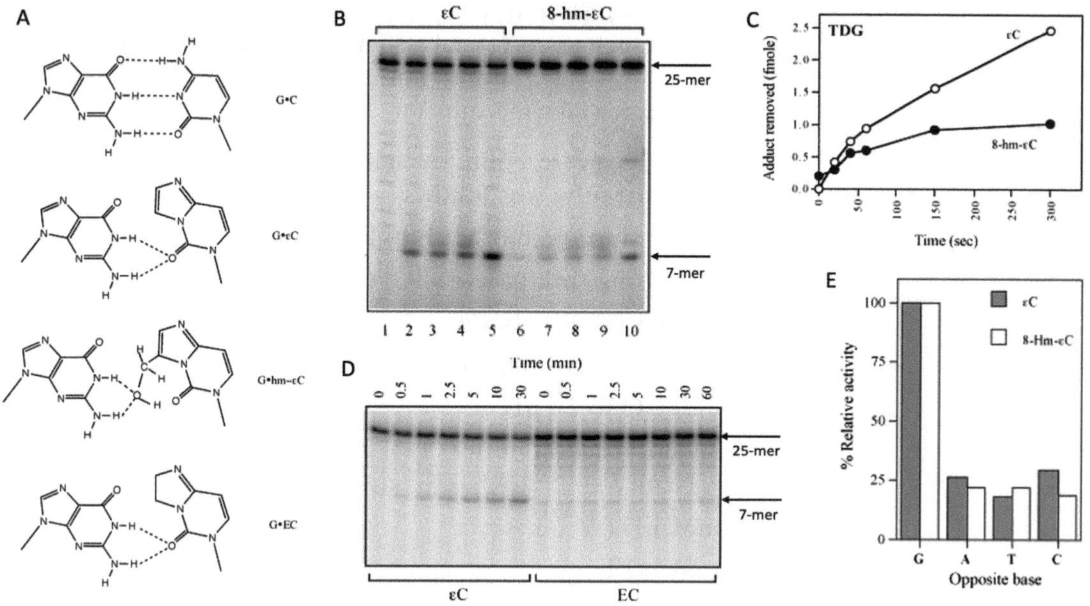

Fig. 3 BER excision assay on human thymine-DNA glycosylase (TDG) substrate specificity and enzyme efficiency. (**a**) Structures of the three exocyclic cytosine analogs paired with G based on data obtained from MD simulations. (**b**) Cleavage by TDG of the 25-mer containing either an εC or 8-hm-εC with an increasing amount of TDG. (**c**) Time course of cleavage of the 25-mer with εC or 8-hm-εC. (**d**) Time course of cleavage of 25-m34 with εC versus EC. (**e**) Effect of opposite bases on TDG activity towards εC or 8-hm-εC. The rate of excision of the modified cytosine base when paired with G was treated as 100%. (These figures are adapted from Ref. [36] with permission)E

6. Put the Eppendorf tube rack in a glass beaker with hot water (>90 °C) and slowly cool to room temperature (<35 °C) over a period of 30–60 min to allow sufficient annealing. Store the final product at −20 °C until use.

3.2 Enzymatic Reaction

Purified repair enzymes, partially purified fractions, and CFEs can be used for a cleavage activity against the lesion of interest in oligomer substrates. Many purified DNA glycosylases and AP endonucleases are available commercially. As mentioned above, many repair activities were identified and characterized in early years using semi-purified protein samples or CFEs. As an example for purified enzymes, Fig. 3 demonstrates human thymine-DNA glycosylase (TDG)-mediated excision of three exocyclic cytosine adducts, including both concentration- and time-dependent repair of εC, 8-hm-εC, and ethano C (EC), and the effect of the opposite base to the adduct on excision [33]. As shown by the structural studies using molecular dynamics simulations, these three structurally analogous adducts (Fig. 3a), formed by unrelated chemical compounds, are processed by TDG with a different specificity, i.e., the substituted etheno adduct 8-hm-εC or ring-saturated adduct EC is less efficiently excised or not excised by the same glycosylase (Fig. 3b to d) [33]. Such studies, combined with structural studies such as co-crystal structures and computational modeling, may provide more understanding of the molecular requirements for enzyme substrate specificity.

1. Set up repair reaction. A standard reaction mixture in an Eppendorf tube is as follows (a total volume 10 μL):

   0.5–2.5 μL protein or CFE to be tested

   0.5 μL 1 mg/mL poly[dI·dC] (*see* **Note 2**)

   1 μL 10× reaction buffer

   1 μL ^{32}P-labeled oligo duplex

   × μL H_2O to 10 μL,

   Standard reaction buffer (1×): 20 mM Tris–HCl (pH 8.0) (*see* **Note 3**).

   1 mM EDTA

   1 mM DTT

   100 ng/μL BSA
2. Incubate the reaction at a 37 °C water bath for a various length of time, depending on the experimental design.
3. Terminate the reaction by heating the Eppendorf tube at 95–100 °C for 3 min and then slowly cooling down to room temperature (*see* **Note 4**).
4. Add 1–5 ng of purified APE1 (depending on the activity) to the reaction and incubate at 37 °C for 20 min, to further cleave the AP site formed by the glycosylase action. For DNA glycosylases with an AP lyse activity, this step is not needed. It is important to make sure that APE itself has no effect on the lesion of interest. This can be done by setting up a control with APE alone (*see* **Note 5**).
5. Add 5 μL of 3× alkaline solution (300 mM NaOH, 90% formamide, 0.05% bromophenol blue), followed by heating the samples at 95 °C for 5 min to denature the duplex. The samples may now be stored frozen for later analysis on the denaturing gel.

3.3 Detection of Oligomer Cleavage on a Denaturing PAGE

1. Prepare a 7 M urea 12% denaturing PAGE.
2. Load 5 μL of the above reaction product per well on the gel, with a 5′-^{32}P-labeled single-stranded oligomer marker(s) (*see* **Note 6**). No purification step is needed even when a CFE is used.
3. Run the gel with 1× TBE buffer on a "Poker Face" sequencing gel apparatus at constant 2000 V for 1 h (*see* **Note 7**).

4. The gel is dried onto a 3MM paper in a gel drier according to the manufacturer's instructions.
5. The gel is scanned with a Phosphorimager (*see* **Note 8**).
6. For band quantitation, a specialized software is used according to the manufacturer's instructions.

This assay is quantitative enough to measure the effect of neighboring base context or opposite base on cleavage efficiency [34]. As shown in ref. [35], eight 15-mer synthetic AP (tetrahydrofuran)-containing oligomer duplexes were used to test the cleavage efficiency of *E. coli* endonuclease IV. Change of the position and composition of the triplet flanking bases adjunct to the AP site caused significant differences in repair rates. Interestingly, the rates also paralleled the thermal stability of the duplexes, as endonuclease IV requires a double-stranded DNA substrate. Another quantitation example is the human TDG cleavage of the oligomer duplex containing an 8-hm-εC (a new substrate then) in comparison with εC (a known substrate) [33] (Fig. 3e), in which the opposite base to either 8-hm-εC or εC is G, A, T, or C. The result showed that the TDG activity had the highest level when these two cytosine adducts paired with G.

Using the same radiolabeled oligonucleotide duplex, the binding assay with the same protein samples can be performed and analyzed on a non-denaturing PAGE. Our group has performed various gel bandshift assays for DNA glycosylase binding to the 25-mer duplex containing an adduct of interest, which provided valuable information in specific recognition, biochemical characterization, and protein purification [10, 35–37].

3.4 NIR and NER Assays

This assay is also suitable for testing a nucleotide incision repair (NIR) activity, which is a specialized form of repair, as exemplified by the processing of *p*BQ-derived exocyclic DNA adducts and oxidative DNA lesions [5, 6]. In Fig. 4, we first used a *p*BQdC-containing 25-mer duplex to explore its potential cleavage by HeLa cell-free extracts [5] and observed a robust activity indicating 5′ incision of the substrate using the cleavage assay (Fig. 4a). This activity was further purified from HeLa cell extracts to a single polypeptide band on silver staining using multiple-column chromatography (Fig. 4b) which was later confirmed as human major AP endonuclease with protein microsequencing [5]. Further experiments with cleavage fragment end-group studies found that the enzyme incises the oligomer 5′ to the adduct and generates 3′--hydroxyl and 5′-phosphoryl termini but leaves the adduct on the 5′ terminus of the cleaved fragment as a "dangling base" (Fig. 4c) [5].

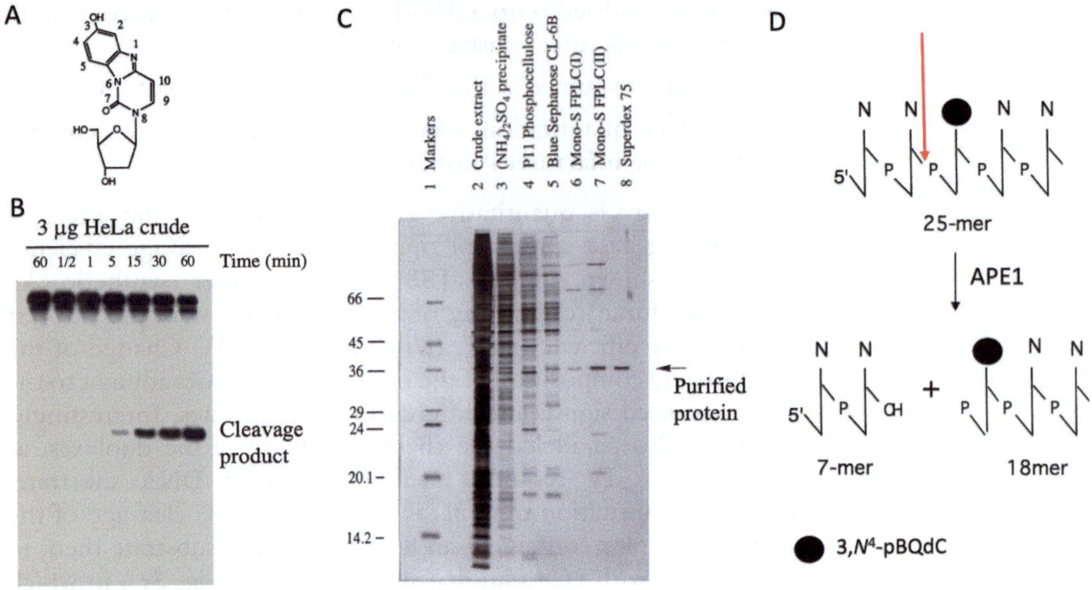

Fig. 4 Identification of the human nucleotide incision activity against the *p*BQdC adduct and a new repair mechanism. (**a**) chemical structure of 3,N^4-benzetheno-2′-deoxycytidine (*p*BQdC). (**b**) Cleavage activity towards the *p*BQdC-containing 25-mer in HeLa CFE. (**c**) Purification of the *p*BQdC activity using multiple column steps (silver staining). (**d**) Proposed mechanism of *p*BQdC activity and APE1 in the processing of the adduct-containing 25-mer. The red arrow indicates the incision site of the enzymes

This assay can be used to study an NER activity against the lesion of interest, if the length of the oligomer duplex is long enough. We previously directly used synthetic oligomers or ligated products (e.g., 40, 60, 90, and 132-base pairs) to test NER incision of various DNA lesions [5, 10, 38, 39] and obtained results in consistence with NER incision patterns reported previously.

3.5 Primer Extension Assay with a DNA Polymerase

Although this is not a repair assay, with the availability of an oligonucleotide template containing a single, site-specifically placed lesion, it can be performed easily to assess the potential mutation of a new derivative, in the presence of classic DNA polymerases or Y-family DNA polymerases specializing in translesion DNA synthesis (TLS) [40]. The primers with appropriate sizes can be synthesized and purified commercially. It is important that they are purified by both HPLC and PAGE to ensure all the oligomers have the exact size before annealing. Technically, many steps such as the 5′-^{32}P labeling and gel electrophoresis are the same as the repair cleavage assay. With this assay, we successfully investigated miscoding features of various DNA adducts with most of the currently identified Y-family of DNA polymerases (pols η, ι, κ, and Rev1), which play major roles in the TLS in coordination with the B-family polymerase, pol ζ [41–43].

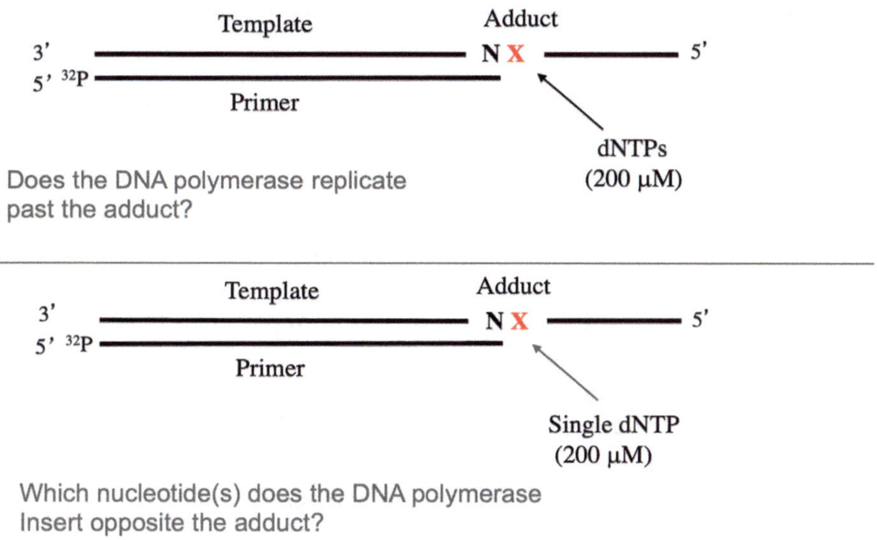

Fig. 5 Assay design for in vitro translesion synthesis (TLS) studies. (**a**) Full replication assay. (**b**) Insertion specificity assay. Note that in both cases, the standing start condition is presented. For a running start experiment, one or more running start dNTP(s) is present at a final concentration of 5–10 μM

As shown in Fig. 5, there are two types of DNA replication assay based on the dNTPs added to the replication reaction:

1. Full replication assay: This is to examine whether a DNA polymerase replicates past the lesion. As shown in Fig. 5, the oligomer to be tested is annealed to its $5'$-^{32}P-labeled primer, terminating 1, 2, or more bases on the $3'$ side of the bases to be tested. The elongation of the primer in the presence of all four dNTPs, if it occurs, is seen as additional bands on the denaturing PAGE (see refs [41–43] for examples).
2. Insertion specificity assay: A single dNTP is placed in the reaction mixture at a time to ask the question: which base is incorporated opposite the adduct, and what are the misincorporation frequencies for each of dNTP? Dr. Myron Goodman's laboratory has established a gel assay for determining the kinetic constants, K_m and V_{max}, to calculate the misincorporation frequency of nucleotide insertion (V_{max}/K_m) opposite a base, which is the basis for assessing mutagenic potential [44]. The assay can be used to directly measure DNA polymerase fidelity, which is defined as the ratio of right (R) to wrong (W) nucleotide incorporations when dRTP and dWTP substrates compete at equal concentrations for primer extension at the same site [45].

4 Notes

1. There is an excess of complementary strand in the annealing reaction, to ensure the template oligonucleotide is fully annealed.
2. The poly[dI·dC] is only used for CFE, to knock down nonspecific binding proteins to the oligomer substrate.
3. Reaction conditions could be optimized based on the repair protein properties. For example, the pH of the reaction buffer can be changed. Also the final salt concentration may be adjusted when a CFE or column fractions are tested, as salt concentrations in such samples typically are high.
4. This step is to make sure the denatured oligomers due to heating anneal again with slow cooling.
5. Another fact is that for monofunctional glycosylase TDG, which is known as a single turnover enzyme, addition of APE stimulates its activity against substrates [36].
6. The size of a marker oligomer depends on the position of the base lesion. Normally the size of the marker is identical to the 5′-^{32}P-labeled cleavage end product of the template. It needs to be ^{32}P end-labeled the same way as the 25-mer oligomer. Dilute the oligo 1:20 to 1:100, mix with the loading buffer, and load 5 μL with other samples.
7. The running buffer (1×TBE) must be collected into the liquid radioactive waste.
8. An X-ray film can also be exposed to the gel in a cassette and then developed to show the bands and cleavage products analyzed.

Acknowledgments

I thank Dr. B. Singer of Lawrence Berkeley National Laboratory for her instrumental support. I also thank Dr. A. Chenna for his long-term collaboration and invaluable help. The work included in this chapter was supported by the National Institutes of Health (NIH) grants CA72079 (to B.H) and CA47723 (to B.S. and B.H.) and was administered by the Lawrence Berkeley National Laboratory under the Department of Energy contract DE AC02-05CH11231.

References

1. Friedberg EC, Walker GC, Siede W, Wood RD (2005) DNA repair and mutagenesis, vol 2. ASM Press, p 1118
2. Chatterjee N, Walker GC (2017) Mechanisms of DNA damage, repair, and mutagenesis. Environ Mol Mutagen 58(5):235–263
3. Hang B (2006) In: Wei Q, Li L, Chen D (eds) Base excision repair in DNA repair, genetic instability, and cancer. World Scientific Publishing Co. Pte. Ltd., Singapore, pp 23–64
4. Singer B, Hang B (1999) Mammalian enzymatic repair of etheno and para-benzoquinone exocyclic adducts derived from the carcinogens vinyl chloride and benzene. IARC Sci Publ 150:233–247
5. Hang B, Chenna A, Fraenkel-Conrat H, Singer B (1996) An unusual mechanism for the major human apurinic/apyrimidinic (AP) endonuclease involving 5' cleavage of DNA containing a benzene-derived exocyclic adduct in the absence of an AP site. Proc Natl Acad Sci U S A 93(24):13737–13741
6. Gros L, Saparbaev MK, Laval J (2002) Enzymology of the repair of free radicals-induced DNA damage. Oncogene 21(58):8905–8925
7. Gros L, Ishchenko AA, Ide H, Elder RH, Saparbaev MK (2004) The major human AP endonuclease (Ape1) is involved in the nucleotide incision repair pathway. Nucleic Acids Res 32(1):73–81
8. Delaney JC, Essigmann JM (2008) Biological properties of single chemical-DNA adducts: a twenty year perspective. Chem Res Toxicol 21(1):232–252
9. Wood RD, Robins P, Lindahl T (1988) Complementation of the xeroderma pigmentosum DNA repair defect in cell-free extracts. Cell 53: 97–106
10. Chenna A, Hang B, Rydberg B, Kim E, Pongracz K, Bodell WJ, Singer B (1995) The benzene metabolite p-benzoquinone forms adducts with DNA bases that are excised by a repair activity from human cells that differs from an ethenoadenine glycosylase. Proc Natl Acad Sci U S A 92(13):5890–5894
11. Biggerstaff M, Wood RD (2006) Repair synthesis assay for nucleotide excision repair activity using fractionated cell extracts and UV-damaged plasmid DNA. Methods Mol Biol 314:417–434
12. Ten Kornberg A (2003) Commandments of enzymology, amended. Trends Biomed Sci 28(10):515–517
13. Singer B, Hang B (1997) What structural features determine repair enzyme specificity and mechanism in chemically modified DNA? Chem Res Toxicol 10(7):713–732
14. Hang B (2004) Repair of exocyclic DNA adducts: rings of complexity. BioEssays 26(11):1195–1208
15. Hang B (2010) Formation and repair of tobacco carcinogen-derived bulky DNA adducts. J Nucl Acids 2010:709521
16. Hang B, Sarker AH, Havel C, Saha S, Hazra TK, Schick S, Jacob P 3rd, Rehan VK, Chenna A, Sharan D, Sleiman M, Destaillats H, Gundel LA (2013) Thirdhand smoke causes DNA damage in human cells. Mutagenesis 28(4):381–391
17. Sarker AH, Chatterjee A, Williams M, Lin S, Havel C, Jacob P 3rd, Boldogh I, Hazra TK, Talbot P, Hang B (2014) NEIL2 protects against oxidative DNA damage induced by sidestream smoke in human cells. PLoS One 9(3):e90261
18. Chenna A, Singer B (1995) Large scale synthesis of p-benzoquinone-2′-deoxycytidine and p-benzoquinone-2′-deoxyadenosine adducts and their site-specific incorporation into DNA oligonucleotides. Chem Res Toxicol 8(6): 865–874
19. Chenna A, Singer B (1997) Synthesis of a benzene metabolite adduct, 3″-hydroxy-1,N2-benzetheno-2′-deoxyguanosine, and its site-specific incorporation into DNA oligonucleotides. Chem Res Toxicol 10(2):165–171
20. Chenna A, Gupta RC, Bonala RR, Johnson F, Hang B (2008) Nucleosides Synthesis of the fully protected phosphoramidite of the benzene-DNA adduct, N2-(4-Hydroxyphenyl)-2′-deoxyguanosine and incorporation of the later into DNA oligomers. Nucleot Nucl Acids 27(8):979–991
21. Chenna A, Perry A, Singer B (2000) Synthesis of 8-(hydroxymethyl)-3,N^4-etheno-2′-deoxycytidine, a potential carcinogenic glycidaldehyde adduct, and its site-specific incorporation into DNA oligonucleotides. Chem Res Toxicol 13(3):208–213
22. Hang B, Downing G, Guliaev AB, Singer B (2002) Novel activity of Escherichia coli mismatch uracil-DNA glycosylase (Mug) excising 8-(hydroxymethyl)-3,N^4-ethenocytosine, a potential product resulting from glycidaldehyde reaction. Biochemistry 41(7):2158–2165
23. Chenna A, Maruenda H, Singer B (1999) Synthesis of para-benzoquinone and 1,3-bis (2-chloroethyl)nitrosourea adducts and their incorporation into oligonucleotides. IARC Sci Publ 150:89–101

24. Maruenda H, Chenna A, Liem LK, Singer B (1998) Synthesis of 1,N6-ethano-2′-deoxyadenosine, a metabolic product of 1,3-bis(2-chloroethyl)nitrosourea, and its incorporation into oligomeric DNA. J Org Chem 63:4385–4389
25. Zhang W, Rieger R, Iden C, Johnson F (1995) Synthesis of 3,*N*[4]-etheno, 3,*N*[4]-ethano, and 3-(2-hydroxyethyl) derivatives of 2′-deoxycytidine and their incorporation into oligomeric DNA. Chem Res Toxicol 8:148–156
26. Bonala RR, Rieger RA, Shibutani S, Grollman AP, Iden CR, Johnson F (1999) 3,N^4-ethano-2′-deoxycytidine: chemistry of incorporation into oligomeric DNA and reassessment of miscoding potential. Nucleic Acids Res 27(24): 4725–4733
27. Bonala RR, Torres MC, Attaluri S, Iden CR, Johnson F (2005) Incorporation of N2-deoxyguanosine metabolic adducts of 2-aminonaphthalene and 2-aminofluorene into oligomeric DNA. Chem Res Toxicol 18(3):457–465
28. Takeshita M, Chang CN, Johnson F, Will S, Grollman AP (1987) Oligodeoxynucleotides containing synthetic abasic sites. Model substrates for DNA polymerases and apurinic/apyrimidinic endonucleases. J Biol Chem 262: 10171–10179
29. Schnetz-Boutaud NC, Mao H, Stone MP, Marnett LJ (2000) Synthesis of oligonucleotides containing the alkali-labile pyrimidopurinone adduct, M_1G. Chem Res Toxicol 13:90–95
30. Fujimoto J, Tran L, Sowers LC (1997) Synthesis and cleavage of oligodeoxynucleotides containing a 5-hydroxyuracil residue at a defined site. Chem Res Toxicol 10:1254–1258
31. Guerniou V, Gasparutto D, Sauvaigo S, Favier A, Cadet J (2003) New synthesis of 5-carboxy-2′-deoxyuridine and its incorporation into synthetic oligonucleotides. Nucleosides Nucleotides Nucleic Acids 22:1073–1075
32. Laxmi YRS, Suzuki N, Dasaradhi L, Johnson F, Shibutani S (2002) Preparation of oligodeoxynucleotides containing a diaste-reoisomer of α-(N^2-2′-deoxyguanosinyl)tamoxifen by phosphoramidite chemical synthesis. Chem Res Toxicol 15:218–225
33. Hang B, Guliaev AB (2007) Substrate specificity of human thymine-DNA glycosylase on exocyclic cytosine adducts. Chem Biol Interact 165(3):230–238
34. Singer B, Hang B (2000) Nucleic acid sequence and repair: role of adduct, neighbor bases and enzyme specificity. Carcinogenesis 21(6):1071–1078
35. Sági J, Hang B, Singer B (1999) Sequence-dependent repair of synthetic AP sites in 15-mer and 35-mer oligonucleotides: role of thermodynamic stability imposed by neighbor bases. Chem Res Toxicol 12(10):917–923
36. Hang B, Chenna A, Rao S, Singer B (1996) 1,N^6-ethenoadenine and 3,N^4-ethenocytosine are excised by separate human DNA glycosylases. Carcinogenesis 17(1):155–157
37. Hang B, Medina M, Fraenkel-Conrat H, Singer B (1998) A 55-kDa protein isolated from human cells shows DNA glycosylase activity toward 3,N4-ethenocytosine and the G/T mismatch. Proc Natl Acad Sci U S A 95(23):13561–13566
38. Rodriguez B, Yang Y, Guliaev AB, Chenna A, Hang B (2010) Benzene-derived N2-(4-hydroxyphenyl)-deoxyguanosine adduct: UvrABC incision and its conformation in DNA. Toxicol Lett 193(1):26–32
39. Kumaresan KR, Hang B, Lambert MW (1995) Human endonucleolytic incision of DNA 3′ and 5′ to a site-directed psoralen monoadduct and interstrand cross-link. J Biol Chem 270(51):30709–30716
40. Vaisman A, Woodgate R (2017) Translesion DNA polymerases in eukaryotes: what makes them tick? Crit Rev Biochem Mol Biol 52(3): 274–303
41. Hang B, Chenna A, Guliaev AB, Singer B (2003) Miscoding properties of 1,N^6-ethanoadenine, a DNA adduct derived from reaction with the antitumor agent 1,3-bis(2-chloroethyl)-1-nitrosourea. Mutat Res 531(1–2): 191–203
42. Singer B, Medina M, Zhang Y, Wang Z, Guliaev AB, Hang B (2002) 8-(Hydroxymethyl)-3,N^4-etheno-C, a potential carcinogenic glycidaldehyde product, miscodes in vitro using mammalian DNA polymerases. Biochemistry 41(6):1778–1785
43. Xie Z, Zhang Y, Guliaev AB, Shen H, Hang B, Singer B, Wang Z (2005) The p-benzoquinone DNA adducts derived from benzene are highly mutagenic. DNA Repair (Amst) 4(12): 1399–1409
44. Boosalis MS, Petruska J, Goodman MF (1987) DNA polymerase insertion fidelity. Gel assay for site-specific kinetics. J Biol Chem 262(30): 14689–14696
45. Bertram JG, Oertell K, Petruska J, Goodman MF (2010) DNA polymerase fidelity: comparing direct competition of right and wrong dNTP substrates with steady state and pre-steady state kinetics. Biochemistry 49(1): 20–28

Chapter 6

In Vitro Reconstitutive Base Excision Repair (BER) Assay

Aruna S. Jaiswal, Elizabeth A. Williamson, Arunima S. Jaiswal, Kimi Kong, and Robert A. Hromas

Abstract

The mammalian cell genome is continuously exposed to endogenous and exogenous insults that modify its DNA. These modifications can be single-base lesions, bulky DNA adducts, base dimers, base alkylation, cytosine deamination, nitrosation, or other types of base alteration which interfere with DNA replication. Mammalian cells have evolved with a robust defense mechanism to repair these base modifications (damages) to preserve genomic stability. Base excision repair (BER) is the major defense mechanism for cells to remove these oxidative or alkylated single-base modifications. The base excision repair process involves replacement of a single-nucleotide residue by two sub-pathways, the single-nucleotide (SN) and the multi-nucleotide or long-patch (LP) base excision repair pathways. These reactions have been reproduced in vitro using cell free extracts or purified recombinant proteins involved in the base excision repair pathway. In the present chapter, we describe the detailed methodology to reconstitute base excision repair assay systems. These reconstitutive BER assay systems use artificially synthesized and modified DNA. These reconstitutive assay system will be a true representation of biologically occurring damages and their repair.

Key words DNA damage, Base excision repair, AP-site, dRP lyase activity, AP endonuclease activity, Flap endonuclease activity, Long-patch repair

1 Introduction

The mammalian genome is continuously exposed to both endogenous and environmental insults that result in base modification (Fig. 1). About 100,000 modifications occur every day on the genome and these modifications are continuously repaired to preserve the genomic integrity. Genomic integrity depends on efficient removal and repair mechanisms [1]. Cells have evolved with repair mechanisms to repair oxidative and DNA alkylation damage from both endogenous compounds and exogeneous environmental agents. Not only oxidative stress (endogenous sources) causes DNA damage but various other events in the metabolic pathways of the cell can also result in base damage of the genome [2, 3]. Exogenous sources can be therapeutic agents (methylating agents) used

Kishor K. Bhakat and Tapas K. Hazra (eds.), *Base Excision Repair Pathway: Methods and Protocols*, Methods in Molecular Biology, vol. 2701, https://doi.org/10.1007/978-1-0716-3373-1_6,

for the clinical intervention, such as nitrogen mustards (e.g., melphalan, cyclophosphamide, and chlorambucil), methyl nitrosourea (e.g., carmustine and lomustine), platinum-based drugs (e.g., cisplatin, oxaliplatin), triazine family (temozolomide, dacarbazine, procarbazine) [4], tobacco-specific nitrosamines, and drugs like streptozotocin that form adducts at N- and O-atoms in DNA bases. Base lesions generated after exposure to these agents are mainly repaired by direct base repair, base excision repair (BER), and, to some extent, nucleotide excision repair (NER) [5]. Most alkylating drugs used in chemotherapy cause DNA lesions at guanine such as O^6-methylguanine (O^6-MeG), N^7-methylguanine (N^7-MeG), and N^3-methylguanine (N3-MeG) and at adenine as N^3-methyladenine (N^3-MeA). The O^6-MeG lesion is repaired by the O6-methylguanine-DNA-methyltransferase (MGMT) pathway. The N^7-MeG and N^3-MeA lesions are repaired by the BER pathway [6]. Another endogenous agent, nitric oxide, also induces AP sites that must be detected and repaired. Inflamed epithelial cells release reactive oxygen (ROS) and reactive nitrogen species (RNS). Nitric oxide also causes DNA damage in the form of 8-oxo-7,8-dihydro-2′-deoxyguanosine and 8-nitroguanine [7]. Nitrosating agents have two major effects on DNA, (i) alkylation of DNA and (ii) conversion of secondary amines into alkylating agents similar to N-methyl-N nitrosourea (MNU) which leads to DNA alkylation and causes genomic instability. Cytosine deamination is most prevalent and converted to uracil upon the hydrolytic attack of the amine group on the outermost 10 bases of the ends of DNA fragments [8]. Deamination of nucleobases is catalyzed by the formation of unstable N-nitroso derivatives of exocyclic amines which converts some of guanine bases in the DNA into xanthine, adenine bases into hypoxanthine, and cytosine bases into uracil (Fig. 1). Deaminated cytosine residues can be removed enzymatically by uracil DNA glycosylase (UDG) [9]. The removal of deaminated bases from DNA is the initial stage of base excision repair which can be impaired by the presence of a nitrosative deaminated base on the DNA strand. This impaired repair of the modified base will be translated to either mutation induction or a pause in replication. If these nitrosative deaminated bases are not removed in a timely fashion, then persistent mutation could trigger the process of tumorigenesis.

BER is considered to be the primary mechanism for repairing most forms of spontaneous hydrolytic, alkylative, and oxidative DNA damage. BER has been subdivided into two major pathways based on the number of nucleotides incorporated during repair synthesis, that is, single nucleotide (SN) and multi-nucleotide, or long patch (LP), which involves incorporation of 2–13 nucleotides and strand displacement polymerization (Fig. 2) [10]. In most cases, excision of a damaged base is initiated by one of the 11 DNA glycosylase enzymes which generates a potentially

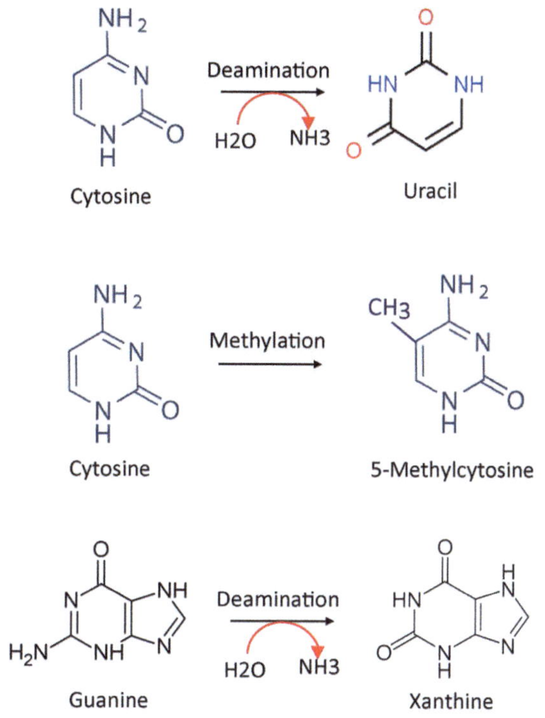

Fig. 1 DNA modification by deamination or alkylation reactions. Deamination and alkylation of cytosine occur very frequently in the cells and modified bases are removed by the BER pathway

cytotoxic apurinic or apyrimidinic site (AP site) intermediate. AP sites are very unstable (labile) and highly mutagenic (Fig. 3) and spontaneously degrade into DNA strand breaks by β-elimination [11]. These unstable AP sites can also result in non-template DNA and RNA synthesis. AP sites are recognized and cleaved by apurinic/apyrimidinic endonuclease 1 (APE1) at 5′- to abasic site to generate a strand break. If AP sites are not repaired efficiently, they can lead to tumor initiation and progression [12]. There are two important enzymatic activities of Pol β in the mammalian single-nucleotide BER (SN-BER): first is DNA resynthesis or insertion of the correct base at the repair site and second is the removal of the 5′-deoxyribose phosphate (dRP) residue [13]. If the 5′-dRP residue is not removed before the ligation step, then DNA ligase I-mediated ligation will not be completed. If this dRP residue is resistant to the dRP lyase activity of Pol β, then the nucleotidyl transferase activity of Pol β will continue adding nucleotides to the 3′-end past the nicking site resulting in the generation of a flap structure of the downstream nicked strand. These flap structures are removed by the structure-specific nuclease FEN1. After the flap structure is removed, the resulting gap is sealed by DNA ligase I or III [10–12]. In SN-BER, the XRCC1/DNA ligase 3 (LIG3) complex catalyzes the DNA ligation step, while LP-BER uses DNA ligase 1 (LIG1) [14].

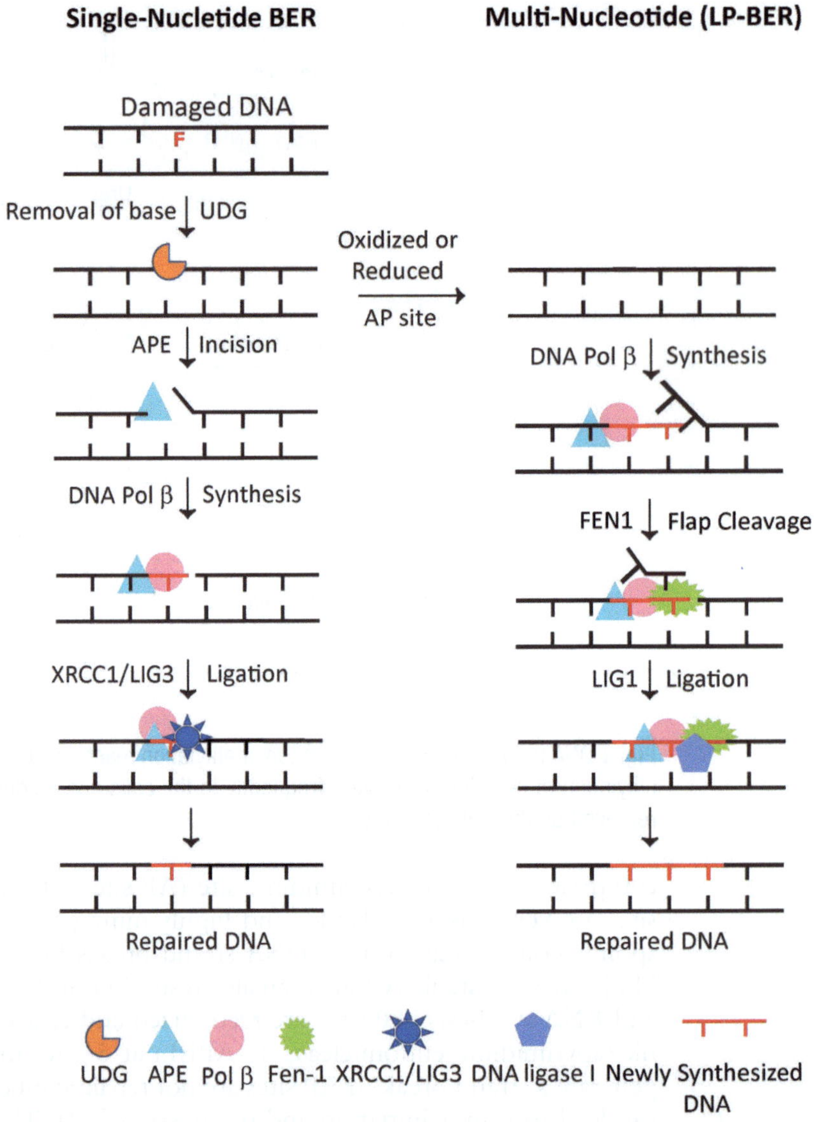

Fig. 2 Stepwise representation showing the sequential steps of BER. There are two sub-pathways for BER, single-nucleotide (SN) BER and multi-nucleotide or long-patch (LP) BER. There is some overlap on the steps in SN-BER and LP-BER. The first step in BER is executed by one of the 11 DNA glycosylases, in which uracil DNA glycosylase removes uracil residue from the DNA and creates an abasic site. DNA glycosylases can be either mono- or bifunctional. Once the base is removed by DNA glycosylases, the AP endonuclease 1 (APE1) incises 5′- to the abasic site and generates a strand break. This single-strand break generates a 3′-hydroxyl group and a unique 5′-deoxyribose phosphate (dRP). This dRP moiety is removed by dRP lyase activity of DNA polymerase β (Pol β) during SN-BER. In SN-BER, the Pol β nucleotidyl transferase activity of Pol β incorporates the correct base at the site of the damaged base, and finally, the XRCC1/DNA ligase III complex seals the gap. However, if the AP site is oxidized or reduced, it becomes resistant to the dRP lyase activity of Pol β. In this situation SN-BER is not a preferred option; rather it opts for the LP-BER pathway. In LP-BER, polynucleotidyl transferase activity of Pol β incorporates 2–13 nucleotides which generates a flap structure. The resulting flap structure can be removed by 5′- flap endonuclease activity of structure-specific flap-endonuclease 1 (FEN1). Finally, the gap is sealed by DNA ligase 1 (LIGI)

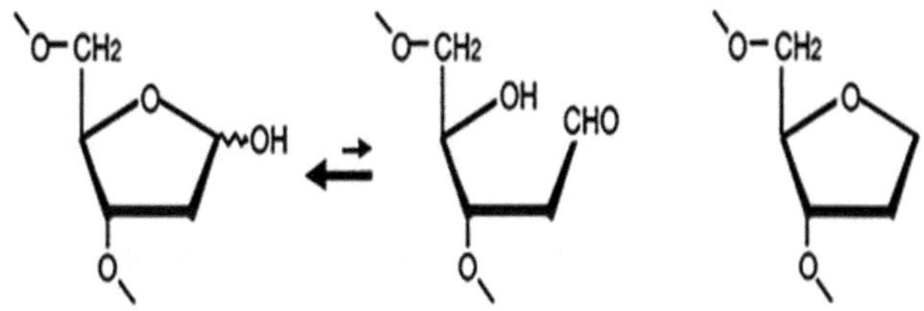

Fig. 3 Structure of naturally occurring abasic sites and chemically synthesized sites

In the present chapter, we provide a detailed assay procedure to reconstitute in vitro BER assay using recombinant purified proteins thus revealing a sequential enzymatic step of a biochemical pathway of DNA repair. This will help us understand the functionality and biological activity of these proteins, deduce the dynamics and spatial interaction of these proteins in the microenvironment of repairosome, and determine how the interacting partner of these proteins can affect repair efficiency. This method can be further explored to define how these proteins assemble at the replication fork and repair the damage to restore the normally replicating forks. In addition, these reconstitutive assays can also be used to screen the drugs which target BER proteins at various steps of the BER pathway.

2 Materials

2.1 Laboratory Equipment

1. Gilson pipets (P2, P10, P20, P100, P200, and P1000) from Gilson Inc. (WI, USA).
2. Pipet aid from Drummond (PA, USA).
3. DNA gel electrophoresis apparatus model S2 from Life Technologies.
4. Power Pac HV (5000v/500 mA/400w) from Bio-Rad (CA, USA).
5. Gel drier from Bio-Rad (CA, USA).
6. Microfuge Centrifuge model 5417R (Eppendorf).
7. Accumet Basic AB15 pH meter from Fisher Scientific (MA, USA).
8. Glass plates (GBP-419-333-5; GBP-419-394-5) from Moliterno Inc. (CA, USA).
9. Accu Block Digital Dry Bath or water bath from Labnet Inc. (NJ, USA).
10. Typhoon Trio Biomolecular Imager (GE Healthcare Bio-Science Corporation, NJ, USA).

11. 4 °C refrigerator.
12. −20 °C freezer from GE.
13. −80 °C freezer from Fisher Scientific.

2.2 Laboratory Reagents

1. Gel preparation glass plates, spacers, and comb.
2. Mini-PROTEAN gel running apparatus.
3. Electrophoresis tank.
4. 37 °C incubator.
5. Boiling water bath/heat block and 37 °C water bath.
6. Saran wrap.
7. Blotting sheets from Thermo Fisher Scientific (cat. no. 057163F).
8. 1 M Tris–HCl pH 7.5 from Sigma (Trizma base cat. no. T1503; Trizma HCl cat. no. T3253).
9. 0.5 M EDTA pH 8.0 from Sigma (cat. no. D0632).
10. Sigmacote from Sigma (cat. no. SL2).
11. 1 M DTT from Sigma (cat. no. D0632).
12. 1 M NaCl from Sigma (cat. no. S3014).
13. 1 M KCl from Sigma (cat. no. P9541).
14. 4 M NaCl.
15. 1 M $MgCl_2$ from Sigma (cat. no. M2670).
16. 10% NP40 from US Biologicals (cat. no. N3500).
17. 100% glycerol from Sigma (cat. no. G6279).
18. 10X TBE buffer from Sigma (cat. no. T4415).
19. 1 M HEPES from Sigma (cat. no. T4415) (pH 7.5).
20. 3 M sodium acetate (pH 5.2).
21. 70% and 100% ethanol.
22. t-RNA from Sigma (cat. no. H3375).
23. Glycogen from Sigma (cat. no. G1767).
24. Bromophenol blue from Cytiva (cat. no. 45-500-11).
25. Xylene cyanol from Sigma (cat. no. X4126).
26. RNase A solution (Sigma, cat. no. R6148).
27. Proteinase K (Sigma, cat. no. P5568).
28. Acrylamide (40% solution) from Bio-Rad (cat. no 161-0144).
29. TEMED from Bio-Rad (cat. no. 161-0800).
30. Sodium borohydride from Sigma (cat. no. 452882).
31. Urea from Bio-Rad (cat. no. 161-0731).
32. 70% ethanol.

33. Formamide from Invitrogen.
34. RNase A solution from (cat. no. R6148).
35. Proteinase K from Sigma (cat. no. P5568).
36. [γ-^{32}P] ATP (3000 Ci/mmol) from Perkin Elmer, Inc. (Boston, MA) (*see* **Note 1**).
37. dATP, dGTP, dTTP, and dCTP (individually or mix from Sigma (cat. no. D7295).
38. 1 mg/mL bovine serum albumin (BSA Sigma, cat. no. A7030).
39. Purified recombinant APE1 protein.
40. Purified recombinant DNA polymerase β protein.
41. Purified recombinant FEN1 protein.
42. Purified recombinant DNA ligase 1 protein.
43. Purified recombinant uracil DNA glycosylase protein from New England Biolab (cat. no. M0372).
44. T4 polynucleotide kinase from New England Biolab (cat. no. M0201S).

Most of the reagents listed above were either directly procured from the vendors listed or prepared in the laboratory to the desired molarity. All the solutions were steam sterilized by autoclaving at 121 °C and 15 psi of pressure or filter sterilized for heat-sensitive solutions by filtering through 0.22 μm filters. Utmost precaution was taken to appropriately store the reagents under nuclease-free condition.

2.3 Oligonucleotides

All oligonucleotides for in vitro nuclease and reconstitutive SN- and LP-BER assays were custom synthesized from Sigma-Genosys (Woodlands, TX). The nucleotide sequence of these oligonucleotides contained a uracil residue or an AP site analog, 3-hydroxy-2-hydroxymethyltetrahydrofuran (designated as F in the oligomers), both positioned at 24-nt. All the oligomers used were PAGE (polyacrylamide gel electrophoresis) purified and reconstituted in TE buffer pH 8.0.

Sense 63-mer U-DNA

5′- CTAGATGCCTGCAGCTGATGCGC **U** GTACGGATC CACGTGTACGGTACCGAGGGCGGGT

CGAGA

Sense 63-mer F-DNA

5′- CTAGATGCCTGCAGCTGATGCGC **F** GTACGGATC CACGTGTACGGTACCGAGGGCGGGTCGAGA

Anti-Sense 63-mer F-DNA

5′- TCTCGACCCGCCCTCGGTACCGTACACGTGGATC CGTACGGCGCATCAGCTGCAGGCAT

CTAG

dRP Lyase Sense 43-mer DNA

5′- TAGACTAGATGCCTGCAGCTGATGUCGCCGTACG GATCCACGT Fam

dRP Lyase Anti-Sense 43-mer DNA

5′- ACGTGGATCCGTACGGCGGCATCAGCTGCAGGCA TCTAGTCTA

Fen1 UPS Sense 24-mer DNA

5′-TAGACTAGATGCCTGCAGCTGATG

Fen1 DWS Sense 28-mer DNA

5′-Fam AAATTGGGTTCGCCGTACGGATCCACGT

3 Methods

3.1 Oligonucleotide Radiolabeling

Polyacrylamide gel electrophoresis (PAGE)-purified oligonucleotides were reconstituted in TE buffer pH 8.0 before labelling. Oligos for BER assays were labeled on sense strands and then annealed to their corresponding complementary oligos. Nucleotide sequences for SP-BER and LP-BER substrates contained one uracil residue or THF (THF is 3-hydroxy-2-hydroxymethyltetrahydrofuran), positioned at 24-nt as described in the oligonucleotide section. For BER assays the 5′-end of the sense strand is labelled with T4-polynucleotide kinase (PNK). After labelling, sense oligo was annealed to the corresponding complementary oligo. All the oligonucleotide sequences used in this study are provided in the reagent section. The reaction for labelling of an oligonucleotide was constituted on ice in a total volume of 25 μL by aliquoting water, followed by addition of 2.5 μL of 10× T4 kinase buffer and 1.5 pMol of sense strand, namely, 63-mer F DNA or U-DNA, to a microcentrifuge tube. Enzymatic transfer of the label was initiated by addition of ^{32}P-γ-ATP and T4 polynucleotide kinase using a standard protocol described by the manufacturer [15]. The reaction was incubated at 37 °C for 20 min. After 20 min, the T4-polynucleotide kinase was inactivated by addition of 1/10 volume of 0.5 M EDTA or heating the reaction content to 65 °C for 10 min to stop the reaction. The labelled sense strand oligo was annealed to an unlabeled antisense oligo (cold) as described below. The labeled annealed oligomer was purified using Nick column as described by the manufacturer (GE Healthcare Biosciences, Pittsburgh, PA), quantitated and stored at −20 °C until use (*see* **Note 1**). Nucleotide sequences for SP-BER and LP-BER substrates contained one uracil residue or THF, positioned at 24-nt as described [15] (*see* **Notes 1** and **2**).

Reaction for 25 μL volume:

Sense 63-mer F-oligonucleotide = 1–2 μL

10× polynucleotide kinase buffer = 2.5 μL

^{32}P-γ-ATP (10 μCi/μL) = 5.0 μL

T4-polynucleotide kinase (10,000 μ/mL) = 1.0 μL

Water (double distilled) = q.s. to 25 μL

The reaction was incubated at 37 °C for 20 min in dry bath (*in a radioactive work area*).

Alternatively, we can use nonradioactive assays. While fluorescence is a very sensitive technique, weak signals often limit visualization and quantitation of low-abundance molecules in the context of gene expression analysis, biomarker detection, and cellular imaging. It is important to note that signals for fluorophore can be detected with a Typhoon Trio Biomolecular Imager (GE Healthcare Bio-Science Corporation, Piscataway, NJ, USA) instrument. Accordingly, we synthesized 5′-fluorophore-labelled oligo for the nonradioactive assay procedure. We used fluorescein (6-FAM) as a fluorophore at either the 5′-end or 3′-end of oligonucleotides. It is important to note that FAM is protonated below pH 7.0 and will have reduced fluorescence so utmost precaution was taken to not let FAM-labeled oligos be exposed below pH 7.0. There are multiple fluorophores available, and based upon their characteristics, we can select with optimal stability and signal to noise ratio. What works best for one can be checked with the vendors for oligosynthesis and capability to detect the signal.

3.2 Oligonucleotide Annealing

Equimolar amounts of complementary oligonucleotides were annealed in 0.05 mL containing 10 mM Tris–HCl, pH 7.5, 1 mM EDTA, and 100 mM KCl. Solutions were heated to 85 °C for 10 min and allowed to gradually cool to room temperature. Annealed oligos were stored at −20 °C (*see* **Note 2**).

3.3 Recombinant Protein Purification

All human recombinant hexahistidine-tagged APE1, Pol β, FEN1, and LIG1 proteins were overexpressed in BL21 (DE3) cells and purified to homogeneity according to our published protocols [15–18]. We acquired purified human Pol β from the late Samuel Wilson (NIEHS, Research Triangle Park, NC, USA) and purified XRCC1/LIG3 complex from Alan Tomkinson (University of New Mexico, Albuquerque, NM, USA).

3.4 Base Excision Repair Assay Procedures

In this section, we describe an assay procedure for most of the enzymatic steps involved in BER. We begin with APE1's incision activity followed by other activities. APE1 is a multifunctional enzyme and exhibits several activities such as AP site incision activity (AP endonuclease activity), exonuclease activity, redox activity, nucleotide incision repair activity, and RNA processing activity. AP endonuclease activity is a key activity for BER and APE1 performs this activity with utmost precision by orienting DNA to the appropriate cleavage site in an optimal position for nucleophilic attack within its compact protein active site. However, its exonuclease activity has a significantly different salt and pH requirement and will not be discussed. It remains unclear how other activities of APE1 are coordinated within the cell.

3.4.1 Nuclease Activity Assay of APE1

All incision reactions of recombinant APE1 were conducted under steady state and within a linear range of assay. The reactions were assembled on ice in a standard reaction mixture of 25 μL as described below.

1. A master mix containing 30 mM Tris–HCl, pH 7.5, 30 mM KCl, 5 mM $MgCl_2$, 2 mM DTT, 5% glycerol, 0.01% Nonidet P-40, and 0.1 mg/mL bovine serum albumin was prepared.
2. 10 μL of master mix was added to each tube.
3. 1 pMol to 1 nMol of recombinant purified APE1 was added to each tube leaving the first tube without APE1 which will serve as the uncleaved substrate control. The amount of APE1 depends upon the specific activity of the protein. However, the optimal concentration of the protein was empirically determined.
4. The reaction was initiated by the addition of the ^{32}P-labeled 63-mer F-DNA substrate (approximately 100,000 cpm). The reaction contents were gently mixed and the reaction mixture was incubated at 37 °C for 30 min (*see* **Note 1**).
5. After 30 min, an equal volume of stop buffer containing 0.4% (w/v) SDS, 5 mM EDTA, and 1 μg of proteinase K was added to each tube to stop the reaction.
6. DNA was extracted from the reaction mixture by addition 50 μL of phenol/chloroform/isoamyl alcohol (25:24:1; v/v). The contents of each tube were mixed by vortexing. Microfuge tubes were centrifuged at 12000 rpm for 10 min in a tabletop centrifuge to separate the aqueous phase from the organic phase. The upper phase containing the APE1 cleaved and uncleaved DNA was quantitatively collected in a fresh tube. DNA was ethanol precipitated by addition of 1/10 volume of 3 M sodium acetate, pH 5.2, and two volumes of ethanol.
7. 1–2 μg of glycogen or tRNA was added to the samples as a carrier and mixed well to facilitate efficient precipitation of DNA. A 2.5× volume of ethanol was added to each tube and mixed well. The tubes were incubated at −20 °C overnight preferably or − 80 °C for 1–2 h.

8. The DNA was recovered by centrifugation at 14,000 rpm for 20 min. The ethanol was removed leaving behind the DNA pellet. The recovered DNA was washed once with 2× volumes of cold 70% ethanol and centrifuged at 14,000 rpm for 20 min.
9. Ethanol was carefully removed with a pipet or syringe without disturbing the DNA pellet, and the pellet was briefly dried at room temperature or till residual ethanol has completely evaporated.
10. The pelleted DNA was resuspended in 5–10 μL of sample loading dye solution (90% formamide, 1 mM EDTA, 0.1% xylene cyanol, and 0.1% bromophenol blue).
11. Denaturing polyacrylamide gel (7 M urea and 15% acrylamide) was prepared skillfully without trapping any air bubbles in the gel specifically in the wells. Care should be taken while taking out the comb so that wells are not damaged. Combs and spacer used for casting the gels were of 0.4 mm thickness. The recipe for the gel is described below (*see* **Notes 4** and **5**).
12. Samples were heat denatured at 85 °C for 5 min and cooled quickly on ice. Samples were kept on ice until loaded on the gel. Samples were centrifuged before loading to bring every droplet to the bottom.
13. Samples (4 μL) were loaded onto a 7 M urea and 15% denaturing polyacrylamide gel for electrophoresis and run at 2000 volts for about an hour. Leftover samples can be stored at −20 °C (*see* **Note 4**).
14. After completion of the run, the gel plates were pry-opened and the gel was transferred to a blotting sheet. The gel was dried under vacuum (80 °C for 2 h) and then subjected to autoradiography. Signals were captured on X-ray films.
15. For nonradioactive APE1-cleaved products, samples were resolved on a 7 M urea and 15% denaturing polyacrylamide gel. DNA on the gel was visualized by scanning in the Typhoon Trio Biomolecular Imager (GE Healthcare Bio-Science Corporation, Piscataway, NJ, USA), at excitation of 526 nm and emission at 488 nm. Digital images were acquired and analyzed for quantitation using ImageQuant TL (GE Healthcare Bio-Science Corporation, Piscataway, NJ, USA) software (*see* **Notes 2–4**).

   Recipe for 7 M urea/15% acrylamide gel:

   DDW = 12 mL

   10 × TBE = 6 mL

   7 M urea = 25.22 gm

   15% acrylamide (40%) = 22.5 mL

   APS (10%) = 0.25 mL

   TEMED = 0.025 mL.

3.4.2 dRP Lyase Activity of DNA Polymerase β

DNA Pol β is a 39-kDa multifunctional protein. Its N-term domain consists of an 8-kDa dRP lyase activity domain, while the carboxyl-terminal domain of 31 kDa has nucleotidyl transferase activity. dRP lyase activity of Pol β is essential for SN-BER, while nucleotidyl transferase activity is required for LP-BER [19]. The dRP lyase activity of Pol β was analyzed under steady-state condition and linear range. Reactions for dRP lyase activity of Pol β were assembled on ice in a standard reaction mixture of 25 μL as described below. The substrate used for dRP lyase activity was 3′-^{32}P-labelled or 3′-FAM-labelled U-DNA. This reaction was carried out in several stages.

1. A master mix containing 30 mM Tris–HCl, pH 7.5, 30 mM KCl, 5 mM $MgCl_2$, 2 mM DTT, 5% glycerol, 0.01% Nonidet P-40, and 0.1 mg/mL bovine serum albumin was prepared.
2. 10 μL of master mix was aliquoted to each tube.
3. One unit of recombinant UDG was added to each tube containing the reaction mixture while leaving the first tube without UDG. This tube will serve as a negative control. The amount of UDG depends upon the specific activity of the protein. The final volume of reaction mixture was adjusted to 15 μL (*see* **Notes 5** and 7).
4. Add 1 unit of recombinant UDG to the reaction mixture. Optimal concentrations of UDG were empirically determined (*see* **Notes 5** and 7).
5. The reaction was initiated by the addition of ^{32}P-labeled 63-mer U-DNA substrate (approximately 100,000 cpm). The reaction mixtures were incubated at 37 °C for 20 min. Alternatively, we can use a nonradioactive assay procedure using 3'-FAM-labelled U-DNA substrate (*see* **Note 5**).
6. 10 μL of reaction buffer containing 30 mM Tris–HCl, pH 7.5, 30 mM KCl, 8 mM MgC12, 2 mM DTT, 0.01% Nonidet P-40, 5% glycerol, and 0.1 mg/mL bovine serum albumin was added to each tube.
7. 1 μL (~25–100 pMol) of recombinant APE1 was added to the reaction mixture and incubated for an additional 30 min at 37 °C in a dry bath. The optimal concentration of APE1 and time should be empirically determined (*see* **Notes 6–8**).
8. 10–25 fMol of recombinant Pol β was added to each tube and incubated at 37 °C for 20 min. The optimal concentration of Pol β and time should be empirically determined (*see* **Notes 4** and **6**).
9. Cold sodium borohydride (340 mM NaBH4, freshly prepared) was added to the each tube and was incubated on ice for 30 min.

10. DNA was extracted from the reaction mixture by the addition of 50 μL of phenol/chloroform/isoamyl alcohol (25:24:1; v/v). The contents of each tube were thoroughly mixed by vortexing. Microfuge tubes were centrifuged at 12000 rpm in a tabletop centrifuge to separate the aqueous phase from the organic phase. The upper phase containing the dRP lyase activity product and uncleaved DNA substrate was quantitatively collected in a fresh tube.
11. 1–2 μg of glycogen or tRNA was added to the samples as a carrier and mixed well to facilitate efficient precipitation of DNA. DNA was ethanol precipitated by addition of 1/10 volume of 3 M sodium acetate, pH 5.2, and 2.5 volumes of ethanol. Samples were mixed well and tubes were incubated at −20 °C overnight preferably or −80 °C for 1–2 h.
12. The DNA was recovered by centrifugation at 14,000 rpm for 20 min. The ethanol was removed leaving behind the DNA pellet. The recovered DNA was washed once with 2× volumes of cold 70% ethanol and centrifuged at 14,000 rpm for 20 min.
13. Ethanol was carefully removed with a syringe without disturbing the DNA pellet, and the pellet was briefly dried at room temperature or till residual ethanol evaporated.
14. The pelleted DNA was resuspended in a 5–10 μL of sample loading dye solution (90% formamide, 1 mM EDTA, 0.1% xylene cyanol, and 0.1% bromophenol blue).
15. Denaturing polyacrylamide (7 M urea and 15% acrylamide) gel was prepared skillfully without trapping any air bubbles in the gel specifically in the wells. Care should be taken while taking out the comb that wells were not damaged. Combs and spacer used for casting the gels were of 0.4 mm thickness. The recipe for the gel is described below (*see* **Notes 4** and **5**).
16. Samples were heat denatured at 85 °C for 5 min and cooled quickly on ice. Samples were kept on ice till loaded on the gel. Samples were centrifuged before loading to bring every droplet to the bottom.
17. Samples (4 μL) were loaded onto a 7 M urea and 15% denaturing polyacrylamide gel for electrophoresis and run at 2000 volts for about an hour. Leftover samples can be stored at −20 °C (*see* **Note 4**).
18. After completion of the run, the gel plates were pry-opened and the gel was transferred to a blotting sheet. The gel was dried under vacuum (80 °C for 2 h) and then subjected to autoradiography. Signals were captured on X-ray films.
19. For nonradioactive APE1-cleaved products, samples were resolved on a 7 M urea and 15% denaturing polyacrylamide gel. DNA on the gel was visualized by scanning in the Typhoon

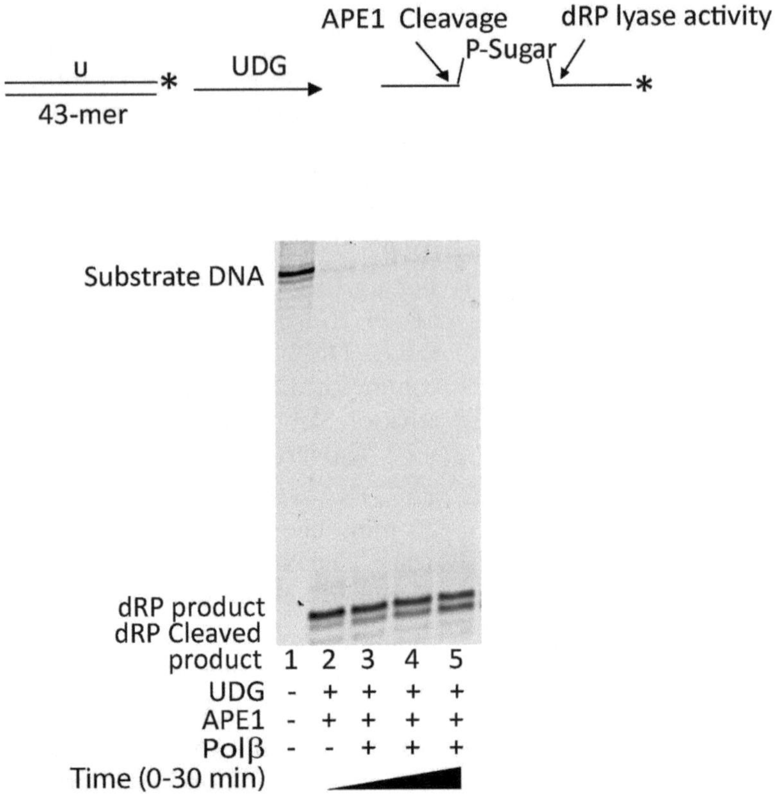

Fig. 4 Analysis of dRP lyase activity of DNA polymerase β. Schematic representation for the reconstitution dRP lyase of DNA polymerase β. All the components of the assay system were sequentially added, and the reaction was carried out at 37 °C at different time points. The dRP lyase activity of Pol β was analyzed using the 3′-labelled U-DNA substrate as discussed in the procedure. Lane 1 represents substrate only without any cleavage, lane 2 shows 24-mer incised product in the presence of UDG and APE1, and lanes 3–5 show the time-dependent dRP lyase activity of Pol β. The incubation time was 0–30 min at 37 °C

Trio Biomolecular Imager (GE Healthcare Bio-Science Corporation, Piscataway, NJ, USA), at excitation of 526 nm and emission at 488 nm. Digital images were acquired and analyzed for quantitation using ImageQuant TL (GE Healthcare Bio-Science Corporation, Piscataway, NJ, USA) software. An example of dRP lyase activity of Pol β is shown in Fig. 4 (*see* **Notes 2–4**).

Recipe for 7 M urea/15%acrylamide gel remains the same as in Subheading 3.4.1.

3.4.3 FEN1 Activity Assay

The reaction for in vitro FEN1 activity was carried out in a standard reaction mixture of 25 μl. The reaction was assembled on ice as described below [11, 15, 20]. The DNA substrate used for FEN1 activity was 5′-labelled downstream sense oligo. All oligonucleotides including 5′-upstream sense oligonucleotide, 5′--labeled upstream sense oligonucleotide, and antisense oligonucleotide were mixed in equimolar quantity and annealed at 85 °C for 10 min followed by slow cooling to room temperature.

1. A master mix containing 30 mM Tris–HCl, pH 7.5, 30 mM KCl, 5 mM $MgCl_2$, 2 mM DTT, 5% glycerol, 0.01% Nonidet P-40, and 0.1 mg/mL bovine serum albumin was prepared.
2. 10 μL of master mix was added to each tube.
3. 0.1–0.2 pMol of purified recombinant FEN1 was added to each tube leaving the first tube without FEN1. The amount of FEN1 was optimized before proceeding for complete LP-BER. The optimal concentration of FEN1 protein was empirically determined.
4. The reaction was initiated by the addition of ^{32}P-labeled 51-mer Flap-DNA substrate (approximately 100,000 cpm). The reaction contents were gently mixed and the reaction mixture was incubated at 37 °C for 30–60 min depending upon the specific activity of FEN1 (*see* **Note 1**).
5. Reaction was terminated by the addition of an equal volume of stop buffer (0.4% [w/v] SDS, 5 mM EDTA, 1 μg of proteinase K).
6. The cleaved flap structure of DNA was recovered by extracting in phenol/chloroform/isoamyl alcohol (25:24:1; v/v). 50 μL of phenol/chloroform/isoamyl alcohol (25:24:1; v/v) was added and mixed by vortexing. Tubes were centrifuged at 14000 rpm in a tabletop centrifuge to separate the aqueous phase from the organic layer. The upper phase containing the cleaved flap structure and uncleaved substrate was collected in a fresh tube.
7. 1–2 μg of glycogen or tRNA was added to the sample as a carrier and mixed well to facilitate efficient precipitation of DNA. DNA was ethanol precipitated by the addition of 1/10 volume of 3 M sodium acetate, pH 5.2, and 2.5 volumes of ethanol. Samples were mixed by vortexing and the tubes were incubated at −20 °C overnight preferably or −80 °C for 1–2 h.
8. The DNA was recovered by centrifugation at 14,000 rpm for 20 min. The ethanol was removed leaving behind the DNA pellet. The recovered DNA was washed once with 2× volumes of cold 70% ethanol and centrifuged at 14,000 rpm for 20 min.
9. Ethanol was carefully removed with a syringe without disturbing the DNA pellet, and the pellet was briefly dried at room temperature or till residual ethanol evaporated.

10. The pelleted DNA was resuspended in a 5–10 μL of sample loading dye solution (90% formamide, 1 mM EDTA, 0.1% xylene cyanol, and 0.1% bromophenol blue).
11. Denaturing polyacrylamide gel (7 M urea and 15% acrylamide) was prepared without trapping any air bubbles in the gel specifically in the wells. Care should be taken while taking out the comb so that wells were not damaged or deformed. Combs and spacer used for casting the gels were of 0.4-mm thickness. The recipe for the gel is described in Sect. 3.4.1 (*see* **Notes 4** and **5**).
12. Samples were heat denatured at 85 °C for 5 min and cooled quickly on the ice. Samples were kept on ice till loaded on the gel. Samples were centrifuged before loading to bring every droplet to the bottom.
13. Samples (4 μL) were loaded onto a 7 M urea and 15% denaturing polyacrylamide gel for electrophoresis and run at 2000 volts for about an hour. Leftover samples can be stored at −20 °C (*see* **Note 4**).
14. After completion of the run, the gel plates were pry-opened and the gel was transferred to a blotting sheet. The gel was dried under vacuum (80 °C for 2 h) and then subjected to autoradiography. Signals were captured on X-ray films.
15. For nonradioactive FEN1-cleaved products, samples were resolved on a 7 M urea and 15% denaturing polyacrylamide gel. DNA on the gel was visualized by scanning in the Typhoon Trio Biomolecular Imager (GE Healthcare Bio-Science Corporation, Piscataway, NJ, USA), at excitation of 526 nm and emission at 488 nm. Digital images were acquired and analyzed for quantitation using ImageQuant TL (GE Healthcare Bio-Science Corporation, Piscataway, NJ, USA) software. An example of FEN1 activity is shown in Fig. 5 (*see* **Notes 2–4**).

Recipe for 7 M urea/15%acrylamide gel remains the same as in Subheading 3.4.1.

3.4.4 DNA Ligase 1 or XRCC1/Ligase 3 Activity Assay

Similar reactions were assembled as described elsewhere for recombinant DNA ligase 1 or XRCC1/DNA ligase 3 protein to determine the ligase activity [1, 15, 21]. DNA ligase assay was assembled similar to the above experimental protocol. Master mixture was supplemented with 0.5 mM ATP. The reaction was initiated by the addition of ^{32}P-labeled or FAM-labeled appropriate nicked substrate for ligation. The reaction was allowed to proceed at 37 °C for 60 min and terminated by the addition of stop buffer. The DNA was recovered by phenol/chloroform extraction, ethanol precipitation at −80 °C, followed by washing with 70% cold ethanol. The recovered DNA was resuspended in sample loading dye (90% formamide, 1 mm EDTA, 0.1% xylene cyanol). The

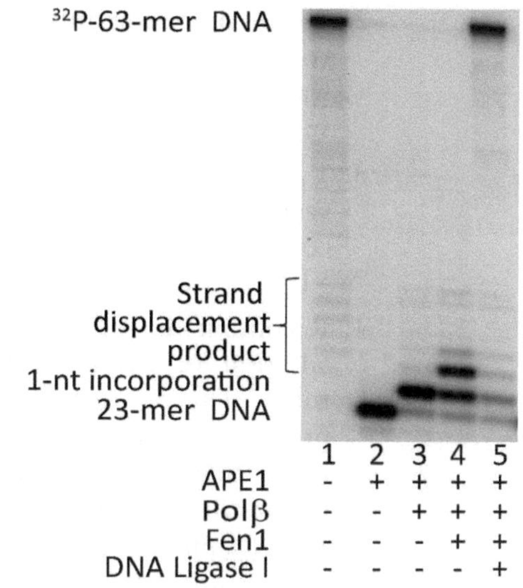

Fig. 5 In vitro reconstitution long-patch (LP) BER. APE1-mediated repair of a THF abasic site is mimicked via LP BER. Ladders in the image represent displacement products generated by DNA polymerase β (denoted as Pol β) creating a flap and usually cleaved by FEN1. Image shows cleaved products at the sequential BER steps and the last lane shows a completely repaired product in the presence of LIG1. Lane 1 represents substrate without any cleavage, lane 2 shows 23-mer incised product after APE1 activity, lane 3 shows addition of 1-nt, and lane 4 shows FEN1-stimulated Pol β displacement activity. Lane 5 shows repair to the 63-mer DNA. The incubation time for the reaction was 60 min at 37 °C

samples were heated at 85 °C for 5 min and cooled quickly on ice. The samples (4 μL) were loaded onto a denaturing gel (15% polyacrylamide and 7 M urea gel) for electrophoresis for separation of ligated products. The gel was scanned in the Typhoon Trio Biomolecular Imager (GE Healthcare, excitation at 526 nm and emission at 488 nm). The digital images were acquired and analyzed for quantitation using ImageQuant TL (GE Healthcare Bio-Science Corporation, Piscataway, NJ, USA) software.

3.4.5 In Vitro Reconstitution of SP- and LP-BER Assay

In vitro SN- and LP-BER assays were carried out as described previously [15, 21, 22]. The reaction mixture for SP-BER was essentially the same as that for LP-BER except that FEN1 was omitted in SP-BER while reaction was initiated with U-DNA substrate. The following is the stepwise protocol to assemble a reconstitutive LP-BER assay:

1. A master mix was prepared by adding the following to the final concentration of 30 mM Tris–HCl, pH 7.5, 30 mM KCl, 8 mM $MgCl_2$, 2 mM DTT, 0.01% Nonidet P-40, 5% glycerol, 0.5 mM ATP, and 0.1 mg/mL bovine serum albumin.

2. 10 μL of master mix was aliquoted to each tube. The final volume was adjusted to 25 μL of a reaction with DNAse-free water after the addition of recombinant proteins and substrate DNA.
3. 1 pMol to 1 nMol of recombinant purified APE1 was added to each tube leaving the first tube without APE1. The protein concentration was empirically determined optimized to achieve ~100 cleavage.
4. Recombinant Pol β (0.2–0.5 pMol) was added followed by addition of FEN1 (0.10–0.200 pMol) to the designated tubes. The optimal concentration of both Pol β and FEN1 proteins required to achieve the optimal generation of 2–13 nucleotide displacement product after FEN1 cleavage was empirically determined.
5. Each dATP, dCTP, dGTP, and dTTP to a final concentration of 20 μM was added to each tube. Alternatively, 10 mM stock solution of dNTP mixture can be used.
6. Recombinant DNA ligase 1 (50–100 pMol) was added to each tube designated as purified recombinant DNA ligase 1.
7. The reaction was initiated by the addition of the ^{32}P-labeled 63-mer F DNA substrate (approximately 100,000 cpm). After initiating the reaction, the contents of the tubes were gently mixed and samples were incubated at 37 °C for 60 min. Alternatively, FAM-labeled oligos can be used (*see* **Note 1**).
8. After 60 min of the reaction, the reaction was terminated by the addition of an equal volume of stop buffer containing 0.4% (w/v) SDS, 5 mM EDTA, and 1 μg of proteinase K to each tube.
9. DNA was carefully extracted from the reaction mixture by the addition of 50 μL of phenol/chloroform/isoamyl alcohol (25: 24:1; v/v). The contents of each tube were thoroughly mixed by vortexing. Microfuge tubes were centrifuged at 12,000 rpm for 10 min in a tabletop centrifuge to separate the aqueous phase from the organic phase. The upper phase containing the repaired and unrepaired DNA was quantitatively transferred to a fresh tube. DNA was ethanol precipitated by the addition of 1/10 volume of 3 M sodium acetate, pH 5.2, and two volumes of ethanol.
10. 1–2 μg of glycogen or tRNA was added to the samples as a carrier and mixed well to facilitate efficient precipitation of DNA. The tubes were incubated at −20 °C overnight preferably or −80 °C for 1–2 h.
11. DNA was recovered by centrifugation at 14,000 rpm for 20 min. Ethanol was removed leaving behind the DNA pellet. The recovered DNA was washed once with 2× volumes of cold 70% ethanol and centrifuged at 14,000 rpm for 20 min.

12. Ethanol was carefully removed with a pipet or syringe without disturbing the DNA pellet, and the pellet was briefly dried at room temperature or till residual ethanol completely evaporated.
13. Pelleted DNA was resuspended in a 5–10 μL of sample loading dye solution (90% formamide, 1 mM EDTA, 0.1% xylene cyanol, and 0.1% bromophenol blue).
14. The gel was prepared (7 M urea and 15% denaturing polyacrylamide) skillfully as per the recipe described in Sect. 3.4.1 without trapping any air bubbles in the gel and in the wells. Also, extreme care was taken while taking out the comb that wells were in good shape. Combs and spacer used for casting the gels were of 0.4 mm thickness. The gel was pre-run before loading the samples for 25–30 min at 2000 volts (*see* **Notes 4** and **5**).
15. Samples were heat denatured at 85 °C for 5 min and cooled quickly on ice. Samples were kept on ice till loaded on the gel and centrifuged before loading.
16. Samples (4 μL) were loaded onto a 7 M urea and 15% denaturing polyacrylamide gel for electrophoresis and run at 2000 volts for about 1.5–2 h. Leftover samples were stored at −20 °C (*see* **Note 4**).
17. After the completion of run, the gel plates were pry-opened and the gel was transferred to a blotting sheet. The gel was dried under vacuum (80 °C for 2 h) and then subjected to autoradiography. Signals were captured on X-ray films.
18. For nonradioactive APE1-cleaved products, samples were resolved on a 7 M urea and 15% denaturing polyacrylamide gel. DNA on the gel was visualized by scanning in the Typhoon Trio Biomolecular Imager (GE Healthcare Bio-Science Corporation, Piscataway, NJ, USA), at excitation of 526 nm and emission at 488 nm. Digital images were acquired and analyzed for quantitation using ImageQuant TL (GE Healthcare Bio-Science Corporation, Piscataway, NJ, USA) software. An example of LP-BER is shown in Fig. 5 (*see* **Notes 2–4**).

Recipe for 7 M urea/15%acrylamide gel remains same as in Subheading 3.4.1.

4 Notes

1. All radioactive work should be carried out in the radioactive area approved by the institution to avoid unnecessary exposure to other co-workers. All radioactive wastes should be appropriately disposed off and segregated according to institutional policies. Radioactive waste cannot be mixed with regular

biohazard waste. Extreme caution should be taken while working with radioactive nuclide.

2. During the oligonucleotide annealing process, cooling of the annealed oligo should proceed gradually from 95 °C to room temperature. It is highly recommended to use a dry heat block for annealing rather than using a PCR machine. The whole process in the dry heat block takes about 2–2.5 h, whereas the temperature in a PCR machine drops more rapidly in comparison which does not allow proper annealing of the sense and antisense oligos.
3. With respect to the nonradioactive procedure, the FAM-labelled oligo reaction mixture should be at a pH of >7. A pH of <7 will result in poor quality image or potentially no image being detected. The signal can be enhanced by increasing the energy level in a photomultiplier tube (PMT values on a program) or directly exposing the gel to a laser beam. Although, FAM-labelled oligos give equally good results, the sensitivity of the assay will be comparatively lower compared to a radioactive probe.
4. While preparing the 15% polyacrylamide gel containing 7 M urea, care should be taken to dissolve the 7 M urea homogeneously by heating mildly during mixing. Once the solution cools down, degas the solution and add APS and TEMED. Add acrylamide/urea solution to the preassembled plates for polymerization. Wait for at least an hour for complete polymerization. Sometimes a longer time may be needed for polymerization. Pre-running of the gel for at least 30 min is essential after washing the wells to remove unpolymerized acrylamide and crystalized urea. After the pre-run, carefully wash the wells by flushing with the running buffer to completely remove urea from the wells. Urea in the wells will cause altered density of the samples and hence contorted movement of the samples and diffusion.
5. A 7 M urea and 15% denaturing polyacrylamide gel should be prepared skillfully without trapping any air bubbles in the gel. Utmost care should be taken while taking out the comb that no wells are damaged or deformed. Combs and spacer used for casting the gels should be 0.4 mm thickness.
6. When working with FAM-labeled oligonucleotides, the gel should be scanned immediately after completion of gel. Leaving the gel without scanning for a long period of time may lead to increased diffusion and decreased fluorescence of the bands.
7. Optimize concentration of UDG and use only the appropriate amount of UDG. If commercial UDG is used in the assay, care should be taken not to use excessive UDG. UDG is also a bifunctional enzyme so it may exhibit AP-lyase activity which

will cleave AP site also. We need abasic lesions generated not a cleaved substrate and not a reduced AP site substrate with NaBH4 which will be resistant to dRP lyase activity of Pol β, which may not be a good substrate for dRP-lyase activity of Pol β.

8. Optimize the concentration of APE1 and time of incubation. Do not use an excessive amount of recombinant APE1. If commercial recombinant APE1 is being used in the assay, then its activity should be verified first. We have observed that commercial recombinant APE1 is poorly active and therefore may require a high concentration of APE1 to achieve nearly complete cleavage.

Acknowledgments

Robert Hromas is supported by the National Institutes of Health (NIH) R01 CA139429 grants.

References

1. Iyama T, Wilson DM 3rd. (2013) DNA repair mechanisms in dividing and non-dividing cells. DNA Repair (Amst) 12:620–636
2. Drabløs F, Feyzi E, Aas PA, Vaagbø CB, Kavli B, Bratlie MS, Peña-Diaz J, Otterlei M, Slupphaug G, Krokan HE (2004) Alkylation damage in DNA and RNA – repair mechanisms and medical significance. DNA Repair (Amst) 3:1389–1407
3. Whitaker AM, Schaich MA, Smith MR, Flynn TS, Freudenthal BD (2017) Base excision repair of oxidative DNA damage: from mechanism to disease. Front Biosci (Landmark Ed) 22:1493–1522
4. Payne MJ, Pratap SE, Middleton MR (2005) Temozolomide in the treatment of solid tumours: current results and rationale for dosing/scheduling. Crit Rev Oncol Hematol 53: 241–252
5. Klaunig JE (2018) Oxidative stress and cancer. Curr Pharm Des 24:4771–4778
6. Jaiswal AS, Banerjee S, Panda H, Bulkin CD, Izumi T, Sarkar FH, Ostrov DA, Narayan S (2009) A novel inhibitor of DNA polymerase beta enhances the ability of temozolomide to impair the growth of colon cancer cells. Mol Cancer Res 7:1973–1983
7. Kawanishi S, Ohnishi S, Ma N, Hiraku Y, Murata M (2017) Crosstalk between DNA damage and inflammation in the multiple steps of carcinogenesis. Int J Mol Sci 18:1808
8. Kunkel TA, Diaz M (2002) Enzymatic cytosine deamination: friend and foe. Mol Cell 10:962–963
9. Briggs AW, Stenzel U, Meyer M, Krause J, Kircher M, Pääbo S (2010) Removal of deaminated cytosines and detection of in vivo methylation in ancient DNA. Nucleic Acids Res 38: e87
10. Narayan S, Jaiswal AS, Law BK, Kamal MA, Sharma AK, Hromas RA (2016) Interaction between APC and Fen1 during breast carcinogenesis. DNA Repair (Amst) 41:54–62
11. Jaiswal AS, Banerjee S, Aneja R, Sarkar FH, Ostrov DA, Narayan S (2011) DNA polymerase beta as a novel target for chemotherapeutic intervention of colorectal cancer. PLoS One 6: e16691
12. Yu Y, Cui Y, Niedernhofer LJ, Wang Y (2016) Occurrence, biological consequences, and human health relevance of oxidative stress-induced DNA damage. Chem Res Toxicol 29: 2008–2039
13. Sobol RW, Prasad R, Evenski A, Baker A, Yang XP, Horton JK, Wilson SH (2000) The lyase activity of the DNA repair protein beta-polymerase protects from DNA-damage-induced cytotoxicity. Nature 405:807–810
14. Wallace SS (2014) Base excision repair: a critical player in many games. DNA Repair (Amst) 19:14–26

15. Jaiswal AS, Williamson EA, Srinivasan G, Kong K, Lomelino CL, McKenna R, Walter C, Sung P, Narayan S, Hromas R (2020) The splicing component ISY1 regulates APE1 in base excision repair. DNA Repair (Amst) 86:102769
16. Balusu R, Jaiswal AS, Armas ML, Kundu CN, Bloom LB, Narayan S (2007) Structure/function analysis of the interaction of adenomatous polyposis coli with DNA polymerase beta and its implications for base excision repair. Biochemistry 46:13961–13974
17. Caglayan M, Batra VK, Sassa A, Prasad R, Wilson SH (2014) Role of polymerase beta in complementing aprataxin deficiency during abasic-site base excision repair. Nat Struct Mol Biol 21:497–499
18. Prasad R, Dyrkheeva N, Williams J, Wilson SH (2015) Mammalian base excision repair: functional partnership between PARP-1 and APE1 in AP-site repair. PLoS One 10:e0124269
19. Prasad R, Beard WA, Chyan JY, Maciejewski MW, Mullen GP, Wilson SH (1998) Functional analysis of the amino-terminal 8-kDa domain of DNA polymerase beta as revealed by site-directed mutagenesis. DNA binding and 5′-deoxyribose phosphate lyase activities. J Biol Chem 273:11121–11126
20. Jaiswal AS, Panda H, Law BK, Sharma J, Jani J, Hromas R, Narayan S (2015) NSC666715 and its analogs inhibit strand-displacement activity of DNA polymerase β and potentiate temozolomide-induced DNA damage, senescence and apoptosis in colorectal cancer cells. PLoS One 10:e0123808
21. Jaiswal AS, Balusu R, Armas ML, Kundu CN, Narayan S (2006) Mechanism of adenomatous polyposis coli (APC)-mediated blockage of long-patch base excision repair. Biochemistry 45:15903–15914
22. Narayan S, Jaiswal AS, Balusu R (2005) Tumor suppressor APC blocks DNA polymerase beta-dependent strand displacement synthesis during long patch but not short patch base excision repair and increases sensitivity to methylmethane sulfonate. J Biol Chem 280: 6942–6949

Part II

Detection and Quantification of Base Lesions, DNA Double-Stand Breaks, DNA Protein Cross-Links, and R Loops

Chapter 7

Detection of Oxidatively Modified Base Lesion(s) in Defined DNA Sequences by FLARE Quantitative PCR

Lang Pan, Yaoyao Xue, Ke Wang, Xu Zheng, and Istvan Boldogh

Abstract

Assessment of DNA base and strand damage can be determined using a quantitative PCR assay that is based on the concept that damage blocks the progression of a thermostable polymerase thus resulting in decreased amplification. However, some of the mutagenic DNA base lesions cause little or no distortion in Watson–Crick base pairing. One of the most abundant such lesion is 8-oxo-7,8-dihydro-2′--deoxyguanosine (8-oxo(d)Gua), although it affects the thermodynamic stability of the DNA, duplex 8-oxo(d)Gua does not inhibit DNA synthesis or arrest most of DNA or RNA polymerases during replication and transcription. When unrepaired, it is a pre-mutagenic base as it pairs with adenine in anti-syn conformation. Recent studies considered 8-oxo(d)Gua as an epigenetic-like mark and along with 8-oxoguanine DNA glycosylase1 (OGG1) and apurinic/apyrimidinic endonuclease1 (APE1) has roles in gene expression via nucleating transcription factor's promoter occupancy. Here, we introduce its identification through fragment length analysis with repair enzyme (FLARE)-coupled quantitative (q)-PCR. One of the strengths of the assay is that 8-oxo(d)Gua can be identified within short stretches of nuclear and mitochondrial DNA in ng quantities. Bellow we describe the benefits and limits of using FLARE qPCR to assess DNA damage in mammalian cells and provide a detailed protocol of the assay.

Key words DNA base damage, Gene regulatory sequences, FLARE-coupled rt-qPCR

1 Introduction

1.1 Importance and Difficulties to Detect Guanine Base Lesions

Reactive oxygen species (ROS) are generated by various oxidoreductases located in the cytoplasmic and mitochondrial membranes due to cellular metabolism and/or in response to a variety of stimuli (growth factors, chemokines, cytokines) or exposure to environmental agents. ROS are needed for the physiological functioning of cells, while in excess resulting in oxidative modification to biological macromolecules. In the case of nucleic acids, ROS induce chemical modifications to DNA and RNA bases and strands. Among oxidatively modified DNA bases, 7,8-dihydro-8-oxo(d)-guanine (8-oxo(d)Gua) is the most abundant [1], which is due to guanine's (Gua) lowest oxidation potential among the four nucleic

Kishor K. Bhakat and Tapas K. Hazra (eds.), *Base Excision Repair Pathway: Methods and Protocols*, Methods in Molecular Biology, vol. 2701, https://doi.org/10.1007/978-1-0716-3373-1_7,

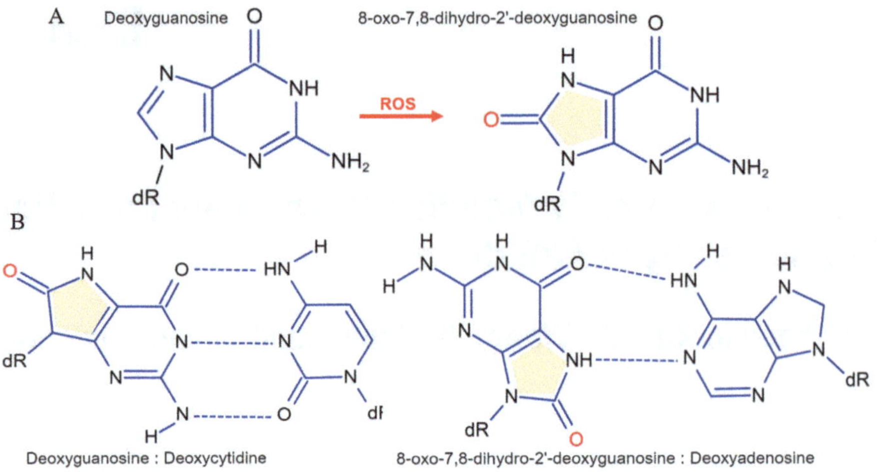

Fig. 1 8-Oxo(d)Gua and its mispairing with adenine. (**a**) Interaction of guanine with ROS primarily hydroxyl radical generates 8-oxo(d)Gua. (**b**) 8-oxo(d)Gua in the anti-conformation forms a Watson–Crick base pair with cytosine (left panel), while in a syn-confirmation, it follows Hoogsteen base pairing and pairs with adenine (right panel). dR, deoxyribose

acid bases [2, 3]. Gua may be oxidized directly or by electron transfer between the bases [3, 4].

8-Oxo(d)Gua differs from guanine only at C8 and N7. C8 harbors an oxygen instead of a hydrogen and a hydrogen on N7 instead of an electron pair (Fig. 1a). While oxygen on C8 changes the thermodynamic stability of the duplex [5], its presence causes no distortion of the DNA structure, nor has effect on the Watson–Crick base arrangement of DNA [6, 7]. Therefore, 8-oxo(d)Gua is not detected during DNA replication, does not induce pausing of the DNA polymerases, nor create a barrier to RNA polymerases during transcription [8, 9]. Because 8-oxo(d)Gua through an anti-syn conformation can pair with an adenine via Hoogsteen base pairing, it is considered a pre-mutagenic lesion [10–12]. In proliferating tissues, a G:C to T:A transversion is fixed, leading to mutations (Fig. 1b). Based on a similar mechanism, the 8-oxo(d)Gua/A Hoogsteen base pair can occur during transcription, leading to transcription-coupled mutagenesis [13].

Removal of oxidatively modified guanine bases is achieved by the DNA base excision repair (BER) pathway, where 8-oxo(d)Gua paired with C is removed by 8-oxoguanine DNA glycosylase1 (OGG1) [14]. In case 8-oxo(d)Gua repair failed and it incorrectly paired with A, the human homolog of MutY DNA glycosylase (MUTYH) removes the A allowing a second chance for a polymerase to insert C opposite 8-oxo(d)Gua for removal by OGG1 [15]. The initial step in the repair is the recognition of the modified base and cleavage of the N-glycosidic bond by a specific DNA glycosylase to generate an abasic (apurinic/apyrimidinic [AP]) site. The AP sites are processed by the downstream enzymes AP

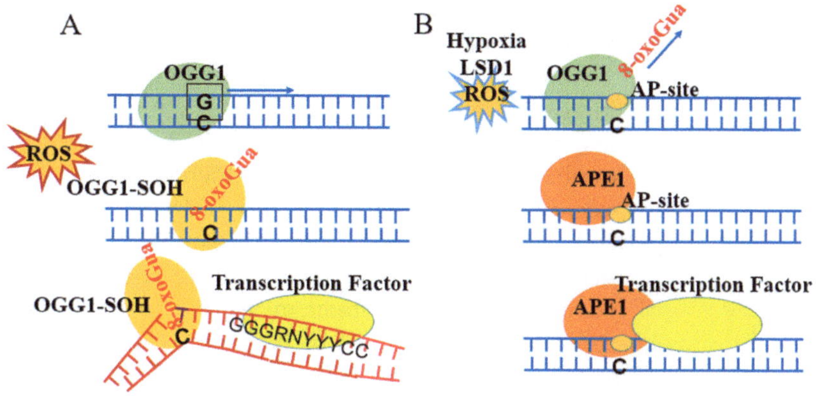

Fig. 2 8-Oxo(d)Gua as an epigenetic-like mark. (**a**), ROS induced by cytokine/chemokine ligand interaction generates 8-oxo(d)Gua and oxidatively disables OGG1' glycosylase activity. OGG1 bound to 8-oxo(d)Gua at gene regulatory regions induces architectural changes in duplex DNA and facilitates binding of TFs to consequently modulate gene expression. GGGRNYYYCC: NFκB binding site; (**b**), under hypoxia or lysine-specific histone, demethylase 1 (LSD1)-generated ROS oxidize guanine to 8-oxo(d)Gua, which is removed by OGG1. The generated AP site is occupied by APE1, which facilitates transcription factor loading to the promoter, for regulation of gene expression. AP site, apurinic/apyrimidinic site; APE-1, apurinic/apyrimidinic endonuclease 1

endoDNase1 (APE1) to cleave the 5′-phosphodiester linkage yielding a nick in the DNA. The repair is completed by DNA polymerase β that removes the sugar fragment at the nick site followed by inserting the correct Gua nucleotide, and DNA ligases seal the gap [14, 16].

Recent studies have shown that 8-oxo(d)Gua (possibly other base modification) is not only an oxidative stress marker, but is implicated in directing cellular transcriptional activities in response to cellular stressors. As such, primarily in nonreplicating cells, 8-oxo (d)Gua can serve as an epigenetic-like mark [17, 18]. Conversely, 8-oxo(d)Gua lesion itself (without or with repair) serves as a signal for transcription to regulate cellular responses to oxidative stress [19–21]. A key enzyme in these processes is OGG1 (Fig. 2a, b). OGG1, via its noncatalytic binding to oxidatively generated guanine lesion in gene regulatory regions (promoter), serves as a nucleation site for transacting factors so that innate inflammation-associated gene expression can occur [17, 19, 22]. A proposed mechanism is that OGG1 flips out the damaged base from the DNA double helix that interacts with cytosine opposite to 8-oxo (d)Gua and alters the structure of adjacent DNA sequences, providing a favorable DNA conformational change for the binding of transcription factors [17, 19, 20]. Under hypoxia-ROS or ROS generated focally on the chromatin by lysine-specific histone, demethylase 1 oxidizes guanine to 8-oxo(d)Gua, which is excised by OGG1 leading to formation of baseless site followed by binding of AP endonuclease 1 (Fig. 2b). Acetylated APE1 bound to a baseless site increases transcription factor loading to regulate gene expression for coordination of multiple cellular processes [18, 21].

Data from the literature indicate significant differences (up to threefold) in histone demethylase 1 abundance of DNA base lesions, particularly 8-oxo(d)Gua [23–25]. The differences may be ascribed to the particular experimental protocols that resulted in over- or underestimation of 8-oxo(d)Gua' abundance [26]. In general, values obtained indirectly by the enzymatic assays are substantially lower than those obtained by using the direct methods of measurement [26]. Direct measurement of the lesion requires isolation of DNA followed by hydrolysis either chemically or biochemically. The released lesions are quantified by sensitive analytical techniques. The bottom line is that during each of the individual steps, DNA can be oxidized [27]. Accordingly, similar levels of 8-oxodGua were identified in DNA using high-performance liquid chromatography (HPLC), HPLC combined with electrochemical detection (HPLC-EC), HPLC–mass spectrometry/MS, and HPLC-gas chromatography–mass spectrometry (GCMS) assays [28–32]. It is noteworthy to mention that the origin of the inaccuracy in 8-oxo(d)Gua levels, at least in part can be explained by artifactual DNA oxidation [33]. Recent technological improvements, including introduction of desferrioxamine and sodium iodine to isolate DNA, have permitted more accurate assessment of intrahelical 8-oxo(d)Gua—levels approximately 1 lesion per 10^6 DNA bases [34–36].

These technologies above provided overall levels of base lesions; however, it is obvious that 8-oxo(d)Gua distribution is not random within the genome, but the highest within guanine-rich transcriptionally active regions of DNA—commonly found in enhancers and promoters as well as key transcription factor binding sites. Moreover, the preferential site of oxidation are the series of guanines, found, e.g., in potential guanine-quadruplex-forming sequences, especially at the 5′-end of guanine runs, resulting from long-distance electron tunneling through the hydrophobic core of DNA (reviewed in [37]). The preferential sites of 8-oxo(d)Gua distribution was demonstrated by using high-throughput sequencing of DNA improved by selective-capture technologies using 8-oxo(d)Gua antibody [38, 39]. Accurate mapping of 8-oxo (d)Gua in the mouse DNA was reported using covalent capture techniques that are based on the selective oxidation of 8-oxo (d)Gua that can be trapped through an imine intermediate [39, 40]. According to recent results, 8-oxo(d)Gua in gene regulatory regions is considered as an epigenetic mark that potentially facilitates transcription factor binding via distinct mechanisms (Fig. 2). To this end comprehensive reviews are published on the roles of 8-oxo(d)Gua-trapped OGG1 in modulating innate and adaptive immunity, pulmonary inflammation [17, 41], and homeostatic/metabolic pathways [20, 42, 43].

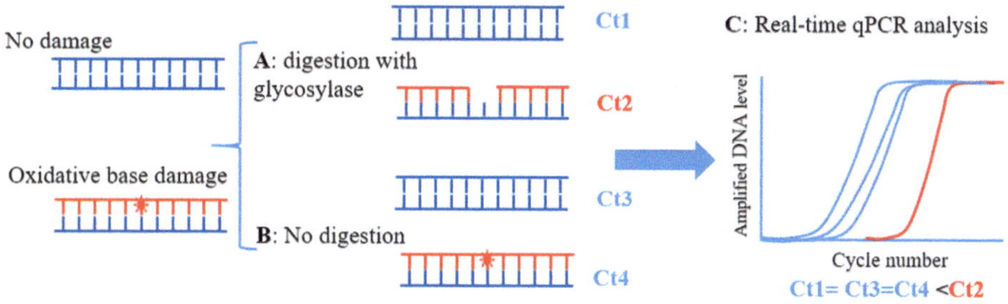

Fig. 3 Graphical depiction of the assay principle of fragment length analysis with repair enzyme (FLARE)-coupled qPCR. Ct, cycle threshold

Below, we provide details of a method that can identify 8-oxo (d)Gua in specific genomic regions with high accuracy in a modest laboratory setting by FLARE-coupled qPCR. In addition, the test can be performed using ng quantities of DNA samples; therefore, the assay is less demanding compared to HPLC-EC, HPLC-MS/MS, and HPLC-GCMS, or tandem liquid chromatography–mass spectrometry, which require microgram quantities of DNA for analysis and highly trained personnel [30–32].

1.2 Principle of the Assay

FLARE-qPCR for assessment of a silent DNA lesion is based on the idea that artifactually generated DNA lesions (e.g., AP sites, Fig. 3) block thermostable DNA polymerase [44–46]. Using the assay one can compare lesion frequencies within a sort starch of DNA based on amplification alone, in which higher amplification corresponds with a lower level of lesion [47, 48]. By assuming that lesion frequency follow Poisson distribution, the amplification of treated samples is compared to amplification of mock treated samples to calculate relative lesion frequency [49]. After damage is repaired, amplification is restored, and thus, this assay is useful in determining the kinetics of DNA repair in nuclear (or mitochondrial) genomes after exposure of cells to chemokines/cytokines or growth factors or even after virus infections [50, 51].

1.3 Advantages of FLARE-Coupled qPCR

Advantages of FLARE-qPCR in assessing 8-oxo(d)Gua (and potentially other silent DNA base lesions) include its sensitivity (ng quantities of DNA required), robustness, and suitability to determine base modification within gene loci (segment of promoter or exon/intron). Moreover, the DNA purification for FLARE-qPCR is the same for each cell type and tissue biopsies as well as for both nuclear and mitochondrial DNA (mtDNA). Because relative short fragments of DNA (80–150 bp) can be tested, FLARE-qPCR is highly sensitive and sufficient to capture physiologically relevant levels of base modification [50, 52]. Using appropriate primer pairs, various gene regions of interest can be examined (Fig. 4). Importantly, this assay allows direct comparison of repair rate (time course) in the same DNA segment or between

A

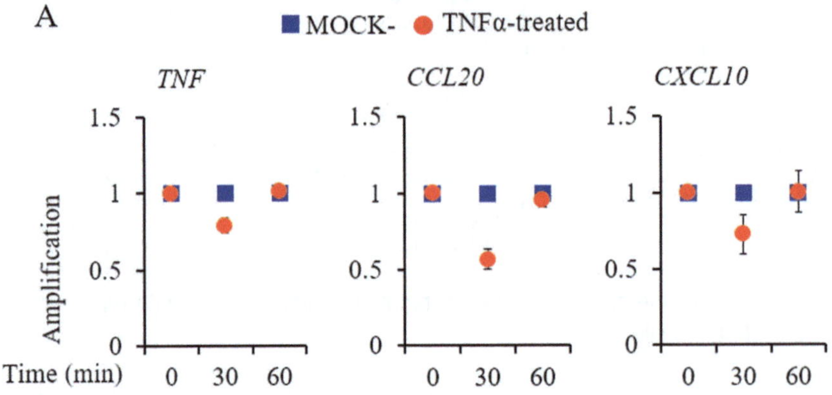

B

TNF (-250 to -120bp)	F: 5′-GGT CCC CAA AAG AAA TGG AGG-3′ R: 5′-TTT ATA TGT CCC TGG GGC GA-3′
CXCL10 (-179 to -71bp)	F: 5′-AGG GAA ATT CCG TAA CTT GGA GGC-3′ R: 5′-TTC ATG GTG CTG AGA CTG GAG GTT-3′
CCL20 (-366 to -170bp)	F: 5′-TGT TCC TGT GTG GGG CTG-3′ R: 5′-TTG CCA CAT GGG GTT TTC C-3′

Fig. 4 Validation of FLARE-qPCR: Generation of physiologically relevant levels of 8-oxo(d)Gua lesion(s) and repair in transcription start-site adjacent regulatory sequences of pro-inflammatory genes. (**a**), A549 cells (human type 2 airway epithelial cells) were exposed to 10 ng per mL tumor necrosis factor alpha (TNFα) and DNA was isolated at 0, 30, and 60 min as described in Subheading 2.2. DNA samples were processed and subjected to real-time-qPCR using SsoAdvanced™ Universal Supermix (BioRad Cat. no. 172–5270). It contains dNTPs, Sso7d fusion polymerase, $MgCl_2$, SYBR Green I, and ROX normalization dyes. rt-qPCR was run on a CFX96 quantitative PCR instrument. Relative quantifications were performed by using the $2^{-\Delta\Delta CT}$ method [64]. (**b**), Primer pairs used in these experiments. *TNF*, tumor necrosis factor, *CCL20*, C-C motif chemokine ligand 20 or macrophage inflammatory protein 3 alpha; *IL6*, interleukin 6 or interferon beta-2. F, forward primer; R, reverse primer, rt-qPCR, real-time qPCR

two genes that are differentially damaged and/or repaired. Because FLARE-qPCR relies on PCR amplification, the test can be performed on ng quantities of DNA samples and do not require expensive instrumentation, such as that needed for liquid chromatography and mass spectrometry.

1.4 Limitations of the Assay

A limitation of the FLARE-coupled qPCR is that polymerase can stall or stop at any baseless site, DNA gaps, or DNA strand lesion, which may not result from enzyme digestion. An additional concern is that the assay is not able to detect base lesion that is not within the amplification region of the primer set. Another potential issue emerges from the fact that primers and thermostable DNA polymerases have limited access to supercoiled covalently closed DNA (e.g., mtDNA) molecule [53]. This issue can be solved by restriction enzymes to linearization of mtDNA [54].

2 Materials

All solutions should be made using DNase-, RNase-free molecular biology grade water and molecular biology grade reagents. Reagents and buffers are stored at 4 °C or as recommended by the manufacturer. When disposing materials, follow waste disposal regulations. *Note*: We do not consistently provide the source of reagents (manufacturer, catalog number) as quality and price of molecular grade reagents (water, buffers, enzymes, and so on) are very competitive. Please use high-quality molecular biology grade reagents according to manufacturer recommendations.

1. Water (RNase-, DNase-free, molecular biology grade).
2. Deferoxamine (DFO) solution: 75 mM deferoxamine mesylate freshly made (49.26 mg DFO per mL) (*see* **Note 1**).
3. Lysis buffer solution: 10 mM Tris–HCl pH 8, 0.5 mM ethylenediamine tetra-acetic acid disodium (EDTA-Na_2), sodium dodecyl sulfate (SDS, 1% w/v). When adjusting pH using acids, add the acid to the aqueous solution for safety.
4. Lysis buffer: 10 mM Tris–HCl (pH 7.5), 5 mM magnesium chloride ($MgCl_2$), Triton X-100 (1%, v/v).
5. Lysis buffer containing DFO: Lysis buffer with 0.15 mM desferrioxamine (freshly made).
6. Sodium iodide (NaI) solution: 7.6 M NaI, 40 mM Tris–HCl pH 8.0, 20 mM EDTA-Na_2 (*see* **Note 2**).
7. RNase: 50 U/mL RNase A and 100 U/mL RNase T; keep all aliquots at −20 °C (or as recommended by manufacturer).
8. Proteinase K: Make 10 mg/mL stock solution of proteinase K (e.g., New England BioLabs, NEB), prepare in digestion dilution buffer, and store at −20 °C.
9. Isopropanol (100%, kept at −20 °C) and ethanol (70%, v/v, kept at −20 °C).
10. Sodium acetate (CH_3COONa) buffer: 20 mM CH_3COONa; store at room temperature.

    CH3COONa buffer containing DFO: Sodium acetate buffer with 0.1 mM DFO (freshly made).
11. QIAGEN Genomic-tips: 20/G for cultured cells (small pellets); small tissue biopsies <20 mg (QIAGEN, cat. no. 10223), 100/G for 100 mg tissues (QIAGEN, cat. no. 10243), and/or tissue kit (QIAGEN, cat. no. 69504).
12. Buffers with Genomic-tip kit (QIAGEN, cat. no. 19060) (buffers G2, QBT, QC and QF, RNase A, and proteinase K).

13. Excision buffer 1×: It is recommended that excision reactions are carried out in 40 mM HEPES–KOH (pH 7.6), 2 mM EDTA, 1 mM DTT, 25 mM KCl, 5% glycerol (all reagents are DNase-, RNase-free), or similar previously validate buffers [55].
14. High-quality recombinant OGG1.
15. Quantitative PCR instrument (e.g., Bio-Rad CFX96™ or compatible instruments such as Applied Biosystems 7500, Applied Biosystems 7900, Stratagene Mx4000™).
16. Thermal Seal RTS™ Sealing Films.
17. DNA template: Use 2 ng to 10 ng DNA.
18. Primers (forward and reverse) diluted to working concentration (10 μM for most assays).
19. qPCR SYBR Green Mix—Examples: SsoAdvanced™ Universal SYBR Green Supermix (BioRad cat no. 172–5270); contains dNTPs Sso7d fusion polymerase $MgCl_2$, SYBRGreenI, and ROX normalization dyes (see instruction manual: https://www.bio-rad.com/webroot/web/pdf/lsr/ literature/10031339.pdf) or LuminoCt®SYBR®Green qPCR ReadyMix™ (Millipore Sigma cat. no. L6544), contains Tris–HCl (pH 8.3), KCl, dNTPs, stabilizers, $MgCl_2$, SYBR Green I, and Jumpstart Taq DNA polymerase (see instruction manual: https://www.sigmaaldrich.com/US/en/product/-sigma/l6544).

3 Methods

3.1 DNA Extraction from Animal Tissues and Cultured Cells

1. Tissue of interest collected from experimental animals placed into 1.5 mL microcentrifuge tubes and snap freeze by carefully submerging the tubes into liquid nitrogen. The snap-frozen tubes can be stored at −80 °C until ready to be used. Cultured cells recommended to be scraped washed in PBS once and stored at −80°C.
2. Perform all steps on ice, all centrifugation steps at 4 °C unless otherwise recommended.
3. Homogenize tissues (organ) in lysis buffer[a] (1.0 mL for 100 mg) containing DFO with polypropylene pellet pestle. If using an ultrasonic homogenizer, perform three 5-s pulses at setting 4 (*see* **Note 3**).
4. Centrifuge samples at 1000 × *g* for 10 min to isolate the nuclear pellet(s). Discard the supernatant.
5. Suspend the pellet in 1 mL lysis buffer containing DFO, and recentrifuge at 1000 × *g* for 10 min. Remove and discard the supernatant.

6. Suspend the pellet in 200 μL of enzyme reaction solution containing DFO. Hand homogenize using polypropylene pellet pestles or Dounce homogenizer. The crude DNA will appear stringy and gel-like; continue homogenizing until the white pellet is gone.
7. Isolate DNA from cultured cells: Suspend cells 1 mL in SDS and DFO-containing lysis buffer and proceed as described below.
8. *Digest RNA* by adding RNA digestion buffer. Mix and pulse-spin the sample by placing it in a microcentrifuge for 5 s.
9. Incubate for 1 h at 50 °C.
10. *Digest protein* by adding 15 μL of buffer containing proteinase K. Mix and pulse-spin in a microcentrifuge for 5 s.
11. Incubate for 3 h at 50 °C and mix sample hourly.
12. If necessary to remove particulates, sample(s) can be centrifuged at 10,000 × *g* for 5 min at 4 °C.
13. Transfer the supernatant into nuclease-free tube.
14. Add 300 μL NaI solution containing deferoxamine (DFO) to each sample and place them on revolving rotator for 1–2 min, or rotate by hand. Precipitate should be visible instantly.
15. Add the same volume of 100% isopropanol (ice cold) to individual sample and rotate them by hand or place samples in revolving rotator for a few minutes. DNA will be visible as a stringy gel-like precipitate.
16. Pellet DNA by centrifugation (10,000 × *g* for 10 min). Remove the supernatant and add 1 mL of cold 70% ethanol.
17. Centrifuge for 5 min at 10,000 × *g* and remove supernatant. Repeat this step three times.
18. Remove ethanol by pipetting. Carefully dry pellets (do not use forced drying as it increases artifactual DNA oxidation and makes resuspension of the DNA difficult).
19. Suspend the purified DNA in 200 μL sodium acetate buffer containing DFO. Solubilize DNA at 4 °C overnight (*see* **Note 4**).

3.2 DNA Extraction Using Qiagen Genomic-Tips Kit (Follow Recommendation of Manufacturer Provided with the Kit)

1. Homogenization of tissues (organs) follow recommendations as described in Subheading 2.1.
2. In cultured cells (10^7) or tissue/organ samples less than 20 mg, the 20/G Genomic-tip DNA isolation protocol is recommended and is described in brief (it is similar to that provided by the manufacturer in each kit) except we recommend supplementing buffers with 0.15 mM DFO (rational is described to Subheading 2.1). There some differences in methods for DNA

isolation based on tissue size; therefore, it is recommended to follow step-by-step protocols for each kit. For example, the link to Qiagen Kit Handbooks is as follows: https://www.qiagen.com/us/service-and-support/learning-hub/search-resources/#pg=1&q=&l=en&uuid=3d1aa7db-4dd3-4be0-87d9-a8186e69dc86 (*see* **Note 5**).

3.3 DNA Quantitation (See Note 6)

3.3.1 NanoDrop Quantification

1. Determine the DNA concentration of each sample by using a NanoDrop apparatus or spectrophotometer (absorbance at 260 nm vs. 280 nm).
2. Blank with 1 μL digestion dilution buffer containing desferrioxamine.
3. To ensure adequate purity, a 260/280 ratio > 1.7 should be obtained.

3.3.2 PicoGreen Quantitation of DNA

1. PicoGreen488 dsDNA quantitation reagent (for example, Lumiprobe; cat. no. 12010). It is recommended to utilize a fluorescence reader capable of measuring fluorescence with 485 nm excitation and 528 nm emission (e.g., BioTek Synergy 2, BioTek Instruments), using black-bottomed 96-well plates.
2. Dilute DNA sample (1:20 or 1:30 or higher in 1× Tris–HCl [pH 8.4]–EDTA), depending on the DNA yield.
3. Pipet 95 μL 1× Tris–HCl (pH 8.4)–EDTA into wells of a black-bottomed 96-well plate in triplicate for standards and samples.
4. Pipet 5 μL of standards into wells (0 ng/μL, 1 ng/μL, 2 ng/μL, 4 ng/μL, 8 ng/μL, 16 ng/μL, and 32 ng/μL).
5. Pipet 5 μL of diluted DNA samples into wells in triplicate.
6. Add PicoGreen solution by pipetting 5 μL PicoGreen reagent to 1 mL 1× Tris (pH 8.4)–EDTA.
7. In subdued lighting, pipet 100 μL PicoGreen solution into each well, cover plate, and place in the dark for 10 minutes to allow for color development.
8. Set excitation at 485 and emission at 528 nM in the fluorescence reader. Check sensitivity of reader (e.g., on Biotek Synergy 5, the sensitivity limit is 75).
9. Place plate in fluorescence reader after 10 min incubation.

3.4 Restriction Digestion of Genomic or Mitochondrial DNA

1. Select a restriction enzyme that does not cut the sequence of interest (within promoter, exon, intron). In case of using NEB enzymes, please find restriction enzyme(s) cutting sites outside sequences of interest (NEBcutter V2.0 tool online (http://nc2.neb.com/NEBcutter2/).

2. Calculate the volume of DNA solution that is needed for restriction digestion (1000 ng). Samples should have an identical amount of DNA for restriction enzyme digestion.
3. Purchase restriction enzyme from, e.g., NEB or similar quality from other sources.
4. NEB buffer stock is 10× and is diluted in molecular biology grade water until the digest is mixed (5 μL). A detailed digest customized protocol for NEB enzymes is available via the following web address: https://www.neb.com/protocols/2012/12/07/optimizing-restriction-endonuclease-reactions
5. In brief, add an appropriate restriction enzyme and place samples in a thermocycler. Set the temperature and digest time in the thermocycler as recommended by the manufacturer (e.g., NEB). After digestion, place samples on ice or at 4 °C.

   Further purification is not necessary prior to QPCR.

   A 50 μL reaction volume is recommended for digestion of 1 μg of DNA substrate.

   Supplement reaction mixture with recommended concentration of S-adenosyl methionine.

   The volume of the enzyme should not exceed 10% of the total reaction volume (50 μL) to prevent "star activity" due to excess glycerol (*see* **Note 7**).

3.5 Generation of DNA Polymerase-Blocking Gaps in DNA by OGG1 (See Notes 8 and 9)

DNA base excision

1. 250 ng DNA in 75 μL digestion (reaction) buffer added to OGG1 (OGG1 quantity is determined in preliminary experiments).
2. Incubate at 37 °C (optimum temperature) for 1 h.
3. After complete digestion, calculate the volume of sample and dilute it in 1× TE to attain 2 ng/μL concentration of DNA required for real-time quantitative PCR.

3.6 Quantitative (q) Polymerase Chain Reaction (qPCR) (See Note 10)

3.6.1 Primer Design for qPCR (Things to Consider; See Note 11)

1. First, be familiar with literature and databases for already existing validated primers.
2. Choose the sequence of interest (e.g., coding, noncoding stand, promoter with defined transcription binding sites, exon, intron).
3. Adjust primer locations such that they are located outside secondary structure within the target sequence. (When choosing the target sequence, avoid regions that have a secondary structure—use online program to assess whether the amplicon will form a secondary structure at the annealing temperature

(https://molbiol-tools.ca/Repeats_secondary_structure_Tm.htm).

4. Plan to amplify a 75–200-bp product (important: product should be 75 bp long minimum to distinguish it from primer dimers).
5. Choose a region that has a GC content of 50–60%.
6. Melting temperature (Tm) must be between 58 and 65 °C (*see* **Note 12**).
7. Repeats of guanines or cytosine longer than three bases should be avoided.
8. Guanines and cytosine being on the ends of the primers are acceptable and advisable.
9. Check the sequences of primers to ensure that there is no 3′ complementarity (avoid primer dimer formation).
10. Design primers using online tools https://primer3.ut.ee/ or https://www.ncbi.nlm.nih.gov/tools/primer-blast/.
11. Use online tools to verify specificity (BLAST: https://blast.ncbi.nlm.nih.gov/Blast.cgi).

*3.6.2 Primer Solubilization (See **Note 13**)*

1. Spin down lyophilized primers by centrifugation (10,000 rpm for 30 s).
2. Resuspend oligos in molecular biology grade (nuclease-free) TE buffer: 10 mM Tris–HCl (pH: 7.6) 0.1 mM EDTA to 100 μM.
3. Mix briefly by vortexing (30 s).
4. Store primer solution at −20 °C until use.
5. Dilute an aliquot to 10 μM from 100 μM with 0.1× TE.

3.6.3 Primer Validation

1. Check primer specificity.
2. Assess primer properties: Melting temperature (Tm), secondary structure, and complementarity.
3. Determine PCR product properties (e.g., size by agarose electrophoresis).
4. Validate the primers and optimize the protocol.

3.6.4 Experimental Steps

1. Prepare enough master mix to run all samples in duplicate (calculate amounts of reagents and add 10% volume to allow for pipetting error).
2. Be sure to include duplicate and no template negative controls.
3. Place all reagents on ice.
4. Vortex and collect contents at the bottom of the tube by brief centrifugation (mix well, avoiding bubbles).

5. Add 10 μL 2× iTaq SYBR Green Supermix.
6. Add forward and reverse primers (300 nM each primer, variable volume).
7. DNA template 2 ng (variable volume).
8. Add DNase-, RNase-free water (mol grade). Total volume should be 20 μL.
9. Set real-time PCR instruments (manufacturer recommendations) and collect the data that can be analyzed by instrument software (*see* **Note 14**).

3.7 Data Analysis

Methods of data analysis commonly include two main scenarios known as absolute and relative quantifications. Although there are qPCR applications that require absolute quantification (e.g., gene copy number or viral load determination), for assessment of DNA base lesion in a given DNA sequence, only relative quantification is required, because base lesions are generated under both physiological and stress conditions. For relative quantifications, two primary methods—the $2^{-\Delta\Delta CT}$ known as the Livak method [56] and Pfaffl method [56]. The Livak method is used when amplification efficiencies of target and reference genes are similar or identical, while if implication efficiencies of the two amplicons are different, the Pfaffl method is used. For assessment of relative changes in DNA lesion levels, the $2^{-\Delta\Delta CT}$ method is recommended.

3.8 Statistical Analysis

- Data analysis represents one of the main challenges in qPCR experiments and the statistical aspects of the analysis are considered key.
- Error values are generated for control samples by comparing each individual control to the average of all controls.
- If only two samples are being compared, a t-test or 1-factor ANOVA can be used to test for statistical significance.
- If more than two samples are being compared, ANOVA should be used.
- If there is more than one variable (e.g., dose, chemical), then an initial (global) multifactor ANOVA should be used, and post hoc comparisons of subsets of the data should only be carried out if warranted by significant global ANOVA results.

4 Notes

1. DFO is to prevent artificial DNA oxidation. Its half-life in solutions is only hours. Cloudy solutions should be discarded and should be made freshly. DFO is an iron chelator, preventing formation of hydroxyl radical by Fenton reactions with transition metals, and its addition to buffers for DNA isolation and digestion protects DNA from further oxidation [57].
2. NaI prevents artifactual oxidation of nucleic acid base(s) and prevents decomposition of 8-oxo(d)Gua of which oxidation potential lower than guanine and is very sensitive to further oxidation [29, 58].
3. In case one is using a volume more than 500 μL, tissue should be transferred to a Dounce homogenizer and homogenize tissues/organ for 15–20 strokes. Avoid bubble formation. Dounce homogenizer should be rinsed thoroughly with sterile ddH_2O between samples. If homogenizing tissue by ultrasonic homogenizer, one can avoid the formation of bubbles by keeping the probe in the buffer for the duration of homogenization.
4. If solution is gel-like, heat it to 60 °C for a few hours. You could also try solubilizing the DNA in digestion dilution or digest stop buffers instead, containing 0.1 mM deferoxamine. However, you will need to adjust the pH to ~5 for the digest with a small volume of sodium acetate buffer.
5. Various kits are recommended for DNA isolation; however, avoid extraction procedures that use phenol due to artifactual DNA oxidation. Kits (e.g., Qiagen, NEB) often used for DNA extraction to isolate long and short DNA provide highly reproducible results. In all cases, supplement buffers with DFO to avoid DNA oxidation. When extracting DNA, vortex the samples well prior to lysis as vortexing does not affect the DNA oxidation in buffer containing DFO.
6. The result of the qPCR is completely dependent on the precise quantification of the DNA samples [59]. Therefore, in case samples are concentrated, we recommend an additional round of dilution and quantitation to ensure equal amounts of DNA for amplification. Use sterile technique for all steps. It is recommended to use disposable gloves to avoid the introduction of nucleases, foreign DNA, or other contaminants. Use a dedicated laminar hood and workstations for individual steps of the procedure. Do not open the PCR tubes as small DNA quantities can volatilize and contaminate other assays. The inclusion of a blank sample (where no DNA is added) helps to assure that no contamination has occurred with foreign DNA or PCR products.

7. "Star activity" is a general property of restriction endonucleases. Restriction endonuclease will cleave sequences, which are similar, but not identical, to their defined recognition sequence under some conditions. To avoid such activities, it is recommended to utilize conditions provided by suppliers. If an enzyme exhibits star activity, it will be indicated in the product data sheet, in the catalog, or our manufacturer website. In addition, additives in the restriction enzyme storage buffer (e.g., glycerol, salt) as well as contaminants found in the substrate solution (e.g., salt, alcohol, EDTA) can be problematic. To avoid such issues, see guidelines at https://www.neb.com/protocols/2012/12/07/optimizing-restriction-endonuclease-reactions.
8. FLARE-qPCR for assessment of DNA lesion such as 8-oxo(d)Gua is based on the idea that removal of a damaged base will generate gaps (AP sites and other kinds of DNA strand lesions) in the DNA strand that block thermostable DNA polymerase [44–46]. Using the assay one can compare the relative level of the lesion (or even lesion frequencies) within a sort stretch of DNA based on amplification alone, in which higher amplification corresponds with a lower level of lesion [47, 48].

   One of the most abundant lesions in DNA and RNA is 8-oxo(d)Gua [60]. The United States and European Standards Committee on Oxidative DNA Damage recommend the use of formamidopyrimidine DNA-glycosylase (FPG) and 8-hydroxyguanine DNA-glycosylase (hOGG1) to detect/excise oxidatively modified guanine [32, 61]. Follow-up studies using the Comet assay showed that hOGG1 recognizes oxidatively modified DNA base lesion 8-oxo(d)Gua with higher specificity than FPG [61]. Therefore, it is recommended that repair enzyme is OGG1 for fragment length analysis coupled to quantitative (q)-PCR for detection of 8-oxo(d)Gua. This method provided superior opportunities to assess 8-oxo(d)Gua and routinely used in our labs [20, 50, 52, 62].
9. Human (h)OGG1 (α isoform) has both N-glycosylase and an apurinic/apyrimidinic (AP)-lyase activity. Via its N-glycosylase activity, OGG1 releases damaged purine(s) from double-stranded DNA, generating an abasic site, than through its AP-lyase activity which cleaves 3′ to the AP site leaving a 5′ phosphate and a 3′-phospho-α, β-unsaturated aldehyde [16, 63].
10. The substrate of hOGG1 includes 8-oxo(d)Gua, 8-oxoadenine, and FapyGua all paired with cytosine (rev in [16]).
11. The primary benefit of real-time (rt)-qPCR over polymerase chain reaction is that it allows you to determine the copy

number of template DNA with precision and high sensitivity over a wide range. rt-qPCR results can either be qualitative (the presence or absence of a sequence) or quantitative (when compared test sample to control). Additionally, rt-qPCR data can be evaluated without gel electrophoresis, resulting in decreased bench time and increased accuracy. Finally, because rt-qPCR reactions are evaluated in a closed-tube qPCR system, chances for contamination are significantly decreased and the need for post-amplification can be eliminated.

12. Various primer design software can be used, as they provide information on annealing temperature, primer length, product length, and GC content. The NCBI primer blast or Primer 3 online software is reliable to design primers for qPCR (https://www.ncbi.nlm.nih.gov/tools/primer-blast/ or https://bioinfo.ut.ee/primer3-0.4.0/primer3/). A reliable qPCR requires highly efficient and specific amplification of product(s). Design a 75–200-bp product. Note: Short PCR products are typically amplified with higher efficiency than longer ones, but the qPCR product must be 75 bp long to distinguish it from primer dimers that could potentially generated. Avoid regions that have secondary structure, when possible. Efficiency of amplification for shorter sequences (75–80 bp) are higher compared to those of longer sequences (200 bp or longer).

13. The melting temperature (Tm) is a key parameter to consider when designing and performing PCR and qPCR. It is defined as the temperature at which 50 percentage (half) of the double-stranded DNA will separate into single stranded. It also indicates the duplex stability. Primer pairs with melting temperatures between 55 and 62°C commonly produce the best results. Use an online Tm calculator, with values of 50 mM for salt concentration and 300 nM for oligonucleotide concentration), for example, online tool: http://biotools.nubic.northwestern.edu/OligoCalc.html.

14. Primers should be suspended in DNase-, RNase-free TE (0.1×) buffer or simply in water. Use of water or 0.1× TE buffer allows you to decrease the amount of EDTA, so it does not interfere PCR through chelation of Mg^{2+}. Make aliquots of stock primer solution in order to avoid multiple freeze/thaw cycles. Once stock (100 μM) primer solution is diluted to a working concentration (e.g., 10 μM), it is kept at 4 °C and is stable for weeks. It is essential to assess the PCR product of your primer set by agarose gel electrophoresis to verify the size of the product and to assure that no other products are generated.

15. To attain correct template quantification in a qPCR assay, each reaction must efficiently amplify a single product, and

amplification efficiency must be independent of template concentration and the amplification of other templates. Mathematical calculation and other considerations in amplification efficiency can be found at https://www.bio-rad.com/webroot/web/pdf/lsr/literature/Bulletin_5279.pdf (only an example). Melting curve analysis is one of the ways for validation of primer pair. From the analysis, it will be obvious that primer-generated nonspecific products and primer dimers that are highly misleading in calculations when the fluorescence of the reporter chemistry depends on the presence of dsDNA, using SYBR Green dye. Formation of secondary nonspecific products and primer dimers* can severely decrease target sequence amplification efficiency and, ultimately, the accuracy of the qPCR results. After completion of the amplification reaction, a melt curve is generated by increasing the temperature in small increments and monitoring the fluorescence signal at each step. As the dsDNA in the reaction is denatured, the fluorescence decreases rapidly. A plot of the negative first derivative of the fluorescence vs. temperature (the rate of change of fluorescence intensity) displays distinct peaks corresponding to the Tm of each product. *A primer dimer is a potential by-product in the polymerase chain reaction. As a result, the DNA polymerase amplifies the primer dimer, leading to competition for PCR reagents, thus potentially decreasing amplification of the DNA sequence targeted for PCR.

Acknowledgements

This work was supported by the National Institute of Allergic and Infectious Diseases (NIAID/AI062885 [I.B.]).

Author Contributions L.P., K.W., X.Z., and Y.X maintained and treated cells with the cytokine TFNα, extracted DNA, perform PCR assays, and quantifications. I.B and L.P wrote the manuscript with help the of X.Y. Artwork was performed by L.P and Y.X. All authors discussed results and approved the content of the manuscript.

References

1. Radak Z, Boldogh I (2010) 8-Oxo-7,8-dihydroguanine: links to gene expression, aging, and defense against oxidative stress. Free Radic Biol Med 49(4):587–596
2. Steenken S (1989) Structure, acid/base properties and transformation reactions of purine radicals. Free Radic Res Commun 6(2–3): 117–120
3. Nunez ME, Hall DB, Barton JK (1999) Long-range oxidative damage to DNA: effects of distance and sequence. Chem Biol 6(2):85–97

4. Hall DB, Holmlin RE, Barton JK (1996) Oxidative DNA damage through long-range electron transfer. Nature 382(6593):731–735
5. Plum GE, Grollman AP, Johnson F, Breslauer KJ (1995) Influence of the oxidatively damaged adduct 8-oxodeoxyguanosine on the conformation, energetics, and thermodynamic stability of a DNA duplex. Biochemistry 34(49):16148–16160
6. Lipscomb LA, Peek ME, Morningstar ML, Verghis SM, Miller EM, Rich A et al (1995) X-ray structure of a DNA decamer containing 7,8-dihydro-8-oxoguanine. Proc Natl Acad Sci U S A 92(3):719–723
7. Oda Y, Uesugi S, Ikehara M, Nishimura S, Kawase Y, Ishikawa H et al (1991) NMR studies of a DNA containing 8-hydroxydeoxyguanosine. Nucleic Acids Res 19(7):1407–1412
8. Kitsera N, Stathis D, Luhnsdorf B, Muller H, Carell T, Epe B et al (2011) 8-Oxo-7,8-dihydroguanine in DNA does not constitute a barrier to transcription, but is converted into transcription-blocking damage by OGG1. Nucleic Acids Res 39(14):5926–5934
9. Pastoriza-Gallego M, Armier J, Sarasin A (2007) Transcription through 8-oxoguanine in DNA repair-proficient and Csb(−)/Ogg1 (−) DNA repair-deficient mouse embryonic fibroblasts is dependent upon promoter strength and sequence context. Mutagenesis 22(5):343–351
10. Grollman AP, Moriya M (1993) Mutagenesis by 8-oxoguanine: an enemy within. Trends Genet 9(7):246–249
11. Kouchakdjian M, Bodepudi V, Shibutani S, Eisenberg M, Johnson F, Grollman AP et al (1991) NMR structural studies of the ionizing radiation adduct 7-hydro-8-oxodeoxyguanosine (8-oxo-7H-dG) opposite deoxyadenosine in a DNA duplex. 8-Oxo-7H-dG(syn).dA (anti) alignment at lesion site. Biochemistry 30(5):1403–1412
12. D'Augustin O, Huet S, Campalans A, Radicella JP (2020) Lost in the crowd: how does human 8-oxoguanine DNA glycosylase 1 (OGG1) find 8-oxoguanine in the genome? Int J Mol Sci 21(21):8360
13. Saxowsky TT, Meadows KL, Klungland A, Doetsch PW (2008) 8-Oxoguanine-mediated transcriptional mutagenesis causes Ras activation in mammalian cells. Proc Natl Acad Sci U S A 105(48):18877–18882
14. Mitra S, Boldogh I, Izumi T, Hazra TK (2001) Complexities of the DNA base excision repair pathway for repair of oxidative DNA damage. Environ Mol Mutagen 38(2–3):180–190
15. Boldogh I, Milligan D, Lee MS, Bassett H, Lloyd RS, McCullough AK (2001) hMYH cell cycle-dependent expression, subcellular localization and association with replication foci: evidence suggesting replication-coupled repair of adenine:8-oxoguanine mispairs. Nucleic Acids Res 29(13):2802–2809
16. Mitra S, Izumi T, Boldogh I, Bhakat KK, Hill JW, Hazra TK (2002) Choreography of oxidative damage repair in mammalian genomes. Free Radic Biol Med 33(1):15–28
17. Ba X, Boldogh I (2018) 8-Oxoguanine DNA glycosylase 1: beyond repair of the oxidatively modified base lesions. Redox Biol 14:669–678
18. Fleming AM, Burrows CJ (2017) 8-Oxo-7,8-dihydroguanine, friend and foe: epigenetic-like regulator versus initiator of mutagenesis. DNA Repair (Amst) 56:75–83
19. Hao W, Wang J, Zhang Y, Wang C, Xia L, Zhang W et al (2020) Enzymatically inactive OGG1 binds to DNA and steers base excision repair toward gene transcription. FASEB J 34(6):7427–7441
20. Hao W, Qi T, Pan L, Wang R, Zhu B, Aguilera-Aguirre L et al (2018) Effects of the stimuli-dependent enrichment of 8-oxoguanine DNA glycosylase1 on chromatinized DNA. Redox Biol 18:43–53
21. Roychoudhury S, Pramanik S, Harris HL, Tarpley M, Sarkar A, Spagnol G et al (2020) Endogenous oxidized DNA bases and APE1 regulate the formation of G-quadruplex structures in the genome. Proc Natl Acad Sci U S A 117(21):11409–11420
22. Pan L, Wang H, Luo J, Zeng J, Pi J, Liu H et al (2019) Epigenetic regulation of TIMP1 expression by 8-oxoguanine DNA glycosylase-1 binding to DNA:RNA hybrid. FASEB J 33(12):14159–14170
23. Cadet J, Douki T, Ravanat JL (2011) Measurement of oxidatively generated base damage in cellular DNA. Mutat Res 711(1–2):3–12
24. Cadet J, Berger M, Douki T, Ravanat JL (1997) Oxidative damage to DNA: formation, measurement, and biological significance. Rev Physiol Biochem Pharmacol 131:1–87
25. Cadet J, D'Ham C, Douki T, Pouget JP, Ravanat JL, Sauvaigo S (1998) Facts and artifacts in the measurement of oxidative base damage to DNA. Free Radic Res 29(6):541–550
26. Collins A (2000) Comparison of different methods of measuring 8-oxoguanine as a marker of oxidative DNA damage. ESCODD (European Standards Committee on Oxidative DNA Damage). Free Radic Res 32(4): 333–341

27. Ravanat JL, Douki T, Duez P, Gremaud E, Herbert K, Hofer T et al (2002) Cellular background level of 8-oxo-7,8-dihydro-2′-deoxyguanosine: an isotope based method to evaluate artifactual oxidation of DNA during its extraction and subsequent work-up. Carcinogenesis 23(11):1911–1918
28. Rodriguez H, Jurado J, Laval J, Dizdaroglu M (2000) Comparison of the levels of 8-hydroxyguanine in DNA as measured by gas chromatography mass spectrometry following hydrolysis of DNA by Escherichia coli Fpg protein or formic acid. Nucleic Acids Res 28(15): E75
29. Dizdaroglu M, Jaruga P, Rodriguez H (2001) Measurement of 8-hydroxy-2′-deoxyguanosine in DNA by high-performance liquid chromatography-mass spectrometry: comparison with measurement by gas chromatography-mass spectrometry. Nucleic Acids Res 29(3):E12
30. Ravanat JL, Turesky RJ, Gremaud E, Trudel LJ, Stadler RH (1995) Determination of 8-oxoguanine in DNA by gas chromatography—mass spectrometry and HPLC—electrochemical detection: overestimation of the background level of the oxidized base by the gas chromatography—mass spectrometry assay. Chem Res Toxicol 8(8): 1039–1045
31. Dizdaroglu M, Coskun E, Jaruga P (2015) Measurement of oxidatively induced DNA damage and its repair, by mass spectrometric techniques. Free Radic Res 49(5):525–548
32. Dizdaroglu M, Jaruga P, Birincioglu M, Rodriguez H (2002) Free radical-induced damage to DNA: mechanisms and measurement. Free Radic Biol Med 32(11):1102–1115
33. Cadet J, Douki T, Ravanat JL (1997) Artifacts associated with the measurement of oxidized DNA bases. Environ Health Perspect 105(10):1034–1039
34. Nakae D, Mizumoto Y, Kobayashi E, Noguchi O, Konishi Y (1995) Improved genomic/nuclear DNA extraction for 8-hydroxydeoxyguanosine analysis of small amounts of rat liver tissue. Cancer Lett 97(2): 233–239
35. Hamilton ML, Guo Z, Fuller CD, Van Remmen H, Ward WF, Austad SN et al (2001) A reliable assessment of 8-oxo-2-deoxyguanosine levels in nuclear and mitochondrial DNA using the sodium iodide method to isolate DNA. Nucleic Acids Res 29(10): 2117–2126
36. Amente S, Bertoni A, Morano A, Lania L, Avvedimento EV, Majello B (2010) LSD1-mediated demethylation of histone H3 lysine 4 triggers Myc-induced transcription. Oncogene 29(25):3691–3702
37. Genereux JC, Barton JK (2010) Mechanisms for DNA charge transport. Chem Rev 110(3): 1642–1662
38. Pastukh V, Roberts JT, Clark DW, Bardwell GC, Patel M, Al-Mehdi AB et al (2015) An oxidative DNA "damage" and repair mechanism localized in the VEGF promoter is important for hypoxia-induced VEGF mRNA expression. Am J Physiol Lung Cell Mol Physiol 309(11):L1367–L1375
39. Ding Y, Fleming AM, Burrows CJ (2017) Sequencing the mouse genome for the oxidatively modified base 8-Oxo-7,8-dihydroguanine by OG-Seq. J Am Chem Soc 139(7): 2569–2572
40. Fleming AM, Ding Y, Burrows CJ (2017) Sequencing DNA for the oxidatively modified base 8-Oxo-7,8-dihydroguanine. Methods Enzymol 591:187–210
41. Vlahopoulos S, Adamaki M, Khoury N, Zoumpourlis V, Boldogh I (2019) Roles of DNA repair enzyme OGG1 in innate immunity and its significance for lung cancer. Pharmacol Ther 194:59–72
42. Sharma P, Sampath H (2019) Mitochondrial DNA integrity: role in health and disease. Cell 8(2):100
43. Sampath H, Lloyd RS (2019) Roles of OGG1 in transcriptional regulation and maintenance of metabolic homeostasis. DNA Repair (Amst) 81:102667
44. Ponti M, Forrow SM, Souhami RL, D'Incalci M, Hartley JA (1991) Measurement of the sequence specificity of covalent DNA modification by antineoplastic agents using Taq DNA polymerase. Nucleic Acids Res 19(11):2929–2933
45. Ayala-Torres S, Chen Y, Svoboda T, Rosenblatt J, Van Houten B (2000) Analysis of gene-specific DNA damage and repair using quantitative polymerase chain reaction. Methods 22(2):135–147
46. Santos JH, Meyer JN, Mandavilli BS, Van Houten B (2006) Quantitative PCR-based measurement of nuclear and mitochondrial DNA damage and repair in mammalian cells. Methods Mol Biol 314:183–199
47. Jennerwein MM, Eastman A (1991) A polymerase chain reaction-based method to detect cisplatin adducts in specific genes. Nucleic Acids Res 19(22):6209–6214
48. Kalinowski DP, Illenye S, Van Houten B (1992) Analysis of DNA damage and repair in murine leukemia L1210 cells using a

quantitative polymerase chain reaction assay. Nucleic Acids Res 20(13):3485–3494

49. Salazar JJ, Van Houten B (1997) Preferential mitochondrial DNA injury caused by glucose oxidase as a steady generator of hydrogen peroxide in human fibroblasts. Mutat Res 385(2): 139–149
50. Ba X, Bacsi A, Luo J, Aguilera-Aguirre L, Zeng X, Radak Z et al (2014) 8-oxoguanine DNA glycosylase-1 augments proinflammatory gene expression by facilitating the recruitment of site-specific transcription factors. J Immunol 192(5):2384–2394
51. Chen KH, Yakes FM, Srivastava DK, Singhal RK, Sobol RW, Horton JK et al (1998) Up-regulation of base excision repair correlates with enhanced protection against a DNA damaging agent in mouse cell lines. Nucleic Acids Res 26(8):2001–2007
52. Pan L, Zhu B, Hao W, Zeng X, Vlahopoulos SA, Hazra TK et al (2016) Oxidized guanine base lesions function in 8-oxoguanine DNA glycosylase-1-mediated epigenetic regulation of nuclear factor kappaB-driven gene expression. J Biol Chem 291(49):25553–25566
53. Hunter SE, Jung D, Di Giulio RT, Meyer JN (2010) The QPCR assay for analysis of mitochondrial DNA damage, repair, and relative copy number. Methods 51(4):444–451
54. Furda AM, Bess AS, Meyer JN, Van Houten B (2012) Analysis of DNA damage and repair in nuclear and mitochondrial DNA of animal cells using quantitative PCR. Methods Mol Biol 920:111–132
55. German P, Szaniszlo P, Hajas G, Radak Z, Bacsi A, Hazra TK et al (2013) Activation of cellular signaling by 8-oxoguanine DNA glycosylase-1-initiated DNA base excision repair. DNA Repair (Amst) 12(10):856–863
56. Pfaffl MW (2001) A new mathematical model for relative quantification in real-time RT-PCR. Nucleic Acids Res 29(9):45e
57. Amente S, Di Palo G, Scala G, Castrignano T, Gorini F, Cocozza S et al (2019) Genome-wide mapping of 8-oxo-7,8-dihydro-2′--deoxyguanosine reveals accumulation of oxidatively-generated damage at DNA replication origins within transcribed long genes of mammalian cells. Nucleic Acids Res 47(1): 221–236
58. Helbock HJ, Beckman KB, Ames BN (1999) 8-Hydroxydeoxyguanosine and 8-hydroxyguanine as biomarkers of oxidative DNA damage. Methods Enzymol 300:156–166
59. Yakes FM, Van Houten B (1997) Mitochondrial DNA damage is more extensive and persists longer than nuclear DNA damage in human cells following oxidative stress. Proc Natl Acad Sci U S A 94(2):514–519
60. Bruner SD, Norman DP, Verdine GL (2000) Structural basis for recognition and repair of the endogenous mutagen 8-oxoguanine in DNA. Nature 403(6772):859–866
61. Smith CC, O'Donovan MR, Martin EA (2006) hOGG1 recognizes oxidative damage using the comet assay with greater specificity than FPG or ENDOIII. Mutagenesis 21(3): 185–190
62. Pan L, Hao W, Zheng X, Zeng X, Ahmed Abbasi A, Boldogh I et al (2017) OGG1-DNA interactions facilitate NF-kappaB binding to DNA targets. Sci Rep 7:43297
63. Boiteux S, Radicella JP (1999) Base excision repair of 8-hydroxyguanine protects DNA from endogenous oxidative stress. Biochimie 81(1–2):59–67
64. Livak KJ, Schmittgen TD (2001) Analysis of relative gene expression data using real-time quantitative PCR and the 2(-Delta Delta C(T)) method. Methods 25(4):402–408

Chapter 8

Isolation and Immunodetection of Enzymatic DNA–Protein Crosslinks by RADAR Assay

Megan Perry and Gargi Ghosal

Abstract

DNA–protein crosslinks (DPCs) are steric hindrances to DNA metabolic processes and the removal and repair of DPCs is a rapidly evolving area of research. A critical component of deciphering this repair pathway is developing techniques that detect and quantify specific types of DPCs in cells. Here we describe a protocol for direct detection of enzymatic DPCs from mammalian cells—the RADAR assay. The method involves isolating genomic DNA and DPCs from cells and binding them to nitrocellulose membrane with a vacuum slot blot manifold. DPCs are detected using antibodies raised against the protein of interest and quantified by normalizing to a DNA loading control. The RADAR assay allows for the detection of specific types of DPCs and the sensitive analysis of the DNA–protein crosslinking activity of various drugs, is adaptable across different cell types and conditions, and requires little specialized equipment.

Key words DNA-protein crosslink, RADAR assay, TOP1-cc, TOP2-cc, Slot blot, DPC repair, DNA repair

1 Introduction

DNA–protein crosslinks (DPCs) are irreversible covalent crosslinks of proteins to the DNA and are steric blockades to virtually all DNA metabolic processes, namely, replication, transcription, recombination, and repair processes. Replication forks stall at DPCs because the DNA polymerase is unable to advance beyond a bulky protein lesion. Similarly, transcription and recombination machinery are stalled or displaced at DPCs [1]. DPCs also present an obstacle to DNA repair processes, as the DNA lesion to be repaired is buried within the crosslinked protein, making it inaccessible to DNA repair enzymes [2]. Unrepaired DPCs lead to single- and double-stranded DNA breaks and genome instability, resulting in tumorigenesis and genetic disease [3]. DPCs are highly variable DNA lesions; the identity of the crosslinked protein, the cell cycle status, and the specific configuration of the DNA—unperturbed duplex

Kishor K. Bhakat and Tapas K. Hazra (eds.), *Base Excision Repair Pathway: Methods and Protocols*, Methods in Molecular Biology, vol. 2701, https://doi.org/10.1007/978-1-0716-3373-1_8,

DNA, unwound single-stranded DNA, or crosslinking to the ends of a single- or double-stranded DNA break—all contribute to the variability of DPCs [4, 5]. Thus, it is not surprising that DPCs can be removed and repaired by the action of multiple enzymes and pathways. DPCs are proteolyzed by proteasome- or protease-dependent repair pathways or cleaved by nucleases [1, 2, 6]. The resulting peptide–DNA adduct is processed by specialized enzymes such as tyrosyl–DNA phosphodiesterases TDP1 and TDP2 and/or canonical DNA double-strand break repair pathways [5–7]. The current understanding of DPC repair is a complex network involving multiple repair pathways that is only beginning to be understood.

DPCs can be classified into two broad categories: nonenzymatic and enzymatic. Nonenzymatic DPCs occur when any protein in the vicinity of damaged DNA becomes crosslinked to the DNA upon exposure to nonspecific damaging agents such as oxygen free radicals, ultraviolet radiation, ionizing radiation, formaldehyde, or platinum-based chemotherapies [2, 8]. Conversely, enzymatic DPCs occur when specific enzymes that functionally bind DNA in a transient manner become covalently trapped. Common enzymatic DPCs include topoisomerase 1 (TOP1), topoisomerase 2 (TOP2), DNA polymerases, poly [ADP-ribose] polymerase 1 (PARP1), and DNA methyltransferase (DNMT1) [2]. While nonspecific DNA damaging agents can induce enzymatic DPCs, these lesions are frequently induced by specific chemotherapeutic agents; for example, camptothecin (CPT) is a potent crosslinking agent of TOP1, etoposide (VP16) crosslinks TOP2, and 5-aza-2'-deoxycytidine induces DNMT1-DNA crosslinks [9–11].

Various techniques have been developed for isolating and investigating the repair dynamics of DPCs. These methods fall into three general categories: (1) repair assays of DPC-containing plasmids in cells or *Xenopus* egg extracts and (2) DPC detection in cells by indirect and (3) direct methods. The first group of assays involves introduction of site-specific DPCs on plasmids to be transfected into cells or used in *Xenopus* egg extract reactions [4, 12–15]. These assays are useful for examining precise DNA replication or DPC proteolysis dynamics; however, these assays are complex, they rely heavily on recombinant proteins and DPC-containing plasmid DNA, and they are not suitable for investigating specific drug activity or total DPC burden in cells. The second group, indirect DPC detection assays, either analyzes the effects of DPC-inducing agents or measure the amount of protein cross-linked DNA [16–19]. The standard indirect DPC assay, the SDS/K^+ precipitation assay, or, its recently developed derivative, the ARK assay precipitates and measures the proportion of global DPCs to total DNA [16, 19–23]. Indirect DPC assays are advantageous for studying nonenzymatic DPC burden and the effects of nonspecific crosslinking agents such as formaldehyde, but do not allow for direct visualization or detection of specific protein–DNA crosslinks. The final category of methods, direct DPC detection, isolates DPCs and uses fluorescent labeling or immunodetection of specific proteins crosslinked to DNA for analysis or quantification [24–27].

In this chapter, we describe a modified protocol for direct detection of specific enzymatic DPCs by the rapid approach to DNA adduct recovery (RADAR) assay [26, 28]. Genomic DNA and DPCs are isolated from cells and free protein is removed from the sample through ethanol precipitation. DPCs are detected and quantified by slot blot analysis (Fig. 1). This method is advantageous for studying drug effects on specific DNA-binding proteins, uses low cell density relative to other methods, and does not rely on ultracentrifugation or fluorescent labeling techniques. The RADAR assay is highly versatile for different proteins of interest and cell types, and it can be modified for immunoprecipitation of crosslinked proteins or detection of nonenzymatic DPCs via SDS-PAGE [23, 29].

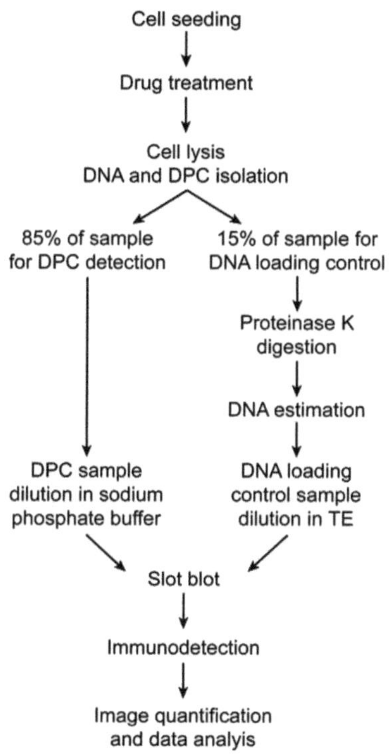

Fig. 1 Schematic of the workflow for the RADAR assay

2 Materials

2.1 Cell Culture and Drug Treatment

1. A549 cells (*see* **Note 1**).
2. Dulbecco's Modified Eagle Medium (DMEM) supplemented with 10% fetal bovine serum (FBS) and 1% penicillin–streptomycin (Pen-Strep).
3. Hemocytometer.
4. Trypan blue.
5. 6-well cell culture plates.
6. DPC-inducing agents: Camptothecin and etoposide, dissolved in DMSO at stock concentrations 50 and 500 μM, respectively.

2.2 DPC Isolation

1. DNAzol® Reagent: Genomic DNA Isolation Reagent (Invitrogen).
2. 1.5 mL microcentrifuge tubes, autoclaved.
3. 200-proof ethanol (EtOH), chilled on ice before use.
4. 70% EtOH, made with sterile deionized water (ddH_2O).
5. 8 mM NaOH, made fresh from concentrated NaOH at the time of use. For 10 mL, add 40 μL of 2 M NaOH to 9.96 mL sterile ddH_2O.

2.3 Slot Blotting, Immunodetection, and Analysis

1. Proteinase K from *Tritirachium album*: stock concentration 10 mg/mL.
2. Heated dry bath or water bath set to 55 °C.
3. Quant-iT™ PicoGreen™ dsDNA assay kit (Invitrogen).
4. 96-well solid white polystyrene microplates.
5. Fluorescence microplate reader.
6. 25 mM sodium phosphate buffer, pH 6.5: Dissolve 0.599 g of Na_2HPO_4 and 0.554 g of NaH_2PO_4 in 200 mL of sterile ddH_2O, adjust the pH, and bring the volume up to 250 mL.
7. Tris–EDTA (TE) buffer, autoclaved: 10 mM Tris–HCl pH 8.0 with HCl, 1 mM EDTA.
8. Positively charged nylon membrane.
9. Nitrocellulose membrane: 0.45 μM.
10. Blotting paper: 0.38 mm thickness.
11. Slot blot blotting manifold.
12. 20× TBST: 121.1 g Tris base, 640 g NaCl; dissolve in 3.2 L of ddH_2O, adjust the pH to 7.6 with concentrated HCl, add Tween-20 to a final concentration of 1%, and bring up the volume to 4 L.

13. 1× TBST: Make from 20× stock solution. Add 0.1 L 20× TBST to 1.9 L of ddH_2O. Final concentrations: 20 mM Tris, 137 mM NaCl, 0.05% Tween-20.
14. 5% milk in 1× TBST: Dissolve 25 *g* of milk powder in 500 mL of 1× TBST.
15. 20× SSC buffer: 175.3 *g* NaCl, 88.2 *g* sodium citrate; dissolve in 700 mL of ddH_2O, adjust the pH to 7.0 with concentrated HCl, and bring up the volume to 1 L.
16. 6× SSC buffer: Make from 20× stock solution. Add 300 mL 20× SSC to 700 mL of ddH_2O. Final concentrations: 900 mM NaCl, 102.5 mM sodium citrate.
17. 2× SSC buffer: Make from 20× stock solution. Add 100 mL 20× SSC to 900 mL of ddH_2O. Final concentrations: 300 mM NaCl, 34.2 mM sodium citrate.
18. Primary antibodies:
    (a) Anti-topoisomerase I (Abcam ab109374; dilution 1: 2000).
    (b) Anti-topoisomerase II-alpha (Cell Signaling Technology 12286S; dilution 1:1000).
    (c) Anti-dsDNA (Abcam ab27156; dilution 1:10,000).
19. 5% BSA in 1× TBST: Dissolve 2.5 *g* bovine serum albumin (BSA) powder, DNase- and protease-free, in 50 mL of 1× TBST.
20. Secondary antibodies:
    (a) Peroxidase-conjugated AffiniPure Polyclonal Goat Anti-Rabbit IgG (Jackson ImmunoResearch Laboratories 111–035-144; dilution 1:10,000).
    (b) Peroxidase-Conjugated AffiniPure Polyclonal Rabbit Anti-Mouse IgG + IgM (Jackson ImmunoResearch Laboratories 315–035-048; dilution 1:10,000).
21. Chemiluminescent substrate for detection of HRP.
22. X-Ray films and film developer or western blot imaging system.
23. ImageJ software.
24. Microsoft Excel.
25. GraphPad Prism software.

3 Methods

3.1 Cell Culture

1. Culture A549 cells at 37 °C in DMEM supplemented with 10% FBS and 1% Pen-Strep.
2. Day 1: Count and seed 1×10^5 cells per well in 6-well plates.
3. Day 3: 48 h after seeding, the cells should be approximately 85% confluent (*see* **Note 1**).

3.2 Drug Treatment

1. Aspirate the culture media from the cells.
2. Refresh each well with 2 mL of new culture media.
3. Treat cells with DPC-inducing drug at the experimental concentrations (*see* **Note 2**).
4. Return the plates to the incubator and incubate the cells in drug-containing media for the desired length of time (*see* **Note 3**).

3.3 Cell Lysis and Sample Preparation

1. Aspirate the media from the cells.
2. Directly lyse the cells in 500 μL of DNAzol® reagent. Pipette up and down approximately five times to ensure that the cells are adequately sheared. Transfer the lysate to a 1.5 mL centrifuge tube (*see* **Note 4**).
3. Add 250 μL of ice-cold 200-proof EtOH to each sample to precipitate the genomic DNA. Thoroughly mix the samples by inversion. Keep the samples at −20 °C for 15 min. After chilling, swirl and invert the tubes by hand slowly to spool the precipitated DNA. Do not vortex. At this stage, the precipitated DNA may or may not be visible as a loose clump floating in the tube.
4. Centrifuge the samples at 4800 × *g* (7,000 rpm) for 5 min at 4 °C. Slowly decant the supernatant into a discard receptacle. Do not aspirate. The precipitated DNA does not always collect at the bottom but may instead be along the side of the tube; take care not to pour it off (*see* **Note 5**).
5. Add 500 μL of 70% EtOH to each sample and vortex at medium-high strength for 2 s. Centrifuge the sample at 7000 rpm for 5 min at 4 °C. Carefully decant the supernatant. Do not aspirate.
6. Repeat the 70% EtOH wash, vortex, and spin as in Subheading 3.3, **step 5**.
7. Decant the supernatant, invert the tube onto a paper towel to absorb the excess EtOH, and air-dry each sample briefly—less than 1 min. The samples do not dissolve if they are left to air-dry for an extended period. A small amount of residual EtOH in the tube will not affect the solubility of the sample.
8. Dissolve each sample in 300 μL of freshly made 8 mM NaOH. Add the NaOH to the tube and let the sample sit undisturbed at room temperature for 10 min. Pipette the sample up and down approximately 15 times with a 200 μL pipette to dissolve the pellet completely (*see* **Note 6**). Once dissolved, store the samples on ice and proceed to DNA quantification and slot blotting (*see* **Note 7**).

3.4 DNA Quantification and Sample Dilution

1. To prepare an aliquot of sample used for DNA quantification, dilute 50 μL of each DPC sample in 97 μL of 8 mM NaOH in a separate tube. Place the undiluted DPC samples back on ice (*see* **step 4b** below).
2. To the diluted samples for quantification, add 3 μL of proteinase K (final concentration of 0.2 mg/mL), vortex briefly, and incubate the samples at 55 °C for 45 min (*see* **Note 8**).
3. Use the Quant-iT™ PicoGreen™ dsDNA assay kit to estimate the DNA concentration of each sample per the manufacturer's instructions (*see* **Note 9**). Reserve the remaining volume of proteinase K-digested samples for immunoblotting the DNA loading control (*see* **step 4a** below). The total DNA yield from seeding 1×10^5 A549 cells is approximately 3 μg.
4. Prepare samples for slot blotting (*see* **Note 10**). Dilute samples for the DNA loading control (set 1) and for DPC immunodetection (set 2) as described below. The sample concentration is determined by the desired amount of DNA to be loaded in a volume of 200 μL (*see* **Note 11**). Vortex diluted samples to mix. Store diluted samples on ice or at 4 °C while assembling the slot blotting apparatus.
   (a) Sample set 1: The reserved proteinase K-digested samples (Subheading 3.4, **step 3**) will be used for the DNA loading control and should be diluted in TE buffer, pH 8. In a clean microcentrifuge tube, prepare 0.5 ng/μL diluted samples to load a total of 100 ng during slot blotting.
   (b) Sample set 2: The DPC samples (Subheading 3.3, **step 8**) should be diluted in 25 mM sodium phosphate buffer, pH 6.5. In a clean microcentrifuge tube, prepare 0.5 ng/μL and 1.25 ng/μL diluted samples to load totals of 100 and 250 ng, respectively, during slot blotting.

3.5 Slot Blotting

1. Begin by slot blotting the proteinase K-digested samples for the DNA loading control. Cut a piece of positively charged nylon membrane large enough to cover the entire surface of the slot blot apparatus and a piece of blotting filter paper of the same size. Wet the membrane and the filter paper in DNase-free water for 5 min (*see* **Note 12**).
2. Assemble the slot blot apparatus per manufacturer's instructions, using a single sheet of wetted blotting paper under the nylon membrane. Ensure that the apparatus is evenly sealed with the provided screws, but do not overtighten. Attach the vacuum line to the bottom reservoir and turn the vacuum on low suction to remove any air bubble or excess liquid. Turn off the vacuum and disconnect the hose from the apparatus.

3. Load all slots with 200 μL of DNase-free water using a multichannel pipette; attach the vacuum and pull the water through a membrane as a wash. The vacuum should be at medium strength. Turn off the vacuum and disconnect the hose from the apparatus when all slots are cleared (*see* **Note 13**).
4. Individually load 200 μL of the diluted proteinase K-digested samples (prepared in Subheading 3.4, **step 4a**) into the slots. Any unused wells should be filled with TE buffer. Attach the vacuum and apply the samples to the membrane under suction as described in Subheading 3.5, **step 3**.
5. Load all slots with 200 μL of TE buffer pH 8.0 using a multichannel pipette and wash the membrane as described in Subheading 3.5, **step 3**.
6. Disassemble the slot blot apparatus and carefully remove the membrane with clean forceps, touching only the edge of the membrane. Block the membrane in 5% milk in 1× TBST for 1 h at room temperature with gentle rocking.
7. Repeat the slot blotting steps (Subheading 3.5, **steps 1–6**) for the sodium phosphate-diluted DPC samples (prepared in Subheading 3.4, **step 4b**) with the following changes:
   (a) Use nitrocellulose membrane (Subheading 3.5, **step 1**).
   (b) Soak the filter paper and membrane in 6× SSC buffer for 10 min (Subheading 3.5, **step 1**).
   (c) Use TE buffer, pH 8.0, for the first wash (Subheading 3.5, **step 3**).
   (d) Use 2× SSC buffer for the second wash (Subheading 3.5, **step 5**).

3.6 Immunodetection

1. Following blocking, rinse each membrane thoroughly with water and submerge each membrane in primary antibody diluted in 5% BSA in 1× TBST.
2. Incubate overnight at 4 °C with gentle rocking.
3. Wash the membranes three times in 1× TBST for 10 min each at room temperature with gentle rocking.
4. Incubate the membranes in secondary antibody diluted in 5% milk in 1× TBST for 1 h at room temperature with gentle rocking.
5. Wash the membranes three times in 1× TBST for 10 min each at room temperature with gentle rocking.
6. After the final wash, briefly rinse each membrane with water. Soak each membrane in chemiluminescent substrate. Develop the membranes using a standard film developer or western blot imaging system (Fig. 2).

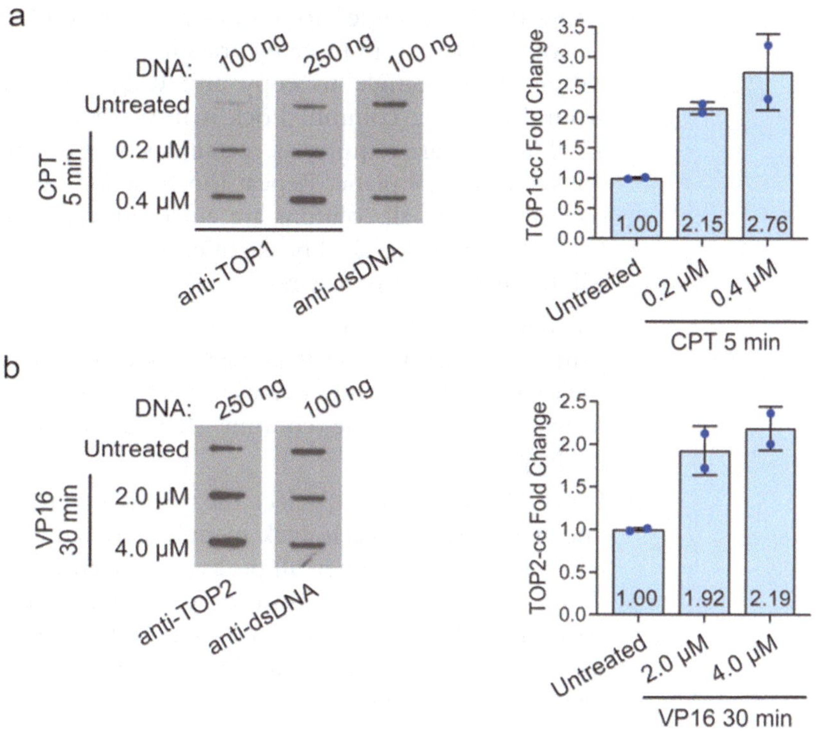

Fig. 2 Isolation and detection of TOP1-ccs and TOP2-ccs. (**a, b**) Left panel, A549 cells were treated for 5 min with CPT (**a**) or 30 min with VP16 (**b**) at the indicated concentrations to generate TOP1-ccs and TOP2-ccs, respectively. DPCs were isolated as described in the protocol. In (**a**), 100 ng and 250 ng of the sample were used for slot blotting and TOP1-ccs were detected by immunoblotting with TOP1 antibody. In (**b**), 250 ng of sample DNA was used for slot blotting and TOP2-ccs were detected using TOP2 antibody. (**a, b**) For loading control, 100 ng of proteinase K-digested sample was used for slot blotting and dsDNA was detected using dsDNA antibody. The right panel shows quantification of the slot blot. TOP1-cc or TOP2-cc fold change was determined from the relative abundance of TOP1 or TOP2 normalized to dsDNA ($n = 2$). Quantification was done with ImageJ and graphed using GraphPad Prism software

3.7 Data Analysis

1. Quantification of the slot blots is critical to the final interpretation and presentation of the results. We recommend first quantifying and calculating the fold change of the DPC of interest, followed by normalization of the results to the DNA loading control (*see* **Note 14**). Begin by opening the scanned image files of the desired DPC exposure in ImageJ.
2. Change the image to grayscale using the taskbar by clicking Image > Type >8-bit. Select Image > Adjust > Brightness and Contrast and use the controls on the pop-up window to adjust exposure as needed. The brightness and contrast should be adjusted to clearly define the borders between the sample signals and the background but should not cause the signals to appear overexposed.

3. Using the *Rectangle* tool, draw a box around a full column of slots and lock it in place by pressing Ctrl +1. Click and drag within the rectangle to copy and drag a second rectangle around the next column. Lock it into place by pressing Ctrl +2. The software requires all rectangles be the same size and in the same vertical plane. Repeat the click and drag and Ctrl +2 command until all columns are enclosed in a rectangle. Press Ctrl +3 to plot the pixel peaks of each selected column, which will be shown in a new window.
4. Standard ImageJ settings will show each rectangle in a separate pane in numerical order from top to bottom, and each plotted peak within those panes corresponds to a single slot, in order from the top of the column (left peak) to the bottom (right peak).
5. If not already selected, select the *Straight* tool and draw a line at the base of each peak to fully close and separate them from one another. When all peaks have been closed, select the Wand tool and click within each peak. This will measure the area under the peak and open a separate Results dialogue box. Be sure to measure the peaks in a uniform order so as to not confuse the samples.
6. When all peaks have been measured, return to the main taskbar and click Analyze > Gels > Label Peaks. This will add a Percent column to the Results window, which corresponds to the percent area of each peak relative to the total area of all peaks measured. Copy and paste the data in the results window into an Excel sheet.
7. To calculate the DPC fold change relative to an untreated control, set the untreated control to one by dividing the percent value by itself. Divide the treated samples' percent values by the untreated percent value.
8. Repeat **steps 3.7.1–3.7.7** for the desired DNA loading control exposure. If the samples are accurately quantified and evenly loaded, the relative values calculated in **step 3.7.7** will all be close to one.
9. Normalize the DPC fold change values by dividing each relative DPC value by its respective relative DNA loading value.
10. Finally, plot the normalized DPC fold change values using GraphPad Prism software or similar graphing software (Fig. 2). Perform the appropriate statistical analyses for the experimental setup.

4 Notes

1. Any mammalian cell line kept under standard culture conditions can be used. The number of cells seeded to achieve 85% confluency should be determined individually for each cell line. The number of wells needed will be determined based on the experimental conditions. It is recommended to include three replicates of each condition if space allows on the slot blot manifold.
2. If multiple drug concentrations or treatment times will be used, it is best to stagger the treatments to minimize variability from one sample to the next, ensuring all samples are processed as close to the desired treatment time as possible.
3. For repair and recovery experiments, remove the drug-containing media after the desired treatment time, wash cells with 1× sterile PBS, refresh with 2 mL of drug-free culture media, and incubate for the desired recovery time.
4. Alternative reagents to the DNAzol® reagent to use for cell lysis—Buffer RLT Plus (Qiagen) or MB Buffer: 6 M guanidinium isothiocyanate (GTC), 10 mM Tris–HCl, pH 6.5, 20 mM EDTA, 4% Triton X-100, 1% Sarkosyl, 1% dithiothreitol [26].
5. Centrifugation of the samples at higher speeds may result in a DNA pellet that is more securely collected at the bottom of the tube, but it will also make the pellet difficult to properly dissolve. High-speed centrifugation is not recommended. A DNA pellet may not be visible after the first spin; this is not necessarily an indication that the precipitated DNA has been lost—the DNA pellet will be visible after the washing steps.
6. If the sample is viscous after dissolving, it is recommended to increase the volume of NaOH.
7. The isolated DPC samples do not contain any protease inhibitors and are prone to degradation; therefore, we recommend the whole protocol, from sample isolation through slot blotting, be done on the same day.
8. Recommended dilution for DNA quantification samples is 1:3, but various experimental conditions may affect final yield and concentration. Any dilution factor can be used as long as it is appropriately factored into the calculations for DNA quantification. If the stock concentration of proteinase K is different than 10 mg/mL, use an appropriate volume for a final proteinase K concentration of 0.2 mg/mL.
9. We recommend the following conditions for using the Pico-Green Kit. DNA standards: 0, 1, 4, 8, 80, 400, 800, and 1400 ng/mL. Use 20 μL of proteinase K-digested sample in

the experimental samples. Be sure to take all dilution steps into account when calculating the total yield and stock concentrations of the samples.

10. This protocol can be modified to analyze posttranslational modifications of DPCs by immunoprecipitation of the RADAR samples and to analyze total DPCs via SDS-PAGE and silver staining, which may be advantageous for studying nonenzymatic DPCs or the effects of nonspecific crosslinking agents [23, 29].
11. Slot blot recommendations provided here are for immunodetection of TOP1-covalent crosslinks (TOP1-ccs) and TOP2-covalent crosslinks (TOP2-ccs), induced by CPT and VP16, respectively. For DNA loading controls, load 50–100 ng of sample. We recommend making at least 250 μL of diluted sample for each technical replicate to compensate for loss of sample when making dilutions or during slot blotting due to standard errors in pipetting. A loading control alternative to immunodetection of DNA is histone H3. Be sure to consider that any drug treatment or cellular manipulation will not affect histone H3 levels in the samples. We recommend loading 200 ng of undigested DPC sample for histone H3. Antibody specifics: anti-histone H3 (Cell Signaling Technology 9715S; dilution 1:2000).
12. All wells must be covered with filter paper and membrane to ensure even vacuum strength across all slots and the filter paper and membrane should be thoroughly wetted in the indicated buffers to maximize sample binding [30, 31].
13. The vacuum must be disconnected from the slot blot apparatus between each wash or sample loading—turning the vacuum off but leaving the hose connected is not sufficient. If one or more wells do not clear on their own even after several minutes, check the seal of the screws and tilt the apparatus toward the vacuum attachment to clear the bottom reservoir. Alternatively, there may be a bubble blocking the well from clearing. Use a clean 200 μL pipette and gently pipette up and down within the well to release the bubble.
14. Various experimental setups as well as technical replicates can affect the details of the quantification. Quantification details for common experimental variations follow:
    (a) For experiments involving technical replicates on the same membrane, we recommend using the average of the percent values of the untreated control samples as the divisor in the calculations.
    (b) If the experiment involves multiple cell lines or conditions, it may be favorable to use the untreated sample of each cell line or condition as a separate reference point,

although it is most common to calculate everything to the untreated negative control cell line or condition. However, it is critical that a single exposure is used to quantify samples that will be graphed together regardless of how many reference points are used.

(c) For experiments examining DPC repair dynamics, it is advisable to use the treated sample as the reference point rather than the untreated. In such a scenario, drug-treated samples are set to one, and recovery time points are calculated relative to the treated sample. The fold change will decrease in the case of DPC recovery.

Acknowledgments

This work was supported by NIGMS COBRE (P20 GM121316; G.G. as project leader), NCI K22 Career Development Award (K22CA188181) to G.G., R01 grants GM141232-01A1 and CA263504-01A1 to G.G.

References

1. Ruggiano A, Ramadan K (2021) DNA-protein crosslink proteases in genome stability. Commun Biol 4(1):11. https://doi.org/10.1038/s42003-020-01539-3
2. Stingele J, Bellelli R, Boulton SJ (2017) Mechanisms of DNA-protein crosslink repair. Nat Rev Mol Cell Biol 18(9):563–573. https://doi.org/10.1038/nrm.2017.56
3. Fielden J, Ruggiano A, Popovic M, Ramadan K (2018) DNA protein crosslink proteolysis repair: from yeast to premature ageing and cancer in humans. DNA Repair (Amst) 71: 198–204. https://doi.org/10.1016/j.dnarep.2018.08.025
4. Tretyakova NY, At G, Ji S (2015) DNA-protein cross-links: formation, structural identities, and biological outcomes. Acc Chem Res 48(6):1631–1644. https://doi.org/10.1021/acs.accounts.5b00056
5. Ide H, Nakano T, Salem AMH, Shoulkamy MI (2018) DNA-protein cross-links: formidable challenges to maintaining genome integrity. DNA Repair (Amst) 71:190–197. https://doi.org/10.1016/j.dnarep.2018.08.024
6. Vaz B, Popovic M, Ramadan K (2017) DNA-protein crosslink proteolysis repair. Trends Biochem Sci 42(6):483–495. https://doi.org/10.1016/j.tibs.2017.03.005
7. Pommier Y, Huang SY, Gao R, Das BB, Murai J, Marchand C (2014) Tyrosyl-DNA-phosphodiesterases (TDP1 and TDP2). DNA Repair (Amst) 19:114–129. https://doi.org/10.1016/j.dnarep.2014.03.020
8. Barker S, Weinfeld M, Murray D (2005) DNA-protein crosslinks: their induction, repair, and biological consequences. Mutat Res 589(2):111–135. https://doi.org/10.1016/j.mrrev.2004.11.003
9. Nitiss JL (2009) Targeting DNA topoisomerase II in cancer chemotherapy. Nat Rev Cancer 9(5):338–350. https://doi.org/10.1038/nrc2607
10. Pommier Y, Marchand C (2011) Interfacial inhibitors: targeting macromolecular complexes. Nat Rev Drug Discov 11(1):25–36. https://doi.org/10.1038/nrd3404
11. Maslov AY, Lee M, Gundry M, Gravina S, Strogonova N, Tazearslan C, Bendebury A, Suh Y, Vijg J (2012) 5-aza-2′-deoxycytidine-induced genome rearrangements are mediated by DNMT1. Oncogene 31(50):5172–5179. https://doi.org/10.1038/onc.2012.9
12. Duxin JP, Dewar JM, Yardimci H, Walter JC (2014) Repair of a DNA-protein crosslink by replication-coupled proteolysis. Cell 159(2): 346–357. https://doi.org/10.1016/j.cell.2014.09.024
13. Chesner LN, Campbell C (2018) A simple, rapid, and quantitative assay to measure repair of DNA-protein crosslinks on plasmids

transfected into mammalian cells. J Vis Exp 133. https://doi.org/10.3791/57413

14. Larsen NB, Gao AO, Sparks JL, Gallina I, Wu RA, Mann M, Raschle M, Walter JC, Duxin JP (2019) Replication-coupled DNA-protein crosslink repair by SPRTN and the proteasome in Xenopus egg extracts. Mol Cell 73(3):574. https://doi.org/10.1016/j.molcel.2018.11.024
15. Sparks JL, Chistol G, Gao AO, Raschle M, Larsen NB, Mann M, Duxin JP, Walter JC (2019) The CMG helicase bypasses DNA-protein cross-links to facilitate their repair. Cell 176(1–2):167. https://doi.org/10.1016/j.cell.2018.10.053
16. Zhitkovich A, Costa M (1992) A simple, sensitive assay to detect DNA-protein crosslinks in intact cells and in vivo. Carcinogenesis 13(8): 1485–1489. https://doi.org/10.1093/carcin/13.8.1485
17. Merk O, Reiser K, Speit G (2000) Analysis of chromate-induced DNA-protein crosslinks with the comet assay. Mutat Res 471(1–2): 71–80. https://doi.org/10.1016/s1383-5718(00)00110-8
18. Morocz M, Zsigmond E, Toth R, Enyedi MZ, Pinter L, Haracska L (2017) DNA-dependent protease activity of human Spartan facilitates replication of DNA-protein crosslink-containing DNA. Nucleic Acids Res 45(6): 3172–3188. https://doi.org/10.1093/nar/gkw1315
19. Hu Q, Klages-Mundt N, Wang R, Lynn E, Kuma Saha L, Zhang H, Srivastava M, Shen X, Tian Y, Kim H, Ye Y, Paull T, Takeda S, Chen J, Li L (2020) The ARK assay is a sensitive and versatile method for the global detection of DNA-protein crosslinks. Cell Rep 30(4):1235–1245. e1234. https://doi.org/10.1016/j.celrep.2019.12.067
20. Trask DK, DiDonato JA, Muller MT (1984) Rapid detection and isolation of covalent DNA/protein complexes: application to topoisomerase I and II. EMBO J 3(3):671–676. https://doi.org/10.1002/j.1460-2075.1984.tb01865.x
21. Costa M, Zhitkovich A, Gargas M, Paustenbach D, Finley B, Kuykendall J, Billings R, Carlson TJ, Wetterhahn K, Xu J, Patierno S, Bogdanffy M (1996) Interlaboratory validation of a new assay for DNA-protein crosslinks. Mutat Res 369(1–2):13–21. https://doi.org/10.1016/s0165-1218(96)90043-9
22. Stingele J, Bellelli R, Alte F, Hewitt G, Sarek G, Maslen SL, Tsutakawa SE, Borg A, Kjaer S, Tainer JA, Skehel JM, Groll M, Boulton SJ (2016) Mechanism and regulation of DNA-protein crosslink repair by the DNA-dependent metalloprotease SPRTN. Mol Cell 64(4):688–703. https://doi.org/10.1016/j.molcel.2016.09.031
23. Vaz B, Popovic M, Newman JA, Fielden J, Aitkenhead H, Halder S, Singh AN, Vendrell I, Fischer R, Torrecilla I, Drobnitzky N, Freire R, Amor DJ, Lockhart PJ, Kessler BM, McKenna GW, Gileadi O, Ramadan K (2016) Metalloprotease SPRTN/DVC1 orchestrates replication-coupled DNA--protein crosslink repair. Mol Cell 64(4): 704–719. https://doi.org/10.1016/j.molcel.2016.09.032
24. Subramanian D, Furbee CS, Muller MT (2001) ICE bioassay. Isolating in vivo complexes of enzyme to DNA. Methods Mol Biol 95:137–147. https://doi.org/10.1385/1-59259-057-8:137
25. Shoulkamy MI, Nakano T, Ohshima M, Hirayama R, Uzawa A, Furusawa Y, Ide H (2012) Detection of DNA-protein crosslinks (DPCs) by novel direct fluorescence labeling methods: distinct stabilities of aldehyde and radiation-induced DPCs. Nucleic Acids Res 40(18):e143. https://doi.org/10.1093/nar/gks601
26. Kiianitsa K, Maizels N (2013) A rapid and sensitive assay for DNA-protein covalent complexes in living cells. Nucleic Acids Res 41(9): e104. https://doi.org/10.1093/nar/gkt171
27. Patel AG, Flatten KS, Peterson KL, Beito TG, Schneider PA, Perkins AL, Harki DA, Kaufmann SH (2016) Immunodetection of human topoisomerase I-DNA covalent complexes. Nucleic Acids Res 44(6):2816–2826. https://doi.org/10.1093/nar/gkw109
28. Perry M, Biegert M, Kollala SS, Mallard H, Su G, Kodavati M, Kreiling N, Holbrook A, Ghosal G (2021) USP11 mediates repair of DNA-protein cross-links by deubiquitinating SPRTN metalloprotease. J Biol Chem 296: 100396. https://doi.org/10.1016/j.jbc.2021.100396
29. Saha S, Sun Y, Huang SN, Baechler SA, Pongor LS, Agama K, Jo U, Zhang H, Tse-Dinh YC, Pommier Y (2020) DNA and RNA cleavage complexes and repair pathway for TOP3B RNA- and DNA-protein crosslinks. Cell Rep 33(13):108569. https://doi.org/10.1016/j.celrep.2020.108569
30. Brown T (2001) Dot and slot blotting of DNA. Curr Protoc Mol Biol Chapter 2 (Unit2):9B. https://doi.org/10.1002/0471142727.mb0209bs21
31. Park J-M, Kang T-H (2015) DNA slot blot repair assay. Bio-Protocol 5(8). https://doi.org/10.21769/BioProtoc.1453

Chapter 9

Slot Blot Assay for Detection of R Loops

Altaf H. Sarker and Priscilla K. Cooper

Abstract

R loops (DNA-RNA hybrid) are three-stranded nucleic acid structures that comprise of template DNA strand hybridized with the nascent RNA leaving the displaced non-template strand. Although a programmed R loop formation can serve as powerful regulators of gene expression, these structures can also turn into major sources of genomic instability and contribute to the development of diseases. Therefore, understanding how cells prevent the deleterious consequences of R loops yet allow R loop formation to participate in various physiological processes will help to understand how their homeostasis is maintained. Detection and quantitative measurements of R loops are critical that largely relied on S9.6 antibody. Immunofluorescence methods are frequently used to localize and quantify R loops in the cell but they require specialized tools for analysis and relatively expensive; therefore, they are not always useful for initial assessments of R loop accumulation. Here, we describe an improved slot blot protocol to detect and estimate R loops and show its sensitivity and specificity using the S9.6 antibody. Since specific factors protecting cells from harmful R loop accumulation are expanding, this protocol can be used to determine R loop accumulation in research and clinical settings.

Key words R loop, Transcription, Replication, RNase H, Senataxin, S9.6

1 Introduction

Transcription is not a continuous process as transcribing RNAPII frequently encounters DNA lesions generated by endogenous or exogenous agents causing RNAPII to pause or stall. Stalled RNAPII at a DNA lesion trigger R loop formation [1]. In mammalian cells transcription–replication conflicts in long genes are associated with R loop accumulation [2]. Replication fork colloids with stalled RNAPII induces strand breaks and unrepaired breaks in the non-template ssDNA of the R loop or the DNA-RNA hybrid formed behind the elongating RNAPII restricts transcription and may block replication forks. Such blocked replication forks generate replication stress and DSBs are potent sources of genomic instability [3]. Cells evolve multiple surveillance mechanisms to restrict R loop accumulation either preventing R loop formation or removing

Kishor K. Bhakat and Tapas K. Hazra (eds.), *Base Excision Repair Pathway: Methods and Protocols*, Methods in Molecular Biology, vol. 2701, https://doi.org/10.1007/978-1-0716-3373-1_9,

R loops once they formed. The factors that prevent R loop formation are THO/TREX complex and TopI. The THO/TREX complex is responsible for packaging of pre-mRNA with RNA binding proteins and prevents R loop formation, thus restricting transcription associated hyper-recombination and DNA damage [4]. Top1 relaxes negative supercoiling behind the elongating RNAPII and suppresses R loop formation [5]. Besides preventing R loop accumulation, the cell uses factors that remove R loops once they are formed. The nuclear protein RNase H1 degrades RNA in the RNA–DNA hybrid regardless of sequence specificity [6]. Overexpressions of RNase H are widely used to remove transcription-associated R loops. Senataxin, an RNA/DNA helicase, also removes R loops and prevents R loop-mediated genomic instability [7]. Mutations in the helicase domain of senataxin are associated with progressive degeneration of motor neurons in the brain leading to neurodegenerative disorders [8].

Despite genomic instability phenotypes associated with unprogrammed R loops, recent evidences emerged that these structures are regulators of gene expression and control several biological processes [9]. For example, R loop structures control class switch recombination (CSR) at the Ig heavy chain locus in activated B cells that are highly repetitive GC-rich sequence and favor R loop formation and stabilization. R loop structures are also implicated in transcription initiation and termination as evidence for R loop formation over the 5′ regions downstream from CpG promoters and also downstream from poly(A) signal at the 3′ region has been reported [10, 11]. Sequence analysis revealed that both promoters and 3′ regions are enriched in G-rich sequences predicted to form the G4 structure that is known to favor R loop formation [12]. Although R loops are a common feature of transcription initiation, however, the mechanism how R loops regulate gene expression remains largely unknown.

Several key DNA repair proteins as BRCA2 and BRCA1 are recently shown to regulate R loops at the promoters and terminators and also at DSBs [13, 14]. Defects in BRCA1 or the nucleotide excision repair (NER) proteins XPG and XPF lead to the accumulation of R loops indicating several DNA repair pathways contribute to R loop regulation [15, 16]. Mutations in R loop resolving senataxin helicase also cause specific neurodegenerative diseases (ALS4, AOA) and it is recently suggested that senataxin is a DNA repair protein that resolves R loops in human cells [17–19]. It is tempting to speculate that these genes may control transcription of long genes by preferentially resolving R loops in the neuronal cells, and thus, they are at the crossroad of transcription, DNA repair, and R loop regulation.

Interest on R loops increases exponentially in recent years to understand how the altered R loop homeostasis is linked to the development of diseases. It is not clear how they maintain dual functions in cells: prevent/resolve R loops to maintain genome stability and promote R loop formation to control gene expression; how do they become pathological? Therefore, abundance of R loops in different conditions needs to be compared in a quantitative way which is often done by genome-wide methods [20] or immunofluorescence with RNA/DNA hybrid S9.6 antibody. However, these methods are always not practical for initial assessments. Although we found a protocol recently published to detect R loops [21], here we describe a slot blot protocol to assess R loop abundance with appropriate controls to ensure specificity.

2 Materials

2.1 Components

(a) Gel-blotting papers (Bio-Rad cat# 1620161).

(b) Slot blot apparatus (Bio-Rad: Bio Dot SF serial no 160BR).

(c) Blotting membrane (Amersham cat# RPN303N).

(d) Blocking solution, 5% w/v dry power low-fat milk in TBS-T.

(e) ECL Select (GE Healthcare cat# RPN2235) or Supersignal West pico (Thermo cat#34580) for detection.

(f) Aspirator.

2.2 Antibodies

Antibodies for the detection of RNA–DNA hybrids (S9.6) and dsDNA as described below:

(a) The stock solution of the primary antibody for dsDNA (abcam cat# ab27156) was 1 mg/mL and diluted to 1:5000 with TBS-T prior to use and the stock stored at 4 °C for short-term storage.

(b) The stock solution of the primary mouse monoclonal antibody for the DNA–RNA hybrid (Kerafast cat# ENH001) was 1 mg/mL and diluted to 1:1000 with TBS-T prior to use and the stock should be stored at 4 °C for short-term storage.

(c) Secondary antibody (for both dsDNA and DNA–RNA hybrid assay), goat anti-mouse immunoglobulin HRP conjugated (diluted to 1:5000 with TBS-T).

3 Methods

For the current protocol we used HEK293T and primary skin fibroblasts (HCA2) as the manipulation protocols are established in our lab, although other cell types can be used. Both cell types were grown in DMEM containing 10% fetal bovine serum.

3.1 Purification of Genomic DNA

(a) Cells were grown in standard tissue culture dish, washed twice with phosphate-buffered saline (PBS), and were collected by trypsinization (0.25% trypsin; Gibco cat#15090-046) followed by centrifugation at a speed of 1200 *g* for 5 min. One 100 × 20 mm dish (Greiner Bio-One cat# 664160) per biological samples was used.

(b) The cell pellet was washed with ice-cold PBS and was collected by centrifugation.

(c) Cells were suspended in 2.5 mL of PBS supplemented with 0.5% SDS and mixed several times to ensure compete lysis. The mixtures were digested with 1 mg of proteinase K (Invitrogen) overnight at 37 °C.

(d) Lysates were viscous at this point and extracted twice with an equal volume of phenol/chloroform/isoamyl alcohol (25: 24:1 pH 8.0) and finally one volume of chloroform.

(e) DNA was precipitated with one volume of isopropanol and by centrifugation at 6500 *g* for 15 min.

(f) Discard the supernatant and the DNA pellets were washed with 1 mL of 70% ethanol and transferred to a sterile Eppendorf tube.

(g) Centrifuge at 12000 *g* for 10 min and discard the supernatant and let the pellet air-dry.

(h) Dissolve the pellet into a buffer containing 1 mM Tris–HCl (pH 8) and 0.1 mM EDTA.

3.2 Digestion of Genomic DNA with Restriction Enzyme to Reduce Viscosity

Dissolved DNA is viscous and digested with restriction enzymes to reduce viscosity (Alternatively performing low speed sonication to reduce viscosity).

(a) Digest the DNA at 37 °C overnight with restriction enzyme cocktail containing EcoRI, HindIII, XbaI, and BsrGI (40 units each).

(b) Measure the DNA concentration after digestion using standard spectrophotometry.

3.3 Ribonuclease Digestion to Determine Specificity of the S9.6 Antibody

To ensure specific binding of the R loop by S9.6 antibody, RNase H digestion should be performed.

(a) A fraction of each sample digested with RNase H (NEB; M0297S). This can be done empirically; however, 5 units RNase H per 10 ug DNA at 37 °C overnight is a starting point.

3.4 Blotting of DNA Samples onto Nylon Membrane and UV Crosslink

(a) Prepare a dilution series of both RNase H digested and undigested DNA in PBS buffer at a concentration of 500, 250, and 125 ng/200 μL.

(b) Cut a nylon membrane exactly the same size of the blotting paper (Bio-Rad cat#1620161) that fit into the slot of the Bio-Rad apparatus (Bio Dot SF serial no 160BR). Sock the membrane in ammonium acetate (1 M solution) for 30″.

(c) Mark the membrane at one corner for orientation and place at the top of two filter papers as per instruction manual (https://www.bio-rad.com/webroot/web/pdf/lsr/literature/M1706542.pdf).

(d) Assemble the apparatus and connect the slot blot apparatus to the vacuum source. Apply 200 μL sample from each dilution to a slot for probing with S9.6 antibody and another for dsDNA antibody. The recommended sample volume in this apparatus is at least 200 μL. Less than 200 μL sample volume may cause uneven loading and corresponding unevenly shaped bands making inaccurate quantification.

(e) Gently apply vacuum sufficient to remove solution from the sample wells. Carefully apply vacuum; otherwise, the membrane may rehydrate and the membrane may cause a hole around the well.

(f) Place the membrane into the center of a UV Stratalinker (Stratagene) and crosslink using the auto-crosslink setting.

3.5 Detection of R Loops Using S9.6 Antibody

(a) Prepare 5% milk solution in TBS-T (Tris-buffered saline +0.1% Tween 20) and incubate the membrane for 1 h in a shaker.

(b) Dilute the S9.6 antibody in TBS-T (1:1000; 1 ug/mL final) and add in one membrane. The dsDNA antibody was diluted in TBS-T (1:5000) and added to another membrane. Both membranes were incubated overnight at 4 °C.

(c) Remove the primary antibody from the membranes and wash 3× with TBS-T (5 min each wash) in a shaker.

(d) Incubate the membrane with secondary antibody (1:5000 dilutions in TBS-T) conjugated with horseradish peroxidase (HRP) for 1 h in a shaker. Both S9.6 and dsDNA antibodies were from mouse and we used mouse secondary antibody (GE Healthcare; NA931V).

(e) Remove the secondary antibody and wash the membrane with TBS-T for three times (5 min each wash).

(f) Develop the membrane using ECL select (GE Healthcare; RPN2235) or Supersignal West pico (Thermo cat#34580) according to the manufacturer's instructions. Use sufficient reagent to cover the membrane.

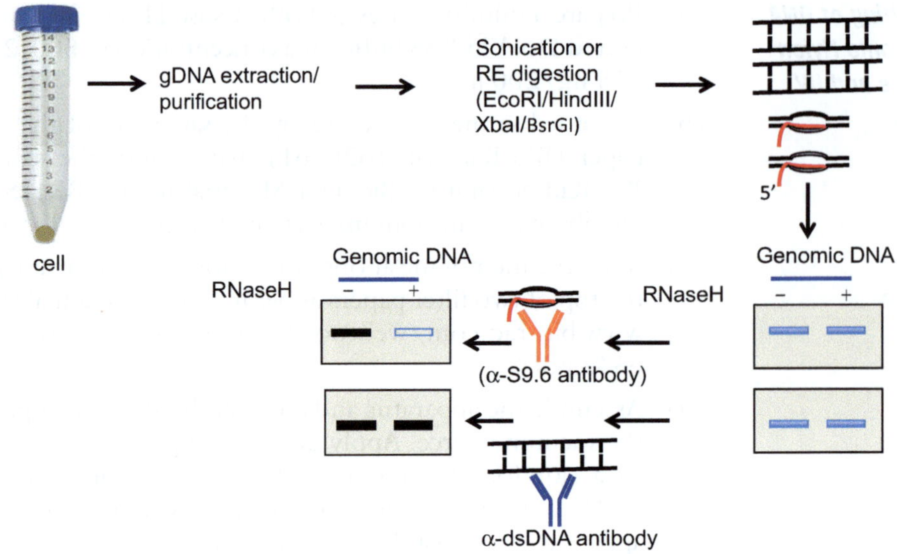

Fig. 1 An overview of the slot blot protocol. Schematics showing different steps performed during the analysis

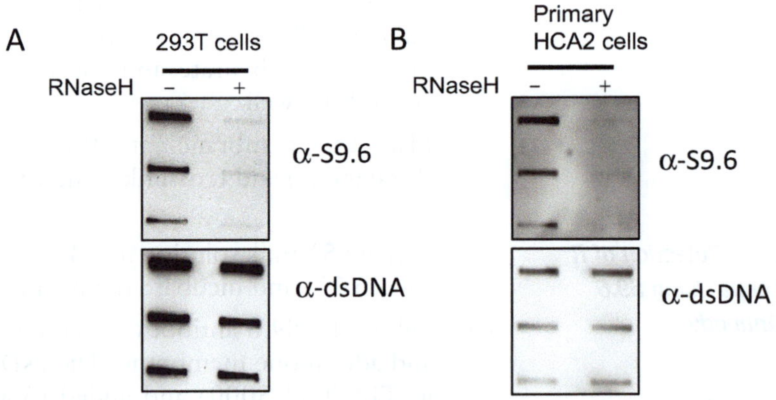

Fig. 2 Specificity of S9.6 antibody in detecting R loops. Genomic DNA were purified from HEK293T (**a**) or primary HCA2 (**b**) cells followed by restriction enzyme (EcoRI/HindIII/XbaI/BsrGI) digestion. The digested genomic DNA was treated with or without RNaseH and 500, 250, and 125 ng of the DNA samples were analyzed by slot blot with the indicated antibodies at the right

(g) Acquire and quantify the signal using the VersaDoc (Bio-Rad) imaging system. An overview of the protocol and a representative image are shown in Figs. 1 and 2.

3.6 Quantification of R Loops

(a) The intensity of each captured band was measured by Quantity One analysis software in the VersaDoc system.

(b) The S9.6 signal intensity was normalized by the corresponding dsDNA signal and plotted. Triplicates were performed to determine the standard error.

4 Notes

The purity of the genomic DNA is critical for the success of the method. The ratio of absorbance at 260 and 280 nm of 1.8 is generally accepted as pure for DNA. A lower ratio indicates the presence of contaminants such as proteins that absorb strongly at 280 nm. As a secondary measure, a ratio of 260 and 230 nm commonly in the range of 2.0–2.2 also indicates pure nucleic acids. A lower ratio may indicate the presence of contaminants which absorb at 230 nm such as EDTA, carbohydrates, and phenol that all have absorbance near 230 nm. Contaminants in the nucleic acid solutions often inhibit RNaseH digestion and also impact on the DNA/RNA immunoprecipitation (DRIP) assay which uses S9.6 antibody for immunoprecipitation.

Acknowledgements

We thank Drs. Susan Tsutakawa and Miaw-Sheue-Tsai (Lawrence Berkeley National Laboratory, Berkeley, California) for scientific discussions, providing reagents, and allowing their lab spaces during the project period. This work was supported by Tobacco Related Diseases Research Program (TRDRP) grant 26IR-0017 to AHS; NIH P01 AG017242 to P.K.C.

References

1. Tresini M, Warmerdam DO, Kolovos P, Snijder L, Vrouwe MG, Demmers JA, van IJcken WF, Grosveld FG, Medema RH, Hoeijmakers JH, Mullenders LH (2015) The core spliceosome as target and effector of non-canonical ATM signalling. Nature 523: 53–58
2. Helmrich A, Ballarino M, Tora L (2011) Collisions between replication and transcription complexes cause common fragile site instability at the longest human genes. Mol Cell 44:966–977
3. Aguilera A, Garcia-Muse T (2012) R loops: from transcription byproducts to threats to genome stability. Mol Cell 46:115–124
4. Salas Armenteros I et al (2017) Human THO-Sin3A interaction reveals new mechanisms to prevent R-loops that cause genome instability. EMBO J 36:3532–3547
5. Tuduri S, Crabbe L, Conti C, Tourriere H, Holtgreve-Grez H, Jauch A, Pantesco V, De Vos J, Thomas A, Theillet C et al (2009) Topoisomerase I suppresses genomic instability by preventing interference between replication and transcription. Nat Cel Biol 11:1315–1324
6. Cerritelli SM, Crouch RJ (2009) Ribonuclease H: the enzymes in eukaryotes. FEBS J 276:1494–1505
7. Mischo HE, Gomez-Gonzalez B, Grzechnik P, Rondon AG, Wei W, Steinmetz L, Aguilera A, Proudfoot NJ (2011) Yeast Sen1 helicase protects the genome from transcription-associated instability. Mol Cell 41:21–32
8. Yüce Ö, West SC (2013) Senataxin, defective in the neurodegenerative disorder ataxia with oculomotor apraxia 2, lies at the interface of transcription and the DNA damage response. Mol Cell Biol 33:406–417
9. Yu K, Chedin F, Hsieh CL, Wilson TE, Lieber MR (2003) R-loops at immunoglobulin class switch regions in the chromosomes of stimulated B cells. Nat Immunol 4:442–451
10. Ginno PA, Lott PL, Christensen HC, Korf I, Chedin F (2012) R-loop formation is a distinctive characteristic of unmethylated human CpG Island promoters. Mol Cell 45:814–825

11. Ginno PA, Lim YW, Lott PL, Korf I, Chedin F (2013) GC skew at the 5′ and 3′ ends of human genes links R-loop formation to epigenetic regulation and transcription termination. Genome Res 23:1590–1600

12. Huppert JL, Bugaut A, Kumari S, Balasubramanian S (2008) G-quadruplexes: the beginning and end of UTRs. Nucleic Acids Res 36: 6260–6268

13. Bhatia V, Barroso SI, García-Rubio ML, Tumini E, Herrera-Moyano E, Aguilera A (2014) BRCA2 prevents R-loop accumulation and associates with TREX-2 mRNA export factor PCID2. Nature 511:362–365. https://doi.org/10.1038/nature13374

14. Hatchi E, Skourti-Stathaki K, Ventz S, Pinello L, Yen A, Livingston DM (2015) BRCA1 recruitment to transcriptional pause sites is required for R-loop-driven DNA damage repair. Mol Cell 57:636–647. https://doi.org/10.1016/j.molcel.2015.01.011

15. Zhang X, Chiang HC, Wang Y, Zhang C, Smith S, Li R (2017) Attenuation of RNA polymerase II pausing mitigates BRCA1-associated R-loop accumulation and tumorigenesis. Nat Commun 8:15908. https://doi.org/10.1038/ncomms15908

16. Sollier J, Stork CT, García-Rubio ML, Paulsen RD, Aguilera A, Cimprich KA (2014) Transcription-coupled nucleotide excision repair factors promote R-loop-induced genome instability. Mol Cell 56:777–785. https://doi.org/10.1016/j.molcel.2014.10.020

17. Rawal CC, Zardoni L, Di Terlizzi M, Galati E, Brambati A, Pellicioli A (2020) Senataxin Ortholog Sen1 limits DNA:RNA hybrid accumulation at DNA double-strand breaks to control end resection and repair fidelity. Cell Rep 31:107603. https://doi.org/10.1016/j.celrep.2020.107603

18. Moreira MC, Klur S, Watanabe M, Németh AH, Le Ber I, Koenig M (2004) Senataxin, the ortholog of a yeast RNA helicase, is mutant in ataxia-ocular apraxia 2. Nat Genet 36:225–227. https://doi.org/10.1038/ng1303

19. Chen YZ, Bennett CL, Huynh HM, Blair IP, Puls I, Chance PF (2004) DNA/RNA helicase gene mutations in a form of juvenile amyotrophic lateral sclerosis (ALS4). Am J Hum Gene 74:1128–1135. https://doi.org/10.1086/421054

20. Crossley MP, Bocek MJ, Hamperl S, Swigut T, Cimprich KA (2020) qDRIP: a method to quantitatively assess RNA-DNA hybrid formation genome-wide. Nucleic Acid Res 8:e84. https://doi.org/10.1093/nar/gkaa500

21. Ramirez P, Crouch RJ, Cheung VG, Grunseich C (2021) R-loop analysis by dot-blot. J Vis Exp 167:e62069. https://doi.org/10.3791/62069

Chapter 10

Assays with Patient-Derived Organoids to Evaluate the Impact of Microbial Infection on Base Excision Repair (BER) Enzymes

Ibrahim M. Sayed, Anirban Chakraborty, and Soumita Das

Abstract

Microbes play an important role in regulating cellular responses and the induction of chronic diseases. Infection and chronic inflammation can cause DNA damage, and the accumulation of mutations leads to cancer development. The well-known examples of cancer-associated microbes are *Helicobacter pylori* in gastric cancer and *Fusobacterium nucleatum (Fn), Bacteroides fragilis, and E.coli NC101* in colorectal cancer (CRC). These carcinopathogens modify the expressions of the base excision repair enzymes and cause DNA damage. This chapter will show how *Fn* can initiate CRC through the downregulation of a critical enzyme of the base excision repair (BER) pathway that subsequently causes accumulation of DNA damage. We used the stem cell-based organoid model and enteroid-derived monolayer (EDM) from the murine and human colon to assess the impact of infection on the expression of BER enzymes on the transcriptional and translational levels and to develop other functional assays. For example, we used this EDM model to assess the inflammatory response, DNA damage response, and physiological responses, where we correlated the level of these parameters to BER enzyme levels.

Key words Colorectal cancer (CRC), *Fusobacterium nucleatum* (*Fn*), Base excision repair (BER), Organoid, Enteroid-derived monolayer (EDM)

1 Introduction

The stem cell-based approaches have recently made significant progress in the cancer field [1–4]. To understand the impact of base excision repair (BER) in microbe-associated cancer, we have developed some new methodology that will be focussed on this chapter. We will use patient-derived organoids, organoid-derived monolayers, and CRC-associated microbes (for example, *Fusobacterium nucleatum* [*Fn*] [5–8]) to model CRC-in-a-dish to study the expressions of base excision repair enzymes and the accumulation of DNA damage.

Kishor K. Bhakat and Tapas K. Hazra (eds.), *Base Excision Repair Pathway: Methods and Protocols*, Methods in Molecular Biology, vol. 2701, https://doi.org/10.1007/978-1-0716-3373-1_10,

The published studies showed the development of stem cell-based 3D organoids from murine and human colon tissues using a basement matrix membrane [9–11]. In addition, other groups and we have developed 3D organoids from CRC models, for example, murine APC (adenomatous polyposis coli)-deficient adenomas and human CRC samples [8, 10]. Three-dimensional organoids can be differentiated to 2D polarized epithelial cells (enteroid-derived monolayers [EDM]) where the colon-specific cells such as enterocytes, enteroendocrine cells, goblet cells, Paneth cells, and tuft cells are proportionally represented [8, 12–14]. The 2D EDM model named "gut-in-a-dish" resembles the physiological gut lining, which can be used to assess the effect of various stressors such as microbes, microbial products, and environmental toxicants, including E-cigarette vapor, on the integrity of the gut barrier [8, 12–17]. Dysbiosis of the gut microbiome is associated with the initiation and progression of CRC [4, 18]. *Fusobacterium nucleatum* (*Fn*) is anaerobic Gram-negative bacteria that drive CRC through the activation of the Wnt/β-catenin signaling pathway and modulation of autophagy [5, 6, 19, 20]. We recently used the "gut-in-a-dish" model and showed that *Fn* infection suppresses the host DNA repair pathway, precisely the level of BER enzyme NEIL2, increases the inflammatory response, and accumulates DNA damage in the colonic epithelial cells [8].

In this method, we will describe the development of the murine and patient-derived 3D organoid, the procedure of 2D colonic EDM generation from 3D organoid, and the infection of the 2D EDM with microbe-associated cancers (for example, *Fn*). We have optimized the EDM models for the following functional assays: (a) evaluation of the expression level of BER enzymes at mRNA and protein levels, (b) determination of the level of cytokines released in the culture supernatants by multiplex array or by specific ELISAs, and (c) assessment of DNA double-strand break by immunostaining of γH2AX following *Fn* infection (Fig. 1). We have also shown the use of EDM in LA (long-amplicon)-PCR and the measurement of DNA damage following infection.

2 Materials

2.1 Culture of *Fusobacterium nucleatum*

(i) *Fusobacterium nucleatum* (*Fn*) subsp. nucleatum Knorr (ATCC-25586).

(ii) Anaerobic jar systems.

(iii) Chopped meat media (Anaerobe Systems, Morgan Hill, CA) that includes lean ground beef w/v (50%), sodium hydroxide v/v (2.5%), pancreatic digest of casein w/v (3%), yeast extract w/v, other components (< 4%), and deionized water.

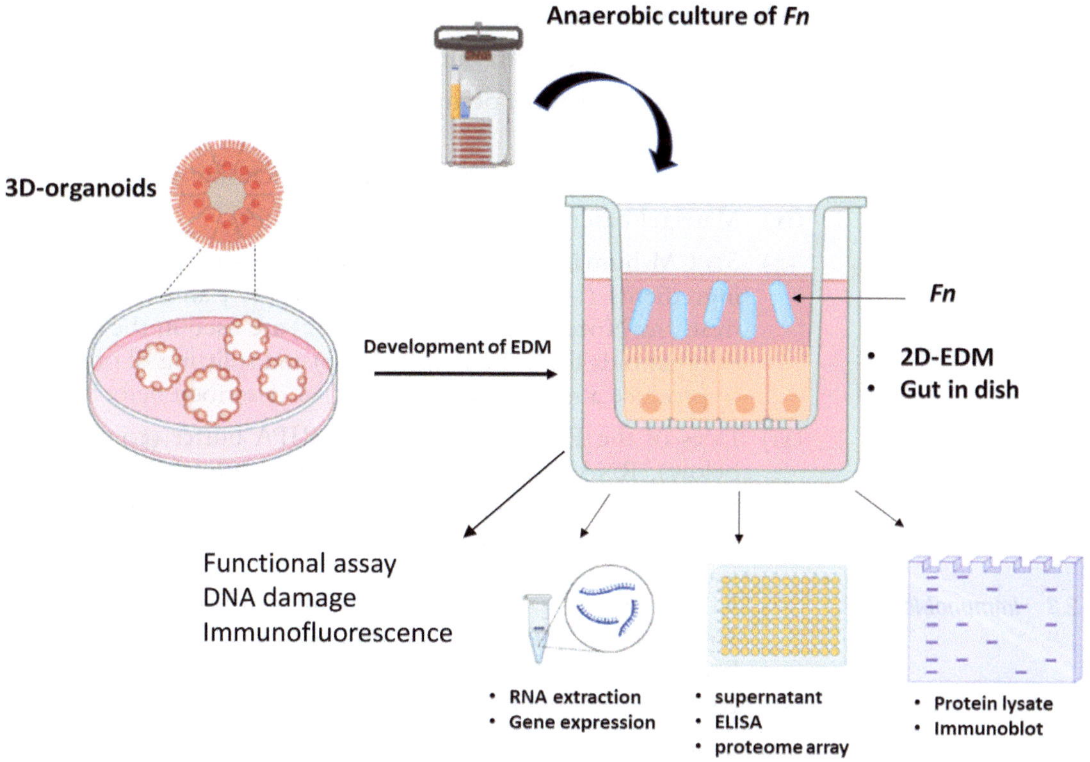

Fig. 1 Schematic flow of the protocol design. Crypt-based stem cells are grown as 3D organoids using Matrigel. Three-dimensional organoids are developed into 2D polarized EDM, and *Fusobacterium nucleatum* (*Fn*) or any other pathogens can be added into the apical part of the EDM. The EDM model (gut in dish) has several functional assays relevant to base excision repair pathways. The RNA lysate from the EDM can be used for the transcriptome assay, and protein lysate can be used for immunoblotting application. Supernatant collected from the apical and/or basolateral side can be used to assess the level of cytokines by ELISA or proteome array. EDM can be used for immunofluorescence staining or measuring DNA damage by LA-PCR

(iv) Anaero Pack, anaerobic cultivation sets (Mitsubishi Gas Chemical [MCG], Japan). The main component of this pack is ascorbic acid (vitamin C).

(v) GeneQuant pro (Biochem LTD, Cambridge, England).

The culture of *Fn* is verified from time to time (*see* **Note 1**). Other microbes such as *H. pylori* and *B. fragilis* are cultured as described in the previous studies [16, 21].

2.2 The Development of Enteroid-Derived Monolayer (EDM)

(i) 3D organoids in culture with 50% conditioned media (CM) (*see* **Notes 2–3**). Fifty percent CM is prepared from L-WRN cells with Wnt3a, R-spondin, and Noggin (ATCC CRL-3276™) as described before [17].

(ii) Advanced DMEM/F12.

(iii) Rho-kinase inhibitor (ROCKi, Y-27632).

(iv) N-acetyl cysteine.

(v) Fetal bovine serum.

(vi) TrypLE™ Select Enzyme (1×).

(vii) Transwell 24-well permeable support with 0.4 μm PET membrane.

(viii) Collagenase, type I (Thermo Fisher Scientific, USA).

(ix) Matrigel basement matrix (Corning, USA).

(x) 5% CM human monolayer EDM media is prepared from the ingredients (i–v) as described before [17] and it should be fresh (*see* **Note 4**). Ingredients (vi–ix) are used in the isolation of organoids from the biopsies (viii) and splitting of organoids to produce a monolayer (vi, vii, ix) as described before [17].

(xi) Lysis of the EDMs is done using RIPA buffer (composition below), RNA lysis buffer provided with Quick-RNA Micro-Prep Kit (Zymo Research, USA), or DNA lysis buffer provided with Gentra Puregene Cell Kit (Qiagen, Hilden, Germany).

2.3 Immunoblotting

(i) RIPA lysis buffer which consists of 50 mM Tris–HCl (pH 7.4), 150 mM NaCl, 1% (v/v) NP-40, 0.25% (w/v) Na deoxycholate, 0.1% (w/v) SDS, and 1 mM EDTA. Protease inhibitor cocktail (1 tablet/10 ml buffer) was added to the RIPA buffer.

(ii) Gradient Bis–Tris gel (4–20%) (Bio-Rad).

(iii) Nitrocellulose membrane 0.45 μm pore size.

(iv) Transfer buffer (1×) (Bio-Rad).

(v) Blocking buffer: Skimmed milk (5% w/v) in TBST buffer (1× Tris-buffered saline and 0.1% Tween 20).

(vi) Primary antibodies: Rabbit anti-NEIL-2, rabbit anti-NEIL-1, rabbit anti-NTH-1, and rabbit anti-OGG-1. All these antibodies are in-house developed.

(vii) Washing buffer: 1× TBST buffer.

(viii) Secondary antibodies: Anti-isotype secondary antibody conjugated with horseradish peroxidase.

(ix) ECL™ Western Blotting Detection Reagents.

2.4 Immunofluorescence

(i) EDM seeded in transwell and infected or not with *Fn*.

(ii) Fixation: 4% paraformaldehyde (PFA).

(iii) Washing: PBS (1×).

(iv) Blocking and permeabilization: 0.1% Triton TX-100, 2 mg/mL bovine serum albumin (BSA) diluted in PBS.

(v) Primary antibody: Mouse anti-p-Histone H2A.X Antibody (Ser 139) (sc-517,348, Santa Cruz biotechnology, dilution 1: 500). Antibody is diluted in blocking buffer.

Table 1
List of primers used for the assessment of the transcriptional level of BER enzymes

Gene	Species	Forward primer (5′ → 3′)	Reverse primer (5′ → 3′)
NEIL1	Mouse	TCGTAGACATCCGTCGCTTT	TGTCTGATAGGTTCCGAAGTACG
NEIL1	Human	CCAGGCAGTGGGAAGTCA	AGGGAGGGTGGCAGAGTC
NEIL2	Mouse	CTGCCGCCTTTCAGTCTCT	TCTGGATCAAACCGAAGGAA
NEIL2	Human	GGGGCAGCAGTAAGAAGCTA	GGAATAATTTCTTTCCATGGACCT
NTH1	Mouse	GCATGAACTCAGGGAAGGAAGA	CCTCACCATTAGCCGCTTCA
NTH1	Human	GACAGCATCCTGCAGACAGA	TTGATGTATTTCACCTTGCTCCT
OGG1	Mouse	TTATCATGGCTTCCCAAACC	GTACCCCAGGCCCAACTT
OGG1	Human	CCAGACCAACAAGGAACTGG	CAAATGCATTGCCAAGGA

(vi) Alexa Fluor 488-conjugated goat antimouse IgG (Invitrogen A11001, dilution 1:500). Antibody is diluted in blocking buffer.

(vii) Antifade aqueous mounting medium (Biotium, CA, USA) and coverslips.

(viii) Leica Automated Inverted Microscope Leica Microsystems Leica CTR4000.

2.5 Functional Assays with the EDM Model

(i) TEER measurement: Epithelial volt/Ohm (TEER) meter and high-throughput automated TEER measurement (REMS AutoSampler) as described before [17].

(ii) DNA/RNA oxidative damage (Cayman, Clone 7E6.9) ELISA kit.

(iii) LA-PCR as described before [8, 22–26].

(iv) Bacterial clearance in the EDM as described before [17].

(v) Cytokine level: ELISA, proteome profiler array, or multiplex cytokine assay.

(vi) Transcriptional level of BER enzymes: qScript™ cDNA SuperMix (Quantabio, USA), 2× SYBR Green qPCR Master Mix (Biomek, USA), QuantStudio 3 Real-Time PCR Systems (Applied Biosystems), and primers listed in Table 1.

3 Methods

3.1 Culture of Fusobacterium nucleatum (Fn)

(i) Culture *Fusobacterium nucleatum* (*Fn*) anaerobically in the chopped meat media in an anaerobic jar system including anaerobic gas kit generating system (Anaero Pack).

(ii) Incubate the anaerobic jar at 37 °C for 48 h under anaerobic conditions.

(iii) For *Fn* subculture, transfer 10 μL from the supernatant of old chopped meat media into a new chopped meat media using the same anaerobic condition as mentioned in **step (a)**.

(iv) On the day of infection, determine the bacterial colony-forming unit (CFU) by measuring the optical density of the media supernatant at a wavelength of 600 nm using GeneQuant pro.

(v) The cultured *Fn* bacteria should be verified from time to time for the expression of *Fn*-specific genes (*see* **Note 1**).

Culture of other microbes such as *H.pylori* and *B. fragilis* can be performed as described in the previous studies [16, 21].

3.2 The Development and Characterization of 3D Organoids

(i) Crypt-derived stem cells were isolated from the colon biopsies of mouse, healthy human, and CRC patients using collagenase digestion as described before [11, 15, 17].

(ii) Growth of stem cells as 3D organoids using basement matrix membrane such as Matrigel.

(iii) The growth of 3D organoids and phenotypic characterization can be performed microscopically (Fig. 2a).

(iv) The viability of 3D organoids can be determined using trypan blue or MTT assay. Also, the phenotypic appearance of the organoids will determine the health status of the organoids (*see* **Note 2**).

(v) The 3D organoids can be processed to formalin-fixed paraffin-embedded blocks for histology preparation using Histogel solution. The processed spheroids can be used for immunofluorescence or immunohistochemistry. The steps of spheroid processing are described below (**steps vi–xii**).

(vi) The 3D organoids are fixed using 4% paraformaldehyde at room temperature for 30 min.

(vii) Scrap the organoids using PBS, and then precipitate the organoid pellets by centrifuge at 300 g for 3 min.

(viii) Ink the spheroids in 15 mL tube using a drop of Gill's hematoxylin, gently agitate, and stand for 5 min.

(ix) Place the Histogel at 65 °C in the water bath till it reaches the melting point.

(x) Dilute the stained spheroids with PBS till the hematoxylin turns pale purple, and then centrifuge to precipitate the organoid pellets.

(xi) Transfer 100–200 μL of Histogel quickly to the precipitated organoids, suspend them well, and then allow to cool down at room temperature for 10–15 min.

(xii) Store the organoids/Histogel preparation in Eppendorf at 4 °C till the time of processing to FFPE blocks.

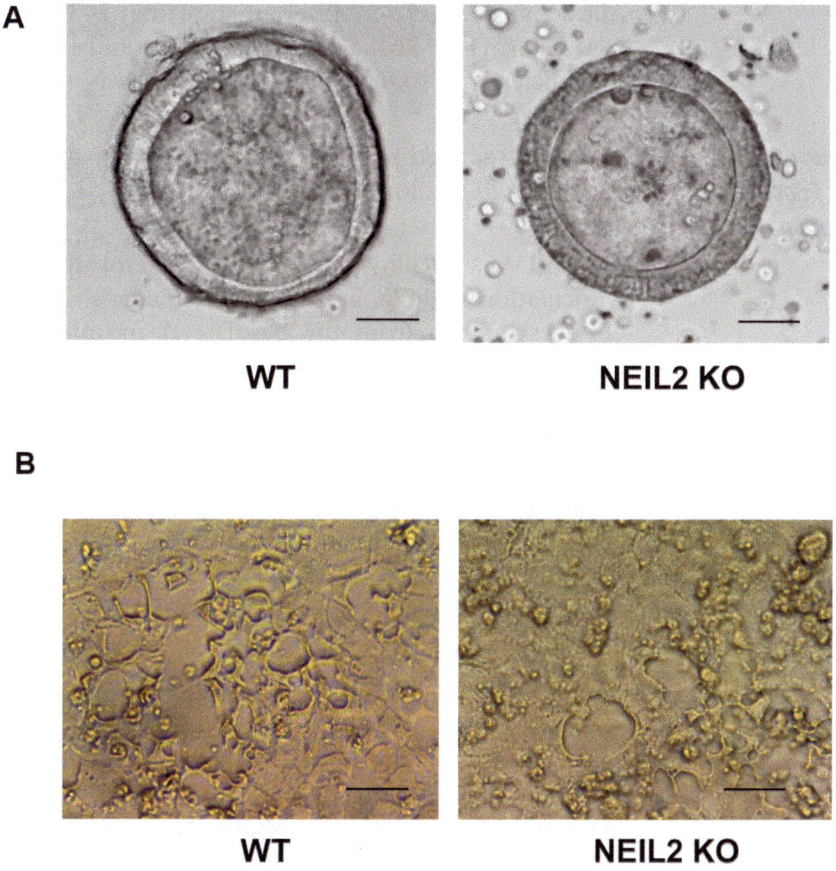

Fig. 2 Morphological characterization of 3D organoids and 2D EDM generated from the colon of wild-type (WT) and NEIL-2 knockout mice. (**a**) Microscopic appearance of WT and NEIL-2 KO 3D organoids. (**b**) Microscopic appearance of WT and NEIL-2 KO 2D EDM

3.3 The Development of 2D Polarized EDMs from 3D Colon Organoids

(A detailed protocol of EDM development is described before [17]). Below are the main steps for the generation of EDM:

(i) Coat the transwell with 50 μL of the PBS-diluted matrigel at a 1:40 ratio. Allow coating the transwell with the PBS-diluted matrigel for 60 min at room temperature.

(ii) Split the 3D organoids (*see* **Note 2**) into single cells by incubating the organoids with TrypLE at 37 °C water bath for 5–10 min. Disrupt the organoids mechanically in the TrypLE mixture by pipetting the cells up and down up to 15–20 times (*see* **Note 3**). Filter the isolated cells using 70 μm cell strainers.

(iii) Count the isolated cells and dissolve them in 5% CM human monolayer EDM media (*see* **Note 4**).

(iv) Seed 2×10^5 cells/transwell and add 200 μL of 5% CM in the apical part and 700 μL of 5% CM in the basolateral part.

(v) Allow the stem cells to differentiate within 2–3 days into polarized EDM (Fig. 2b). Several readouts could be used to assess the efficiency of EDM's differentiation (*see* **Notes 5–10**).

3.4 Infection of EDMs with Fn and Functional Characterization

(i) Enteroid-derived monolayer (EDM) is prepared from healthy growing 3D colonic spheroid as described in the previous section.

(ii) The EDMs are allowed to differentiate for up to 48 h before inoculation with *Fn* infection. Old media should be replaced by fresh media from the basolateral and apical part of the transwell after 24 h of the EDM preparation.

(iii) On the day of the infection, media in the basolateral part and apical part of the transwell are replaced with fresh 5% CM media. EDM is challenged with *Fn* at a multiplicity of infection (moi) 1:100 (*see* **Note 11**). The bacteria are added at the apical part of EDM for 8 h (*see* **Note 12**).

(iv) After 8 h of infection, bacteria is removed, and media is replaced by fresh 5% CM. The TEER is measured to assess the invasiveness of *Fn* to the epithelium.

(v) After 24 h of infection, the supernatants are collected from the basolateral and apical part of the transwell for proteome array and cytokine analysis (*see* **Notes 13–14).** The cells on the transwells are lysed using one of the following buffers (**steps vi–viii**) according to the design of the experiment.

(vi) RNA lysis buffer is provided with Quick-RNA MicroPrep Kit. Lysis of cells is done by adding 150 μL of lysis buffer, incubating at room temperature for 2–5 min, and then pipetting the buffer with cells 10–15 times. The cell lysate is suitable for RNA extraction according to the manufacturer's instruction of the Qucik-RNA MicroPrep Kit (*see* **Note 15**). The MicroPrep Kit gives the best RNA quality and concentration compared to the MiniPrep Kit due to the smaller volume of elution. The extracted RNA is used to assess the transcriptional level of BER enzymes using the primers listed in Table 1 and RT-qPCR methods as published before.

(vii) RIPA lysis buffer as described in Subheading 2.3. Lysis of cells is done by adding 50 μL of RIPA lysis buffer, incubating at room temperature for 2–5 min, and then pipetting the buffer with cells 10–15 times. The cell lysate is suitable for immunoblot to test the translational level of BER enzymes.

(viii) DNA lysis buffer is provided with Gentra Puregene Cell Kit. Lysis of cells is done by adding 50 μL of lysis buffer, incubating at room temperature for 2–5 min, and then pipetting the buffer with cells 10–15 times. The cell lysate is suitable for DNA extraction according to the manufacturer's instruction of

Gentra Puregene Cell Kit. The extracted DNA is used for the analysis of the accumulation of DNA Strand Break using long-amplicon quantitative PCR (LA-qPCR) assay (*see* **Note 16**).

3.5 Assessment of the Transcriptional Level of the BER Enzymes

(i) The extracted RNA (750 ng) is converted into cDNA using the qScript™ cDNA SuperMix in a final volume of 20 μL according to the manufacturer's instructions.

(ii) The synthesized cDNA is diluted to 80 μL using DNA-/RNA-free water. Quantitative PCR (RT-qPCR) is performed using 2× SYBR Green qPCR Master Mix and QuantStudio 3 Real-Time PCR Systems for measuring the level of BER transcripts. The cycle threshold (Ct) of each gene is normalized to the Ct of the housekeeping gene using the ΔΔCt method. The primers used for testing the BER enzymes are listed in Table 1.

3.6 Assessment of the Translational Level of BER Enzymes

(i) Infect the EDMs with *Fn,* and then lyse cells using RIPA buffer.

(ii) Separate the proteins in the whole-cell extracts from EDMs (20 μg protein loaded) onto a Bio-Rad 4–20% gradient Bis–Tris gel, at 125 V until the dye front goes out of the gel for better separation.

(iii) Electro transfer the protein from the Bio-Rad gradient Bis–Tris gel into the nitrocellulose membrane using 1× Bio-Rad transfer buffer (overnight transfer at constant 45 V or 2-h transfer at 100 V both at the cold room).

(iv) Block the membranes using 5% w/v skimmed milk in 1× TBST buffer, and then incubate the membranes with in-house rabbit developed antibodies (NEIL2, NTH1, and OGG1; used in 1: 500 dilutions). (All the antibodies are incubated overnight at 4 °C at cold room/for better blocking incubate with 5% w/v skimmed milk followed by 5% BSA in TBST, or antibody incubation at 5% skimmed milk in TBST).

(v) Wash the membrane with 1× TBST, each wash at least 10 min. Repeat the washing step at least three times.

(vi) Incubate the membrane with anti-rabbit secondary antibody conjugated with horseradish peroxidase in 5% skimmed milk and incubated at room temperature for 1 h.

(vii) Wash the membrane again with 1× TBST (10 min/wash). Repeat the washing step three times.

(viii) Develop the membrane using ECL™ Western Blotting Detection Reagents (RPN2209) (1–2 min) and image immediately using Kwikquant Imager. The Western blot images were quantified using ImageJ automated digitizing system.

3.7 Immunofluorescence of γH2AX Phosphorylation (Fig. 3a)

(i) Infect the EDMs with *Fn,* and then remove the media from the apical and basolateral compartments of the transwell.

(ii) Wash the EDMs gently using PBS at room temperature (*see* **Note 17**) three times.

(iii) Fix the EDMs with 4% paraformaldehyde (PFA) at room temperature for 15 min.

(iv) Remove the PFA, wash with cold PBS (*see* **Note 17**), and then block the EDMs using blocking buffer.

(v) Incubate the permeabilized EDM with the primary antibody—mouse anti-p-Histone H2A.X Antibody (Ser 139) (diluted 1: 500 in blocking buffer)—at 4 °C overnight.

(vi) Remove the primary antibody and wash the EDMs (*see* **Note 18**) with PBS three times (10 min/wash).

(vii) Add the secondary antibody—Alexa Fluor 488-conjugated goat antimouse IgG (diluted 1:500 in blocking buffer) and DAP (diluted 1:500 in blocking buffer)—for 1 h at room temperature in the dark.

(viii) Remove the secondary antibody and wash the EDM three times (10 min/wash) with PBS.

(ix) Dry the membrane and mount it in gel/mount antifade aqueous mounting medium with coverslips.

(x) Images are acquired using a confocal microscope using a 40× and a 63× objective lens (Fig. 3a).

(xi) Z-stack images were acquired by successive 1 μm-depth Z-slices of EDMs in the desired confocal channels. Fields of view that were representative of a given transwell were determined by randomly imaging three different fields. Z-slices of a Z-stack are overlaid to create maximum-intensity projection images; all images were processed using Image J software.

3.8 Measurement of DNA/RNA Oxidative Damage by ELISA

(i) The supernatant from the basolateral part can be used for measuring the level of damaged bases such as 8-hydroxyguanosine, 8-hydroxy-2′-deoxyguanosine, and 8-hydroxyguanine using DNA/RNA oxidative damage ELISA kit.

(ii) For quantitative measurement of the oxidized base, a standard curve is developed from a serial dilution of 8-hydroxyguanosine. The assay ranges at 0.15–20 ng/mL, and the assay's sensitivity (defined as 80% B/B0) is 0.45 ng/mL.

(iii) The standards and samples were prepared using the same culture media of the EDM to minimize the media background as suggested by the manufacturer's instructions.

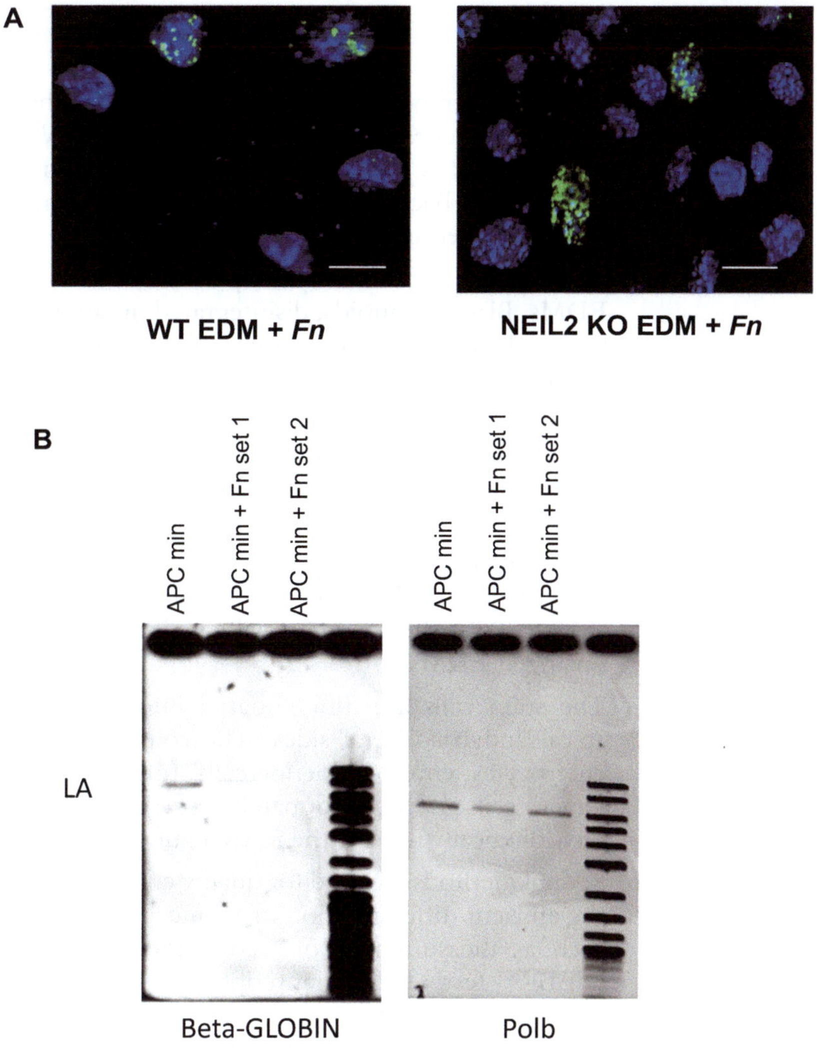

Fig. 3 (**a**) Immunofluorescence of γH2AX phosphorylation: The level of double-strand breaks was determined by immunofluorescence (IF) staining of γH2AX (green); DAPI was used for nuclei staining (blue). (**b**) LA-qPCR assay to measure the level of DNA strand breaks in APC Min/+ EDM following *Fn* infection. Amplification of polβ and β-globin genes was assessed in uninfected and *Fn*-infected EDM

(iv) The concentration of damaged bases is determined by the following:

% B/B0 (% bound/maximum bound): Ratio of the absorbance of a sample and/or standard to the maximum binding (B0) well.

B0 (maximum binding): Maximum amount of the tracer that the antibody can bind in the absence of free analyte

(v) The standard curve is developed by plotting % B/B0 values versus concentrations of a known amount of 8-hydroxyguanosine.

4 Notes

1. *Fn* strains were verified from time to time for the expression of *Fn*-specific genes such as *16srRNA, FadA, FN1527, FN1529, and FN1792* as reported before [27, 28]. Extract DNA from the cultured bacterial pellets and the detection of these genes was performed by PCR.
2. Healthy 3D organoids are required to develop polarized 2D EDMs. Black organoids, disintegrated organoids, or organoids with black centers are not good sources for differentiated EDMs.
3. Mechanical disruption of the organoids should be done with TrypLE and the pipetting should be continued until no cell clumps/aggregates appear to develop good homogenized EDM.
4. The human monolayer media (5% CM) should be prepared fresh. It should be kept at 4 °C for up to 1 week. Longer storage of the media will reduce the efficiency of polarized EDMs.
5. The stem cells are differentiated into polarized EDM with apical and basolateral sides. The colon-specific cells such as enterocytes, enteroendocrine cells, tuft cells, goblet cells, and Paneth cells are proportionately represented. The EDM model physiologically mimics the in vivo gut [8, 12–14].
6. Several methods are used for quality control (QC) and to assess the efficient differentiation of stem cells into polarized EDMs, such as measurement of transepithelial electrical resistance (TEER), testing the transcript levels of stemness (such as Lgr-5) and differentiation markers (such as carbonic anhydrase, mucin2, chromogranin, etc.), and/ or testing the expression of the differentiation markers by immunofluorescence. The differentiated EDMs should show a gradual increase in the TEER value with time compared to undifferentiated stem cells. The increase in TEER value is due to the formation of tight junctions among the epithelial cells. Also, the differentiated EDMs should show an increase in the expression of differentiation markers and a decrease in the stemness markers [8, 9, 12].
7. Poorly differentiated EDMs are not a good option for *Fn* infection, since *Fn* infects the epithelial cells through the interaction with FadA and E-cadherin that is expressed on the epithelial cells, not in the stem cells.
8. For the stem cells to differentiate into polarized EDMs, the amount of stem cell-enriched WRN conditioned media (WNT 3a, R-spondin, and Noggin) should be reduced. Therefore, the 50% CM is diluted to 5% CM and the many growth factors

required for the growth of the stem cells are omitted from the 5% CM EDM monolayer media. The reduction in the WRN and growth factors is crucial to the differentiation process of the stem cell [9–11]. Change in the concentration of WRN and/or growth factors could affect the development of polarized EDM.

9. The 2D EDM model cannot be passaged and propagated. However, 3D organoids can be passaged and cultured up to 30–50 passages. The 3D organoids are the sources of 2D EDMs.
10. The multiplicity of infection (moi) is determined by measuring the CFU of bacteria and counting the number of cells seeded. For *Fn* infection in the colon EDM model, moi 1:100 works well. However, moi can be varied from one cell line to another and according to the *Fn* strains used in the infection experiments. Changing the moi could affect the readouts of the experiments.
11. *Fn* is added to the apical part of the transwell, not the basolateral part. Since the apical part of polarized EDMs represents the side of epithelium facing the internal gut, the basolateral side of EDMs represents the side of gut epithelium facing the bloodstream. To mimic the gut exposure to the microbiome, we challenge the apical part of EDM with *Fn*.
12. Supernatant collected from the basolateral part can be used to assess the level of inflammatory cytokine released and/or used in other functional assays such as measurement of the cell death and evaluation of the level of the damaged oxidative bases. The supernatant should be centrifuged first at 4000 rpm for 10 min to precipitate cell debris and/or bacteria pellets that could interfere with the functional assays.
13. For the determination of the relative levels of selected cytokines and chemokines induced following infection, a membrane-based antibody array/proteome profiler array could be used. This assay is validated to determine the cytokine levels in the cell culture supernatant, cell lysates, tissue lysates, serum, plasma, saliva, urine, and human milk. The detection of cytokines in the basolateral supernatant of EDM should be done following the manufacturer's instructions and as described in our previous study [8].
14. The supernatant from the basolateral part can be used for measuring the degree of cell cytotoxicity by measuring the level of lactate dehydrogenase (LDH) released from the damaged cells using a LDH Glo™ Cytotoxicity Assay kit. Briefly, supernatants need to be diluted using LDH storage buffer, and then, LDH detection reagent is added to each sample and the samples need to be loaded in white opaque 96-well plate. Incubate the plate for 60 min at room temperature, and then

measure the degree of luminescence (relative luminescence unit [RLU]). The standard curve is developed by doing serial dilutions of LDH-positive control (purified lactate dehydrogenase from rabbit muscle) in 5% CM EDM media. Using 5% EDM CM as a diluting agent for the LDH-positive control is to eliminate the cultural background; also the same media is used a medium background. The standard curve ranged from 32 to 0.5 mU/mL.

15. Lyse the cells by adding 150 μL of RNA lysis buffer, and the cell lysate can be saved at −80 °C till the process of RNA extraction. RNA extraction should be done using the RNA MicroPrep Kit, where the elution should be done with a small elution volume (15 μL). Elution of RNA using a larger volume of water leads to low RNA concentration and/or poor-quality RNA.
16. Lyse the cells by adding 50 μL of DNA lysis buffer; the cell lysate can be saved at −80 °C till the process of DNA extraction. The DNA extraction is based on precipitation of the cell protein and then precipitating the DNA from the supernatant using isopropanol. The precipitated DNA is washed using 70% ethanol and precipitated again. The DNA threads are dissolved in 50 μL of DNA hybridizing solution at 65 °C for 1 h The extracted genomic DNA can be used to assess the accumulation of DNA strand break in the transcribed genes (such as polβ and globin) following the protocol described previously (Fig. 3b).
17. In IF of γH2AX phosphorylation, PBS is used for washing (**steps b** and **d**). In **step b**, PBS should be at room temperature since the EDM is incubated at 37 °C. In **step e**, PBS should be cold since this washing step comes after fixation of EDM with cold methanol at −20 °C for 20 min. Changing the temperature of PBS, especially in the first washing step (**b**), can lead to the detachment of cells from the transwell.
18. It is also possible to microinject the gut pathogens into the lumens of 3D organoids, as done previously [29–32].

Acknowledgments

This work was supported by the National Institutes of Health (NIH) grants DK107585 and AG069689 (to S.D.). S.D. is supported by the Leona M. and Harry B. Helmsley Charitable Trust. We thank Amer Ali Abdel- Hafeez, Pradipta Ghosh, and Tapas K Hazra for consultation in some of the immunofluorescence and DNA damage assay. We are grateful to the UC San Diego HUMANOID Center of Research Excellence (CoRE) for providing organoid media.

References

1. Larsen BM, Kannan M, Langer LF et al (2021) A pan-cancer organoid platform for precision medicine. Cell Rep 36(4):109429
2. Azar J, Bahmad HF, Daher D et al (2021) The use of stem cell-derived organoids in disease modeling: an update. Int J Mol Sci 22(14): 7667
3. Idris M, Alves MM, Hofstra RMW et al (2021) Intestinal multicellular organoids to study colorectal cancer. Biochim Biophys Acta Rev Cancer 1876(2):188586
4. Sayed IM, El-Hafeez AAA, Maity PP et al (2021) Modeling colorectal cancers using multidimensional organoids. Adv Cancer Res 151: 345–383
5. Rubinstein Mara R, Wang X, Liu W et al (2013) Fusobacterium nucleatum promotes colorectal carcinogenesis by modulating E-cadherin/β-catenin signaling via its FadA Adhesin. Cell Host Microbe 14(2):195–206
6. Yu T, Guo F, Yu Y et al (2017) Fusobacterium nucleatum promotes chemoresistance to colorectal cancer by modulating autophagy. Cell 170(3):548–563.e16
7. Geng F, Zhang Y, Lu Z et al (2020) Fusobacterium nucleatum caused DNA damage and promoted cell proliferation by the Ku70/p53 pathway in Oral cancer cells. DNA Cell Biol 39(1):144–151
8. Sayed IM, Chakraborty A, Abd El-Hafeez AA et al (2020) The DNA glycosylase NEIL2 suppresses fusobacterium-infection-induced inflammation and DNA damage in colonic epithelial cells. Cell 9(9):1980
9. Sato T, Vries RG, Snippert HJ et al (2009) Single Lgr5 stem cells build crypt-villus structures in vitro without a mesenchymal niche. Nature 459(7244):262–265
10. Jung P, Sato T, Merlos-Suárez A et al (2011) Isolation and in vitro expansion of human colonic stem cells. Nat Med 17(10): 1225–1227
11. Miyoshi H, Stappenbeck TS (2013) In vitro expansion and genetic modification of gastrointestinal stem cells in spheroid culture. Nat Protoc 8(12):2471–2482
12. Sayed IM, Sahan AZ, Venkova T et al (2020) Helicobacter pylori infection downregulates the DNA glycosylase NEIL2, resulting in increased genome damage and inflammation in gastric epithelial cells. J Biol Chem 295(32):11082–11098
13. Sharma A, Lee J, Fonseca AG et al (2021) E-cigarettes compromise the gut barrier and trigger inflammation. iScience 24(2):102035
14. Roodsant T, Navis M, Aknouch I et al (2020) A human 2D primary organoid-derived epithelial monolayer model to study host-pathogen interaction in the small intestine [original research]. Front Cell Infect Microbiology 10: 272
15. Sayed IM, Suarez K, Lim E et al (2020) Host engulfment pathway controls inflammation in inflammatory bowel disease. FEBS J 287(18): 3967–3988
16. Sayed IM, Sahan AZ, Venkova T et al (2020) Helicobacter pylori infection down-regulates the DNA glycosylase NEIL2, resulting in increased genome damage and inflammation in gastric epithelial cells. J Biol Chem 295(32):11082–11098
17. Sayed IM, Tindle C, Fonseca AG et al (2021) Functional assays with human patient-derived enteroid monolayers to assess the human gut barrier. STAR Protocols 2(3):100680
18. Sayed IM, Ramadan HK-A, El-Mokhtar MA et al (2021) Microbiome and gastrointestinal malignancies. Curr Opin Physio 22:100451
19. Kostic AD, Gevers D, Pedamallu CS et al (2012) Genomic analysis identifies association of fusobacterium with colorectal carcinoma. Genome Res 22(2):292–298
20. Kostic AD, Chun E, Robertson L et al (2013) Fusobacterium nucleatum potentiates intestinal tumorigenesis and modulates the tumor-immune microenvironment. Cell Host Microbe 14(2):207–215
21. Chung L, Thiele Orberg E, Geis AL et al (2018) Bacteroides fragilis toxin coordinates a pro-carcinogenic inflammatory Cascade via targeting of colonic epithelial cells. Cell Host Microbe 23(2):203–214.e5
22. Chakraborty A, Wakamiya M, Venkova-Canova T et al (2015) Neil2-null mice accumulate oxidized DNA bases in the transcriptionally active sequences of the genome and are susceptible to innate inflammation. J Biol Chem 290(41): 24636–24648
23. Chatterjee A, Saha S, Chakraborty A et al (2015) The role of the mammalian DNA end-processing enzyme polynucleotide kinase 3′-phosphatase in spinocerebellar ataxia type 3 pathogenesis. PLoS Genet 11(1):e1004749
24. Chakraborty A, Tapryal N, Venkova T et al (2016) Classical non-homologous end-joining pathway utilizes nascent RNA for error-free double-strand break repair of transcribed genes. Nat Commun 5(7):13049
25. Gao R, Chakraborty A, Geater C et al (2019) Mutant huntingtin impairs PNKP and ATXN3,

disrupting DNA repair and transcription. elife 17:8

26. Chakraborty A, Tapryal N, Venkova T et al (2020) Deficiency in classical nonhomologous end-joining-mediated repair of transcribed genes is linked to SCA3 pathogenesis. Proc Natl Acad Sci U S A 117(14):8154–8165
27. Gupta S, Ghosh SK, Scott ME et al (2010) Fusobacterium nucleatum-associated beta-defensin inducer (FAD-I): identification, isolation, and functional evaluation. J Biol Chem 285(47):36523–36531
28. Liu P, Liu Y, Wang J et al (2014) Detection of fusobacterium nucleatum and fadA adhesin gene in patients with orthodontic gingivitis and non-orthodontic periodontal inflammation. PLoS One 9(1):e85280
29. Engevik MA, Engevik KA, Yacyshyn MB et al (2015) Human Clostridium difficile infection: inhibition of NHE3 and microbiota profile. Am J Physiol Gastrointest Liver Physiol 308(6):G497–G509
30. Forbester JL, Goulding D, Vallier L et al (2015) Interaction of salmonella enterica serovar typhimurium with intestinal organoids derived from human induced pluripotent stem cells. Infect Immun 83(7):2926–2934
31. Leslie JL, Huang S, Opp JS et al (2015) Persistence and toxin production by clostridium difficile within human intestinal organoids result in disruption of epithelial paracellular barrier function. Infect Immun 83(1):138–145
32. Williamson IA, Arnold JW, Samsa LA et al (2018) A high-throughput organoid microinjection platform to study gastrointestinal microbiota and luminal physiology. Cell Mol Gastroenterol Hepatol 6(3):301–319

Chapter 11

Characterizing the Repair of DNA Double-Strand Breaks: A Review of Surrogate Plasmid-Based Reporter Methods

Arijit Dutta, Joy Mitra, Pavana M. Hegde, Sankar Mitra, and Muralidhar L. Hegde

Abstract

DNA double-strand breaks (DSBs) are the most lethal genomic lesions that are induced endogenously during physiological reactions as well as by external stimuli and genotoxicants. DSBs are repaired in mammalian cells via one of three well-studied pathways depending on the cell cycle status and/or the nature of the break. First, the homologous recombination (HR) pathway utilizes the duplicated sister chromatid as a template in S/G_2 cells. Second, the nonhomologous end-joining (NHEJ) is the predominant DSB repair pathway throughout the cell cycle. The third pathway, microhomology-mediated/alternative end-joining (MMEJ/Alt-EJ), is a specialized backup pathway that works not only in the S phase but also in G_0/G_1 cells that constitute the bulk of human tissues. In vitro experimental methods to recapitulate the repair of physiologically relevant DSBs pose a challenge. Commonly employed plasmid- or oligonucleotide-based substrates contain restriction enzyme-cleaved DSB mimics, which undoubtedly do not mimic DSB ends generated by ionizing radiation (IR), chemotherapeutics, and reactive oxygen species (ROS). DSBs can also be indirectly generated by reactive oxygen species (ROS). All such DSBs invariably contain blocked termini. In this methodology chapter, we describe a method to recapitulate the DSB repair mechanism using *in cellulo* and in vitro cell-free systems. This methodology enables researchers to assess the contribution of NHEJ vs. Alt-EJ using a reporter plasmid containing DSB lesions with non-ligatable termini. Limitations and challenges of prevailing methods are also addressed.

Key words DNA double-strand breaks, Homologous recombination, Non-homologous end-joining, Microhomology-mediated alternative end joining, Reporter assay

1 Introduction

DNA double-strand breaks (DSBs) constitute the most lethal damage in the mammalian genome, and their repair is essential for the maintenance of genomic stability and prevention of cell death. Nonhomologous end-joining (NHEJ), the predominant DSB repair pathway, is active throughout the cell cycle, while homologous recombination (HR) takes place preferentially in the S/G_2 phase (Fig. 1) [1]. Although the minor alternative end-joining

Kishor K. Bhakat and Tapas K. Hazra (eds.), *Base Excision Repair Pathway: Methods and Protocols*, Methods in Molecular Biology, vol. 2701, https://doi.org/10.1007/978-1-0716-3373-1_11,

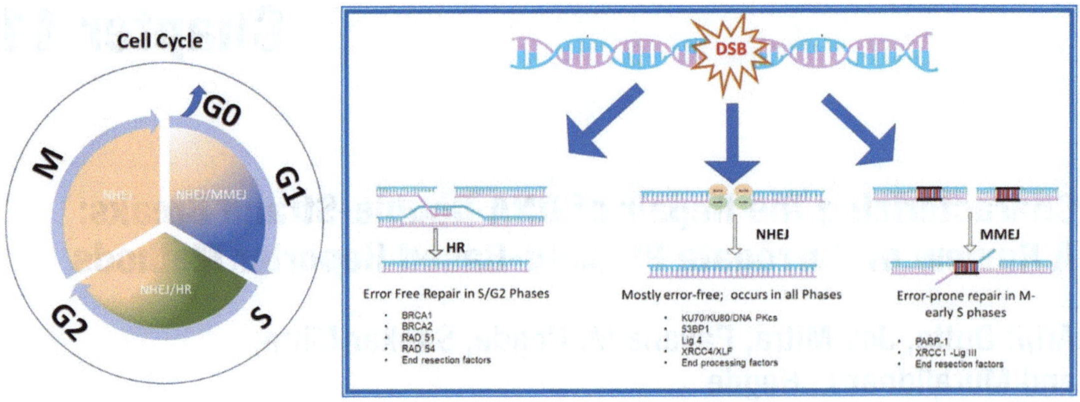

Fig. 1 Schematics of DNA double-strand break (DSB) repair pathways. The image on the left illustrates the relative contributions of the three DSB repair pathways, namely, homologous recombination (HR), nonhomologous end-joining (NHEJ), and microhomology-mediated end-joining (MMEJ), in various cell cycle stages. The HR pathway involves template-dependent repair of DSBs but limited to the S-phase in dividing cells. Microhomology-mediated end-joining (MMEJ) is a form of alternative end-joining (Alt-EJ) that operates in the G0/G1 and early S-phases exploiting short stretches of DNA microhomologies flanking DSB ends. This repair process leads to larger deletion of the genomic regions, particularly at repetitive DNA. The image on the right outlines the key repair steps in each of these three pathways, indicating canonical repair factors, such as BRCA1-BARD1, BARC2, RAD51, and long-range end-resection factors MRN, DNA2, BLM, and EXO1 in the HR, end-processing factors MRN and PNKP, damage sensors like Ku70-Ku80 heterodimer, DNA-PKcs, 53BP1, and repair proteins XRCC4, XLF, and DNA ligase 4 in the NHEJ, and PARP1, MRN, and the ligation complex XRCC1-DNA ligase 3 in the MMEJ/Alt-EJ pathway

(Alt-EJ) pathway was initially considered to function as a backup to NHEJ, current studies underscore its significant role in maintaining genomic integrity both in dividing and nondividing cells [2, 3]. The key components of the classical NHEJ and HR machinery have been linked to the Alt-EJ repair process, based on notable genetic and biochemical studies. For example, Pfeiffer and Vielmetter first showed that linearized plasmids could be recircularized in vitro by incubating with cell-free *Xenopus* egg extract [4]. These results were subsequently repeated by North et al., using human cell extracts [5]. Additional studies documented in vitro DSB repair via end-joining of duplex oligonucleotides and in plasmid DNA containing either enzymatically cleaned DSBs or radiomimetic drug-induced dirty-ended DSBs. Such DSB repairs were carried out using cultured human cells and *Xenopus* egg extracts [6, 7]. The methods for the detection of the repaired products involved Southern blotting with radiolabeled oligonucleotide probes, followed by ethidium bromide (EtBr) staining, autoradiography of radiolabeled substrate, or β-glycosidase-dependent bacterial mutagenesis assays using bacterial transformation of in vitro repaired plasmid in *E. coli*. Other methods for the detection of DSB repair, such as asymmetric field inversion gel electrophoresis (AFIGE) and quantitative PCR assays, have been subsequently developed [8–10]. Furthermore, DSB substrates with chemically incompatible, non-ligatable DSB termini, including 5′-hydroxyl, 3′-phosphate, and 3-′-phosphoglycolate, were generated in vitro mimicking physiologically relevant complex DSBs induced by ionizing radiation (IR) or radiomimetic drugs, revealing the importance of end-processing enzymes like polynucleotide kinase/phosphatase (PNKP) in DSB repair [11, 12].

Although several key factors of the DSB repair pathway were characterized using end-joining assays with cell-free extracts, such approaches could not recapitulate the repair of DSBs generated in chromatin. However, genomic DSBs produced by rare-cutter meganucleases, for example, I-Sce I, not only allowed *in cellulo* monitoring of DSB repair but also led to the identification of several regulatory factors as well as relative contributions of NHEJ and HR factors to the DSB repair at the chromatin level [13–15]. Presently, I-Sce I-based repair assays have been routinely employed to measure the mutagenic NHEJ and Alt-EJ processes [16, 17]. Although this tool provided a significant advantage in analyzing various DSB repair pathways acting on the chromatin, I-Sce I-generated DSBs do not mimic the complexity of IR-induced damages. Due to the technical challenge of generating blocked DSBs at the chromatin level, we designed linearized plasmids with blocked termini, whose in-cell repair could reveal the underpinning of the DSB repair pathway choice in irradiated human cells. Here, we describe the design of the substrate, assay details, methods of readouts, and technical challenges of various DNA repair assays.

2 Materials

1. pEGFPN1 (Addgene).
2. NEB enzymes: EcoRI, XhoI, UDG, FpG, T4 DNA ligase, Nt. BbvCI, Nb.BbvCI.
3. Max Efficiency *E. coli* DH5α competent cells (Invitrogen).
4. XL10 *E. coli* ultracompetent cells (Agilent).
5. Qiagen PCR purification kit.
6. Qiagen plasmid isolation kit.
7. Phenol/chloroform/isoamyl alcohol mixture, 25:24:1 (v/v), Sigma.
8. Agarose.
9. Gel electrophoresis apparatus with power unit.
10. Heat blocks.
11. Temperature-controlled shaker.
12. Lipofectamine 2000.

13. Dulbecco's Modified Eagle Medium (DMEM), Gibco.
14. Fetal bovine serum, Gibco.
15. Dulbecco's phosphate-buffered saline (DPBS), Gibco.
16. Fluorescence microscope.
17. Flag M2 resin, Sigma.

3 Methods

3.1 Reporter Plasmid Preparation

The following protocol generates a linearized reporter plasmid containing a 3′-phosphate (P) terminus that mimics a 3′-P blocked DSB substrate. Introduction of microhomology sequences of desired length flanking the DSB generation site allows for semi-quantitative and comparative estimation of NHEJ vs. Alt-EJ *in cellulo* and in vitro, which will be discussed in the following sections (Subheadings 3.2 and 3.3). A DNA single-strand break with 3′-P and 5′-P termini could be generated at the uracil (U) base by treating with uracil DNA glycosylase (UDG), which excises the U, followed by strand scission of the abasic site by FapyG DNA glycosylase (Fpg). Thus, using this strategy, a DSB with 3′-P and 5′-P termini could be generated in a circular plasmid DNA containing two closely spaced "U"s in opposite strands. We describe below a method to introduce two closely spaced "U"s in a reporter plasmid backbone, such as pEGFP-N1, and then linearizing it with UDG and Fpg to generate a plasmid with 3′-P and 5′-P termini (Fig. 2). Because direct ligation of a U-containing duplex oligonucleotide in a linearized vector is inefficient, a two-step method of nicking followed by ligation of a U-containing oligonucleotide is a highly efficient approach to introducing two closely spaced "U"s in opposite strands, as discussed below.

3.1.1 Generation of a Plasmid Construct, Named pNt.Nb, Containing a Pair of Microhomology Sequences, GTGAGG and CCTCAGC, Site for Nicking Endonuclease (Nt.BbvCI/Nb.BbvCI)

Anneal oligonucleotides NB2 and NT2 to generate a duplex by mixing them in equimolar ratio (1 μM each) in a 20 μL buffer containing 10 mM Tris–HCl (pH 7.5), 50 mM NaCl, and 1 mM EDTA (pH 8.0), followed by incubation at 37 °C for 1 h.

NB2: 5′ P- TCGAGGCTGAGGTACCTCACTTACCTCACGTG CTGAGGG-3′ OH, and

NT2: 5′ P- AATTCCCTCAGCACGTGAGGTAAGTGAGGTAC CTCAGCC-3′ OH.

Linearize 1 μg of pEGFP-N1 with EcoRI and BamHI and gel purify the vector backbone. Ligate linearized pEGFP-N1 and NT2. NB2 duplex by setting a ligation reaction with T4 DNA ligase (NEB) in 1× ligase reaction buffer (NEB), per the manufacturer's protocol. Use 1 μL of the ligation mixture to transform into *E. coli* DH5α competent cells. Screen clones to confirm integration of NT2.NB2 in pEGFP-N1 (named pNt.Nb) by Sanger sequencing.

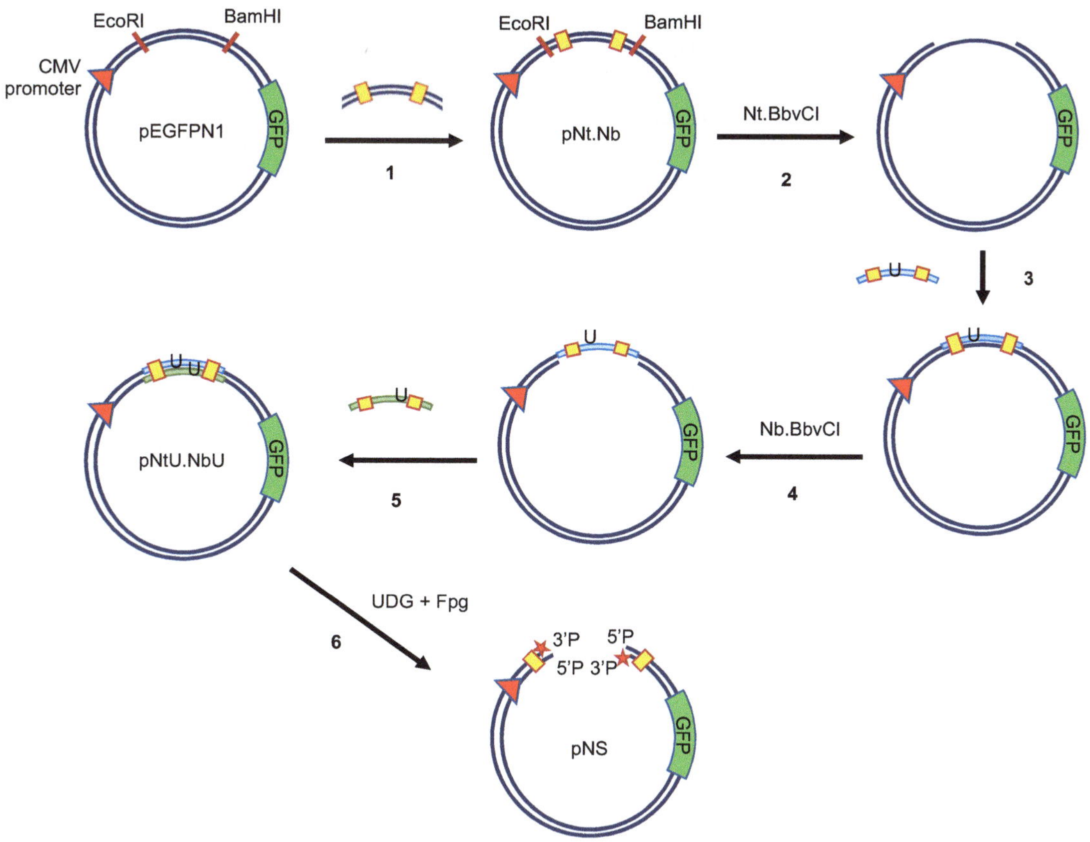

Fig. 2 Schematics of the generation of the reporter plasmid substrate. Step 1: Introduction of microhomology sequence containing duplex DNA oligonucleotide into pEGFPN1 at the EcoRI and BamHI sites by molecular cloning to generate pNt.Nb. Step 2: Generation of a gapped plasmid intermediate via nicking with the enzyme Nt.BbvCI and partial denaturation. Step 3: Annealing and ligation of the first U-containing complementary oligonucleotide with the gapped plasmid. Step 4: Generation of the second gapped plasmid intermediate via nicking with the enzyme Nb.BbvCI and partial denaturation. Step 5: Annealing and ligation of the second U-containing complementary oligonucleotide with the gapped plasmid, to generate pNtU.NbU. Step 6: Generation of pNtU.NbU with UDG and Fpg to generate linearized pNS with 3′-P and 5′-P unligatable DNA ends. The figure is adapted and modified from the original article, by Dutta et al. [18]

3.1.2 Generation of the U-Containing Circular Plasmid, pNtU.NbU

Incubate 100 μg of pNt.Nb with 5 μL of the nicking endonuclease Nt.BbvCI (NEB) in 1× rCutSmart buffer (NEB, USA) in 50 μL reaction volume. Purify the nicked pNt.Nb vector by removing the residual enzyme utilizing the Qiagen PCR purification kit, as per the manufacturer's protocol. Mix 100 molar-fold excess of the U-containing oligonucleotide NtU (5′ P- TCAGCACGTGAG GUAAGTGAGGTACC -3′ OH) with nicked pNt.Nb in 45 μL reaction volume in a buffer containing 10 mM Tris–HCl (pH 7.5) and 50 mM NaCl, and then incubate at 65 °C for 10 min in a heat block, followed by slow cooling to room temperature. Add 1 μL T4 DNA ligase and 5 μL 10× T4 DNA ligase buffer in a final volume of 50 μL and incubate overnight (15 h) at 16 °C to allow ligation of the NtU with pNt.Nb; the product was named as pNtU.Nb. The next day, incubate pNtU.Nb with 5 μL nicking endonuclease Nb.BbvCI (NEB) in 1× rCutSmart buffer (NEB) in 50 μL reaction volume. Purify the nicked plasmid utilizing the Qiagen PCR purification kit, as was done earlier. Anneal 100 molar-fold excess of oligonucleotide NbU (5′ P-TGAGGTACCTCACU TACCTCACGTGC-3′ OH) with nicked pNtU.Nb in 45 μL reaction volume in a buffer containing 10 mM Tris–HCl (pH 7.5) and 50 mM NaCl, as before, by incubating at 65 °C for 10 min in a heat block, followed by slow cooling to room temperature. Add 1 μL T4 DNA ligase and 5 μL 10× T4 DNA ligase buffer in a final volume of 50 μL and incubate overnight (15 h) at 16 °C to allow ligation of NbU. Gel purify the final ligation product, pNtU.NbU, with U's on opposite strands. Check pNt.Nb and the intermediate products, like nicked pNt.Nb, pNtU.Nb, nicked pNtU.NB, and the final product pNtU.NbU via resolution in 1% TAE agarose gel.

3.1.3 Generation of Linearized Reporter Plasmid, pNS with 3′-P and 5′-P Termini

Incubate 10 μg of pNtU.NbU with 2 μL UDG (NEB) in 1× UDG buffer in a 20 μL reaction volume at 37 °C for 1 h. Add 2 μL of Fpg (NEB) enzyme and incubate for another 1 h. Check the product on 1% agarose gel by running along with pNtU.NbU. Gel purify the product, as named pNS, by carefully excising only the band corresponding to the linearized plasmid.

3.2 In-Cell *NHEJ/Alt-EJ Repair Assay*

3.2.1 Transfection of Mammalian Cells with pNS (Fig. 3)

Culture human cells, such as U2OS (osteosarcoma) or A549 (lung carcinoma) in recommended medium, such as DMEM with 10% fetal bovine serum (FBS). Grow cells on a 60 mm tissue culture dish to 40–50% confluence. On the next day, transfect the cells with 100 ng of pNS using 1 μL Lipofectamine-2000 reagent, as per the manufacturer's protocol, and incubate overnight (15 h). Check transfected vs. untransfected cells for GFP fluorescence using a fluorescence microscope (Fig. 3).

3.2.2 Extraction of Recircularized Plasmid Using the Modified Hirt Method

Extract the plasmids from cells using a Qiagen plasmid isolation kit. Add 100 μL of resuspension buffer P1, followed by 100 μL of lysis buffer P2 directly on the 60 mm plates, and finally 100 μL of neutralizing buffer P3. Transfer the entire content to a microcentrifuge tube with an air-lock cap and centrifuge for 10 min at 16,000 *g*. Remove the proteins via phenol/chloroform extraction and finally precipitate the DNA by adding 750 μL of ice-cold ethanol along with 10 μg of glycogen, and then incubate at −80 °C for 1 h. Next, remove the ethanol and wash the precipitate with 70% ethanol. Air-dry the precipitate and resuspend it in 20 μL of TE buffer (10 mM Tris–HCl, pH 8.0, 1 mM EDTA).

3.2.3 Recircularized Plasmid Enrichment and Sequencing

Transform XL10-gold ultracompetent *E. coli* cells (Agilent) with 5 μL of the plasmid extracts, as per the manufacturer's protocol, and spread on LB agar plates supplemented with 50 μg/mL of

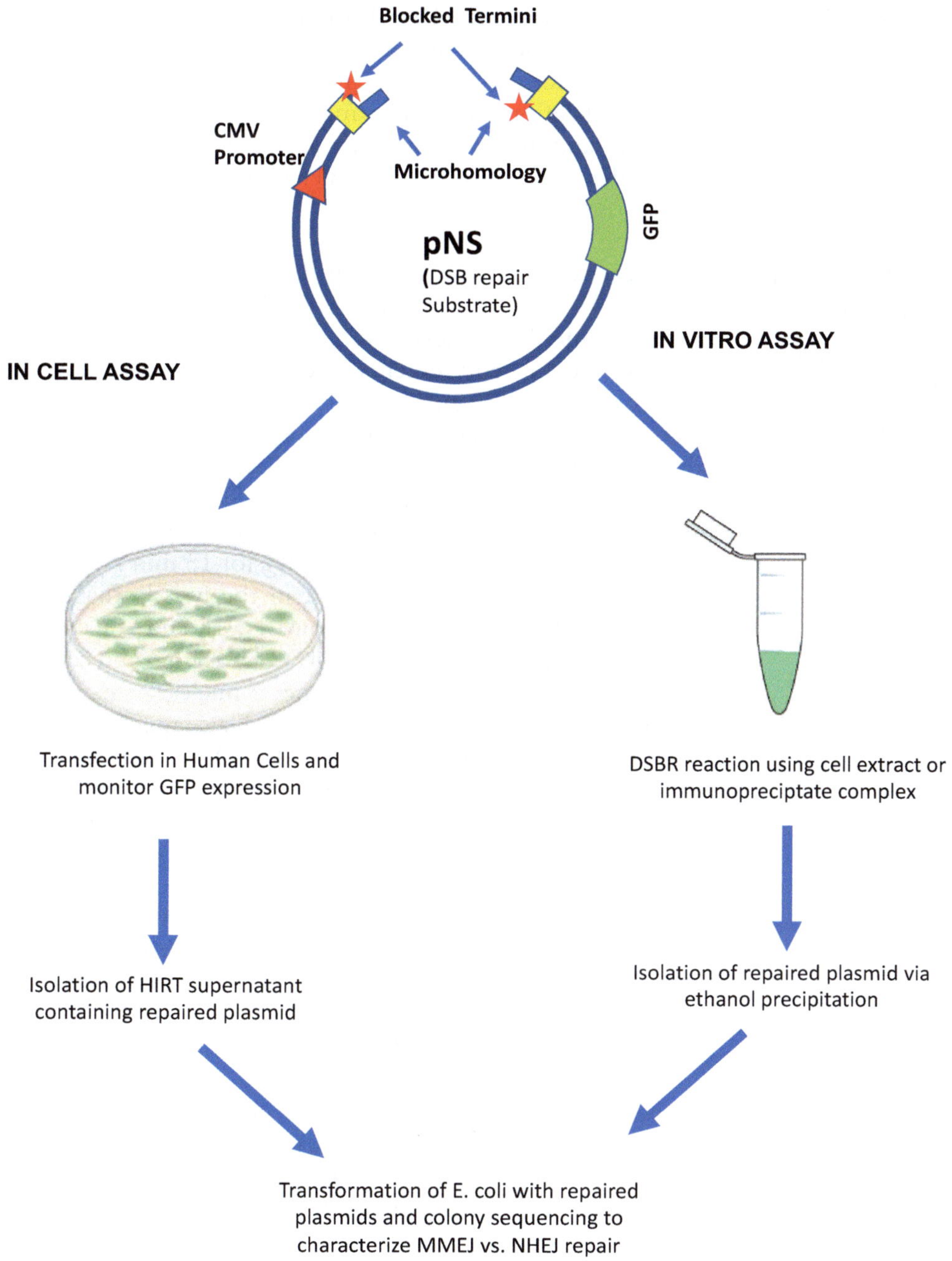

Fig. 3 *In cellulo* and in vitro reporter assay: Linearized reporter substrate pNS could be introduced in human cell lines via transfection, after desired treatment such as oxidative stress, ionizing radiation, etc., in combination with inhibition or siRNA-mediated deletion of specific DNA repair/DNA-damaging signaling factors. Repair of the plasmids could be monitored via GFP expression, only after recircularization of the plasmid, followed by isolation of the episomal DNA. In vitro repair of pNS could be performed via biochemical reconstitution assays with purified proteins, cell extract, or immune complex isolated from human cells. After purification, the repaired plasmids are amplified in *E coli* via transformation and individual colonies are sent for sequencing. Repaired regions of the plasmid DNA were subsequently sequenced for assessing the relative contribution of NHEJ vs. MMEJ/Alt-EJ pathways to this in vitro repair system

kanamycin sulfate. Pick 40–50 colonies for colony sequencing with the CMV-F primer. Analyze the sequencing results by aligning with the original target sequence in pNt.Nb. Mark sequences with 1–4 nt(s) deletion with intact microhomology sites as the NHEJ outcome, and mark those with a single stretch of deletion of nucleotides extending to the microhomology sequences as the Alt-EJ product.

3.3 In Vitro NHEJ/Alt-EJ Repair Assay

3.3.1 Nuclear Extract Preparation and Co-immunoprecipitation

Transfect exponentially growing U2OS cells with Flag-tagged XRCC1 (Alt-EJ factor) or XRCC4 (NHEJ factor) plasmid using Lipofectamine-2000 transfection reagent. After 48 h, treat cells as per experimental requirements, such as with small-molecule inhibitors or DNA-damaging agents (e.g., IR or neocarzinostatin). Harvest the cells by scraping and resuspending them in cytoplasmic extraction buffer containing 10 mM Tris–HCl (pH 7.9), 0.34 M sucrose, 3 mM $CaCl_2$, 2 mM $MgCl_2$, 0.1 mM EDTA, 1 mM DTT, and 0.1% Nonidet P-40 with protease inhibitors. Briefly vortex the cell suspension and centrifuge at 3500 *g* for 15 min at 4 °C. Collect the nuclear pellet and resuspend it in the lysis buffer containing 20 mM Tris–HCl (pH 7.5), 150 mM NaCl, and 1% Triton-X100 with protease inhibitors. Vortex at 4 °C for 15 min and centrifuge at 14,000 rpm to obtain the nuclear extract. For co-immunoprecipitation (co-IP) of Flag-tagged XRCC1 or XRCC4 complex, mix 50 μL of anti-Flag M2 resin beads (Sigma) with 200–300 μL of the nuclear extract at 4 °C for 1–2 h. Add 0.15 U/μL of benzonase (EMD Millipore) during co-IP to prevent DNA binding-mediated nonspecific pulldown of DNA repair protein complexes.

After incubation, centrifuge the tubes (2000 *g*) for 2 min at 4 °C. Discard the supernatant and wash the beads three times with the lysis buffer. After the third wash, resuspend the anti-Flag resin beads in a buffer containing 2 mM $MgCl_2$, 60 mM NaCl, 50 mM HEPES, 2 mM DTT, and 50 μg/mL bovine serum albumin (BSA).

3.3.2 In Vitro Plasmid Recircularization Assay

Incubate 20 μL of resuspended co-IP eluate or anti-Flag M2 resin beads as control with pNS in a reaction mixture containing 1 mM ATP and 1 mM dNTPs with gentle shaking at 30 °C for 1 h. Spin down the resin beads and use 5 μL of the supernatant to transform XL10-gold ultracompetent *E. coli* cells. Pick colonies and perform DNA sequencing as discussed earlier in Subheading 3.2.3.

4 Notes and Final Summary

There are some limitations in the described protocols, such as moderate efficiency in the repaired plasmid recovery and low-throughput quantitative readout. Nevertheless, its biggest advantage over the prevailing methods is that the DSB terminus in the substrate closely represents the endogenously generated broken genome ends under various biological and/or pathological conditions. This approach could further be utilized to generate linearized reporter plasmids with varying lengths of microhomologies and/or different types of blocked ends such as 3-′-phosphoglycolate, peptide-conjugated ends, etc. It is also possible to perform multiplex assays to monitor the repair of different types of blocked ends in the same cells exploiting different fluorescent reporter proteins, including cyan or red fluorescent protein for each type of reporter plasmids [19]. Moreover, this method could be improvised into a high-throughput assay with a DNA barcoding approach and via the use of next-generation sequencing of the cell extracts [20].

Acknowledgments

Research in the authors' laboratory is supported by grants from the National Institute of Neurological Disorders and Stroke (NINDS) and the National Institute of Aging (NIA) of the National Institutes of Health (NIH) under award numbers R01NS088645, RF1NS112719, R03AG064266, and R01NS094535 and the Houston Methodist Research Institute's internal funds. M.L.H. thanks Everett E. and Randee K. Bernal for their support via the Centenial Endowed Directorship of DNA Repair. The content is solely the responsibility of the authors and does not necessarily represent the official views of the funding agencies. The authors thank Dr. Gillian Hamilton at Houston Methodist Research Institute (Houston, TX) for her assistance with editing the document.

References

1. Davis AJ, Chen DJ (2013) DNA double strand break repair via non-homologous end-joining. Transl Cancer Res 2(3):130–143. https://doi.org/10.3978/j.issn.2218-676X.2013.04.02
2. Decottignies A (2013) Alternative end-joining mechanisms: a historical perspective. Front Genet 4:48. https://doi.org/10.3389/fgene.2013.00048
3. Deriano L, Roth DB (2013) Modernizing the nonhomologous end-joining repertoire: alternative and classical NHEJ share the stage. Annu Rev Genet 47:433–455. https://doi.org/10.1146/annurev-genet-110711-155540
4. Pfeiffer P, Vielmetter W (1988) Joining of nonhomologous DNA double strand breaks in vitro. Nucleic Acids Res 16(3):907–924
5. North P, Ganesh A, Thacker J (1990) The rejoining of double-strand breaks in DNA by human cell extracts. Nucleic Acids Res 18(21): 6205–6210
6. Pastwa E, Somiari RI, Malinowski M, Somiari SB, Winters TA (2009) In vitro non-homologous DNA end joining assays – the 20th anniversary. Int J Biochem Cell Biol 41(6):1254–1260. https://doi.org/10.1016/j.biocel.2008.11.007
7. Fairman MP, Johnson AP, Thacker J (1992) Multiple components are involved in the efficient joining of double stranded DNA breaks in human cell extracts. Nucleic Acids Res 20(16): 4145–4152
8. Iliakis G, Mladenov E, Cheong N (2012) In vitro rejoining of double strand breaks in

genomic DNA. Methods Mol Biol 920:471–484. https://doi.org/10.1007/978-1-61779-998-3_32

9. Budman J, Chu G (2005) Processing of DNA for nonhomologous end-joining by cell-free extract. EMBO J 24(4):849–860. https://doi.org/10.1038/sj.emboj.7600563
10. Ma Y, Lieber MR (2006) In vitro nonhomologous DNA end joining system. Methods Enzymol 408:502–510. https://doi.org/10.1016/S0076-6879(06)08031-1
11. Gu XY, Bennett RA, Povirk LF (1996) End-joining of free radical-mediated DNA double-strand breaks in vitro is blocked by the kinase inhibitor wortmannin at a step preceding removal of damaged 3′ termini. J Biol Chem 271(33):19660–19663
12. Datta K, Purkayastha S, Neumann RD, Winters TA (2012) An in vitro DNA double-strand break repair assay based on end-joining of defined duplex oligonucleotides. Methods Mol Biol 920:485–500. https://doi.org/10.1007/978-1-61779-998-3_33
13. Bellaiche Y, Mogila V, Perrimon N (1999) I-SceI endonuclease, a new tool for studying DNA double-strand break repair mechanisms in drosophila. Genetics 152(3):1037–1044
14. Yatagai F, Suzuki M, Ishioka N, Ohmori H, Honma M (2008) Repair of I-SceI induced DSB at a specific site of chromosome in human cells: influence of low-dose, low-dose-rate gamma-rays. Radiat Environ Biophys 47(4):439–444. https://doi.org/10.1007/s00411-008-0179-7
15. Moynahan ME, Jasin M (2010) Mitotic homologous recombination maintains genomic stability and suppresses tumorigenesis. Nat Rev Mol Cell Biol 11(3):196–207. https://doi.org/10.1038/nrm2851
16. Bindra RS, Goglia AG, Jasin M, Powell SN (2013) Development of an assay to measure mutagenic non-homologous end-joining repair activity in mammalian cells. Nucleic Acids Res 41(11):e115. https://doi.org/10.1093/nar/gkt255
17. Kostyrko K, Mermod N (2015) Assays for DNA double-strand break repair by microhomology-based end-joining repair mechanisms. Nucleic Acids Res 44:e56. https://doi.org/10.1093/nar/gkv1349
18. Dutta A, Eckelmann B, Adhikari S, Ahmed KM, Sengupta S et al (2017) Microhomology-mediated end joining is activated in irradiated human cells due to phosphorylation-dependent formation of the XRCC1 repair complex. Nucleic Acids Res 45(5):2585–2599. https://doi.org/10.1093/nar/gkw1262
19. Nagel ZD, Margulies CM, Chaim IA, McRee SK, Mazzucato P et al (2014) Multiplexed DNA repair assays for multiple lesions and multiple doses via transcription inhibition and transcriptional mutagenesis. Proc Natl Acad Sci U S A 111(18):E1823–E1832. https://doi.org/10.1073/pnas.1401182111
20. Hawkins JA, Jones SK Jr, Finkelstein IJ, Press WH (2018) Indel-correcting DNA barcodes for high-throughput sequencing. Proc Natl Acad Sci U S A 115(27):E6217–E6226. https://doi.org/10.1073/pnas.1802640115

Part III

Interactome Profiling and Purification of DNA Damage Repair/Response Proteins

Chapter 12

Interactome Profiling of DNA Damage Response (DDR) Mediators with Immunoprecipitation-Mass Spectrometry

Henry C. -H. Law, Dragana Noe, and Nicholas T. Woods

Abstract

Immunoprecipitation-mass spectrometry (IP-MS) is a versatile tool to probe for global protein–protein interactions (PPIs) in biological samples. Such interactions coordinate complex biological processes, such as the DNA damage response (DDR). Induction of DNA damage activates signaling networks where post-translational modifications cause PPI that facilitate DNA repair and cell cycle coordination. Protein interactome profiling of DDR sensors, transducers, and effectors has the potential to identify novel DDR mechanisms that could advance our understanding and treatment of diseases associated with DDR defects, such as cancer. The protocol described here is a routine PPI analysis procedure that can be performed on samples stimulated with DNA damage. All processes and reagents are optimized for maximum sensitivity on the interactome and minimal contamination for the mass spectrometer.

Key words Immunoprecipitation, Mass spectrometry, DNA damage response, Protein-protein interactions, Posttranslational modifications, BRCA1 C-terminal

1 Introduction

Cancer is a disease that can arise from a failure to accurately repair DNA damage, while at the same time DNA-damaging treatments are the most common form of cancer therapy. Therefore, delineating the molecular mechanisms regulating the DDR is essential to understanding the origins and determining the optimum treatments for this disease. The DDR promotes numerous PPI due, in part, to posttranslational modifications (PTMs) of proteins that act to recruit other proteins into complexes that facilitate DNA repair. For instance, in response to DNA damage, specialized kinases ATM, ATR, and DNAPK phosphorylate target proteins that can act as ligands for phospho-epitope binding domains, such as the BRCA1 C-terminal (BRCT) domain [1, 2] and the forkhead-

---

Henry C.-H. Law and Dragana Noe contributed equally to this work

Kishor K. Bhakat and Tapas K. Hazra (eds.), *Base Excision Repair Pathway: Methods and Protocols*, Methods in Molecular Biology, vol. 2701, https://doi.org/10.1007/978-1-0716-3373-1_12,

associated (FHA) domain [3, 4] which are found in a subset of specialized DDR proteins, such as BRCA1, NBN, and CHK2. Additional PTMs, including ubiquitination, can also act as PPI recruitment and repair signals in the DDR [5, 6]. These protein complexes can form at the sites of DNA damage and promote their processing and repair. In addition, emerging evidence from PPI analysis indicates that DDR proteins and domains interact with additional non-DDR proteins to coordinate repair and identifies additional proteins that function in the DDR that may not have been previously recognized [7–9]. Thus, protein–protein interactions can reveal new features of the DDR that may be important to the etiology and treatment of cancer. This protocol describes an immunoprecipitation (IP) procedure that is compatible with mass spectrometry proteomic analysis. The reagents and cleanup procedure in this protocol are optimized to profile the interactome without compromising the sensitivity of the instrument. There are two workflows included in this protocol. The in-gel digestion protocol is suitable for routine interactome analysis. When there is a limited amount of input lysate, or the study's focus is on posttranslational modifications of proteins, the on-bead digestion protocol is recommended.

2 Materials

1. 293FT cells.
2. Mitomycin C.
3. NP-40 (nonyl phenoxypolyethoxylethanol).
4. BCA assay kit.
5. Dynabeads® Antibody Coupling Kit (Thermo Fisher Scientific). The kit contains 2.8 μm magnetic M-270 Epoxy beads and all the reagents for coupling and washing (solutions C1, C2, HB, LB, and SB).
6. Antibody specific for the target protein for coupling with Dynabeads®.
7. Normal IgG antibody (same species as the specific antibody) for coupling with Dynabeads®. This will be used as the negative control.
8. Glacial acetic acid.
9. Ammonium bicarbonate (NH_4HCO_3; AmBic).
10. Tris(2-carboxyethyl)phosphine (TCEP).
11. Dithiothreitol (DTT).
12. Iodoacetamide (IAA).
13. Trifluoroacetic acid (TFA).

14. Sequencing-grade trypsin.
15. 10% Criterion™ XT Precast Bis–Tris gel (Bio-Rad).
16. XT MOPS (3-(N-morpholino)propanesulfonic acid) running buffer (Bio-Rad).
17. XT sample buffer 4× (Bio-Rad).
18. XT reducing agent 20× (Bio-Rad).
19. Coomassie Brilliant Blue G-250 dye.
20. HPLC MS-grade acetonitrile (ACN, 100%).
21. HPLC MS-grade water.
22. HPLC MS-grade trifluoroacetic acid (TFA, 100%).
23. Pierce® C18 tips, 100 μL bed (Thermo Fisher Scientific).

2.1 Stock Solutions

1. 500 mM ethylenediaminetetraacetic acid (EDTA).
2. 1 M Tris–HCl pH 8.0.
3. 1 M HEPES pH 7.4.
4. 5 M sodium chloride (NaCl).
5. 1 M potassium chloride (KCl).
6. 1 M magnesium chloride ($MgCl_2$).
7. 2.5 M calcium chloride ($CaCl_2$).

2.2 Reagents

1. DNA damage response induction and cell lysis.
   1. Phosphate-buffered saline (PBS).
   2. DMEM with 10% FBS, supplemented with penicillin and streptomycin.
   3. Mitomycin C stock solution (0.4 mg/mL; prepare freshly before the experiment).
   4. NETN buffer (0.5% NP-40, 0.5 mM EDTA, 20 mM Tris–HCl pH 8.0, 150 mM NaCl).
2. Dynabeads conjugation and immunoprecipitation.
   1. SWB-A (sample washing buffer A): 10 mM HEPES pH 7.4, 10 mM KCl, 50 mM NaCl, 1 mM $MgCl_2$, 0.05% NP-40.
   2. SWB-B (sample washing buffer B): 10 mM HEPES pH 7.4, 10 mM KCl, 0.07% NP-40.
3. In-gel digestion.
   1. SEB (sample elution buffer): 0.5 M NH_4OH, 0.5 mM EDTA, pH 11.0.
   2. Destain solution: 10% methanol/5% glacial acetic acid/85% water (v/v/v).
   3. Ammonium bicarbonate (AmBic) solutions: 50 mM and 100 mM.

4. Reducing reagent: 20 mM TCEP in 100 mM AmBic (prepare freshly before experiment).
5. Alkylating reagent: 375 mM IAA in 100 mM AmBic (prepare freshly before experiment).
    6. Digestion enzyme: MS-grade trypsin, 20 ng/μL in water (Promega).
    7. Peptide extraction buffer (XB): 50% ACN/49.9% water/0.1% TFA (v/v/v).

4. On-bead digestion.
    1. SWB-C (sample washing buffer C): NETN buffer supplemented with 20 mM Tris–HCl pH 8 and 2 mM $CaCl_2$.
    2. 20 mM Tris–HCl pH 8.0.
    3. 100 mM DTT in 20 mM Tris–HCl pH 8.0.
    4. 100 mM IAA in 20 mM Tris–HCl pH 8.0.
    5. 1 μg/μL trypsin in Milli-Q water.

1.1.2. Sample cleaning prior to LC-MS/MS.
    1. Wetting solution: 50% ACN/50% water (v/v).
    2. Equilibration solution: 0.1% TFA in water.
    3. Rinse solution: 5% ACN/94.9% water/0.1% TFA (v/v/v).
    4. Peptide elution solution: 60% ACN/39.9% water/0.1% TFA (v/v/v).
    5. 0.1% formic acid in water.

2.3 Consumables

1. Cell scraper.
2. 1.5 mL Safe-Lock tubes (Eppendorf).
3. 2.0 mL Safe-Lock tubes (Eppendorf).
4. pH strips (pH 0–14).
5. 10% Criterion™ XT Bis–Tris Protein Gel.
6. Feather™ Disposable Scalpel Set #11 (Electron Microscopy Sciences).
7. Autosampler 11 mm snap-it caps, 6 mm opening, pre-slit (Thermo Fisher Scientific).

2.4 Equipment

1. MagJET Separation Rack, 12 × 1.5 mL tube (Thermo Fisher Scientific).
2. Sample tip-to-end rotator.
3. Analytical balance.
4. Cell culture incubator.
5. Criterion™ Cell and PowerPac™ Basic Power Supply (Bio-Rad).

6. Criterion staining/blotting trays (Bio-Rad).
7. Lab rocking shaker for gel trays.
8. ThermoMixer (Eppendorf).
9. SpeedVac.
10. NanoDrop.

2.5 Nanoflow HPLC and Mass Spectrometry Analysis

1. Mobile phase A: 0.1% formic acid in water.
2. Mobile phase B: 80% ACN/19.9% water/0.1% formic acid.
3. Nanoflow liquid chromatography: UltiMate 3000 UHPLC system (Thermo Fisher Scientific), with a trap C18 LC column (Acclaim PepMap 100, 75 μm × 2 cm, nanoViper, Thermo Fisher Scientific) and an analytical C18 LC column (Acclaim PepMap RSLC, 75 μm × 50 cm, nanoViper, Thermo Fisher Scientific).
4. Mass spectrometer: Orbitrap Fusion Lumos (Thermo Fisher Scientific) or other high-resolution mass spectrometry system.

2.6 Database Search

1. A standalone workstation with the following minimal specification: 2 GHz processor, 2GB RAM, video card and monitor capable of 1280 × 1024 resolution, screen resolution of 96 dpi, 75 GB available on the C drive, NTFS formatted hard disk, Windows 10 operating system with the latest service pack.
2. Proteome Discoverer 2.4.

3 Methods

3.1 Induction of DDR on 293FT Cells

1. Seed the 293FT at 30% confluency and incubate the cells at 37 °C overnight. Assign one plate of 100 mm plate for each of the treatment and antibodies to be used in the experimental design (*see* **Note 1**).
2. Dissolve 2 mg of mitomycin C in 5 mL PBS (*see* **Note 2**).
3. Add the mitomycin C stock solution to the 293FT cells to a final concentration of 10 μg/mL and incubate the cells at 37 °C for 4 h.
4. Rinse the cells with 10 mL of PBS and add 200 μL NETN buffer.
5. Scrap the plate and collect the lysates.
6. Spin the cells at 16,00 g at 4 °C for 3 min and collect the supernatant.
7. Measure the concentration of the protein lysate with BCA assay kit.
8. Proceed to Subheading 3.2 immediately.

3.2 Dynabeads Conjugation and Immunoprecipitation

1. Weigh out 3 mg of Dynabeads® M-270 Epoxy beads in a 1.5-mL microcentrifuge tube, for each planned reaction.
2. Wash the beads with 1 mL of C1 by pipetting up and down. Make sure to pipette gently and slowly up and down, at least ten times.
3. Place the tube on a magnet, wait for 1 min, and carefully remove the supernatant.
4. Pipette 50 μg of the antibody for the protein of interest into the tube. Additionally, pipette the same amount of plain IgG in a different tube, as this will be used as a negative control (*see* **Note 3**).
5. Add the appropriate volume of C1 to each of the tubes from **step 4**, which includes either the specific antibody or the negative control, to a final volume of 150 μL, and add to the beads from **step 3**.
6. Add 150 μL of C2 to the bead solution from **step 5**.
7. Incubate with rotation at 37 °C overnight.
8. Place the tube on a magnet, wait for 1 min, and carefully remove the supernatant.
9. Wash the beads with HB buffer by adding 800 μL of HB to the beads, pipette up and down gently at least ten times, place the beads on magnet for 1 min, and remove the supernatant.
10. Wash the beads with LB buffer by adding 800 μL of HB to the beads, pipette up and down gently at least ten times, place the beads on magnet for 1 min, and remove the supernatant.
11. Wash the beads with SB by adding 800 μL of SB to the beads, pipette up and down gently at least ten times, place the beads on magnet for 1 min, and remove the supernatant.
12. Add 800 μL of SB to the beads and incubate on a rotator for 15 min at room temperature.
13. Place the tube on a magnet, wait for 1 min, and carefully remove the supernatant. The beads are now covalently coupled with the antibody and ready for Co-IP.
14. Add the cell lysate containing 2.5 mg of the total proteins to the antibody-coupled beads from **step 13**.
15. Incubate overnight with rotation, at 4 °C.
16. Wash the beads four times with SWB-A (10 mM HEPES pH 7.4, 10 mM KCl, 50 mM NaCl, 1 mM $MgCl_2$, 0.05% NP-40). Pipette up and down gently at least ten times, place the beads on magnet for 1 min, and remove the supernatant.
17. Wash the beads with SWB-B for 5 min, with rotation (10 mM HEPES pH 7.4, 10 mM KCl, 0.07% NP-40). Place the beads on magnet for 1 min and remove the supernatant.

3.3 Workflow A: In-Gel Digestion

1. Protein elution: Add 500 μL of the SEB (0.5 M NH_4OH, 0.5 mM EDTA, pH 11.0) to the beads and rotate for 10 min.
2. Collect the supernatant and, using a SpeedVac, concentrate the volume to around 50 μL.
3. Check the pH of the concentrated supernatant containing the protein pulldown and neutralize with acetic acid, if necessary.
4. Mix 40 μL of the protein pulldown sample with 10 μL of XT sample buffer (with added XT reducing agent).
5. Incubate for 7 min at 95 °C using a ThermoMixer.
6. Load 45 μL on a 10% Criterion XT Bis–Tris gel and run the electrophoresis.
7. Run the electrophoresis until the dye front moves around 2 inches from its start, and then stop the electrophoresis.
8. Carefully transfer the gel into the staining/blotting tray filled with HPLC-grade water, do a quick wash, and discard the liquid.
9. Cover the gel with around 50 mL of the Coomassie Brilliant Blue G-250 stain solution and shake for 2 h.
10. Destain overnight.
11. Cut out the 2-inch lane into 3 or more bands with a disposable scalpel, and transfer into 1.5 mL Eppendorf tubes. Wash the scalpel with water, then methanol, and then water again, between each band excision.
12. Cut each excised band into around five equal-size gel squares, as this will increase the total gel surface area.
13. Add 100 μL of HPLC-grade water, shake for 10 min, and remove.
14. Add 100 μL of ACN, to shrink gel pieces, incubate for 5–10 min, remove.
15. For every sample mix 10 μL of 20 mM TCEP with 90 μL of 100 mM AmBic. Add 100 μL of TCEP solution to a sample and shake at 37 °C for 30 min on a ThermoMixer.
16. Add 100 μL of ACN and continue shaking for 15 min. If gel pieces are still blue, continue shaking for another 15 min, and remove.
17. Add 100 μL of ACN, incubate for 5 min, and remove.
18. For every sample mix 10 μL of 375 mM IAA with 90 μL of 100 mM AmBic. Add 100 μL of IAA solution to a sample and shake in the dark at room temperature for 20 min, and remove.
19. Add 100 μL of ACN, incubate for 10 min, and remove.
20. Add 15 ng/μL trypsin to cover gel cuts (20–50 μL) and incubate on ice for 30 min, and remove excess trypsin.

21. Add 100 μL of 50 mM AmBic to gel cuts (it should cover it completely) and incubate overnight at 37 °C. Check after 15 min if AmBic still covers the gel pieces; if not, add more AmBic.
22. Prepare fresh XB (50% ACN/49.9% water/0.1% TFA).
23. Label and clean new tubes with 100 μL of XB. These tubes will collect the peptide extracts that were incubated overnight in the tubes from **step 20**.
24. Transfer peptide extracts to newly labeled tubes. Make sure not to transfer any gel piece to a new tube.
25. Add 50 μL of XB to gel cuts, and incubate with shaking on a ThermoMixer, for 15 min at room temperature.
26. Transfer peptide extracts to corresponding tubes.
27. Repeat **steps 24–25**.
28. Dry samples down using SpeedVac until it looks like a white pellet.

3.4 Workflow B: On-Bead Digestion

1. Using the beads from Subheading 3.2 **step 17**, wash the beads twice with SWB-C (NETN buffer supplemented with 20 mM Tris–HCl pH 8 and 2 mM $CaCl_2$).
2. Resuspend the beads in 9 μL 20 mM Tris–HCl pH 8.0.
3. Add 0.4 μL of 100 mM DTT and incubate at room temperature for 30 min at 500 rpm.
4. Add 0.6 μL of 100 mM IAA at room temperature for 10 min at 500 rpm.
5. Add 500 ng of trypsin and incubate on a tip-to-end rotator at 37 °C overnight.
6. Pellet the beads with the magnetic stand and collect the supernatant.
7. Add another 500 ng of trypsin in 20 mM Tris–HCl pH 8 to the beads and incubate at 37 °C for 4 h.
8. Pool the supernatant from **steps 5** and 7 and add 50% formic acid to a final concentration of 2%.
9. Proceed to Subheading 3.5.

3.5 Sample Cleaning Prior to LC-MS/MS

1. Add 100 μL of the equilibration solution (0.1% TFA) to the peptide pellet.
2. Pipette up and down, vigorously, until the pellet is completely dissolved.
3. Wet the C18 tip by aspirating and discarding 100 μL of the wetting solution. Repeat.

4. Equilibrate the C18 tip by aspirating and discarding 100 μL of the equilibrating solution. Repeat.
5. Slowly aspirate 100 μL of the peptide sample into the C18 tip. Dispense and aspirate the sample around ten times, to increase the binding efficiency.
6. Rinse the C18 tip by aspirating and discarding 100 μL of the rinse solution. Repeat.
7. Label a new set of 1.5 mL Eppendorf tubes.
8. Slowly aspirate 50 μL of the peptide elution solution and dispense into the new tube.
9. Dry samples down using SpeedVac.
10. Resuspend the peptide pellet (by either vortexing or vigorous pipetting up and down) with 0.1% formic acid (same as our mobile phase A for HPLC system) so the final concentration of the peptides is around 500 ng/μL (measure with NanoDrop).
11. Centrifuge for 2 min at 12,000 × *g*, at 4 °C.
12. Transfer the supernatant to an autosampler vial for nano LC-MS/MS.

3.6 Nanoflow HPLC and Mass Spectrometry Analysis

1. Around 1 μg of the peptides is analyzed by UltiMate 3000 UHPLC system equipped with 50-cm-long C18 analytical LC column, directly connected to Orbitrap Fusion Lumos mass spectrometer (Thermo Fischer Scientific). After injection, peptides are loaded onto a trap C18 column and washed with mobile phase A, at a flow rate of 4 μL/min. Next, peptides are separated on the analytical C18 column, using a step gradient of 4–25% mobile phase B from 10 to 100 min and 25–45% mobile phase B from 100 to 130 min, at the flow rate of 300 nL/min and 50 °C.
2. Eluted peptides are then analyzed by a high-resolution mass spectrometer, in a data-dependent acquisition mode (DDA). A survey full-scan MS is set for acquiring in the Orbitrap mode, for a scan range from 350 to 1800 *m/z* and a resolution of 120,000. The automatic gain control (AGC) target for MS1 is set to 4×10^5, with the ion filling time of 100 ms. The most intense ions with charge states 2–6 are selected to be isolated in 3-s cycles, fragmented using HCD fragmentation with 35% normalized collision energy, and detected at a mass resolution of 30,000 at 200 *m/z*. The AGC target for MS2 is set to 5×10^4, with the ion filling time of 60 ms. Dynamic exclusion is set for 30 s with a 10-ppm mass window.

3.7 Database Search

1. Transfer the .raw files from the data acquisition workstation to the database searching workstation.

2. Open Proteome Discover 2.4 and create the study with the specified file location (*see* **Note 4**).
3. Download, install, and activate the protein sequence database of the species corresponding to the sample from the Uniprot website (usually human, mouse, or rat; https://www.uniprot.org/).
4. Select the database search method required. If quantitative analysis is required, specify MS1 quantitation in the method.
5. Specify the database to be searched against, as prepared in **step 3**.
6. If posttranslational modification mapping is required, add the ptmRS node to the processing workflow.
7. In the Sequest node of the processing workflow, set the search the parameters as follows: MS1 precursor mass tolerance, 10 ppm; MS2 fragment ion mass tolerance, 20 mmu; fragmentation method, HCD; and ion series, *b*-, *y*- ions.
8. To look for posttranslational modification on the bait protein, specify the required modifications (e.g., phosphorylation) and the residue (e.g., S, T, and Y) in the data search parameter.
9. Submit the job with the specified search parameters. The search is usually completed within 1–4 h.
10. Proteins with a "high" FDR confidence and two unique mapped peptides are considered as a positive identification from the sample.
11. Export the protein and peptide list to evaluate the biological significance in DNA damage response (*see* **Note 5**).

4 Notes

1. *Experimental*: Setting up a proper control is essential to accurately determine the interactome. If the control serves to identify adventitious proteins that could represent false-positive interactions, an isotype antibody (usually normal mouse or rabbit IgG) should be used as control. Unconjugated beads can also be added as a control. It is important to run the controls in at least triplicate samples to build an adequate library of potential false-positive interactions. Because the expression and/or affinity of adventitious proteins could change between experimental conditions, such as in the presence or absence of DNA damage, appropriate controls should be used for each experimental condition tested.
2. *DNA damage induction*: Mitomycin C stocks can be stored in the dark at 4 °C for no more than 2 weeks. It is best if the mitomycin C stock solution is freshly prepared. Other methods such as ionizing radiation, ultraviolet, and hydroxyurea can also

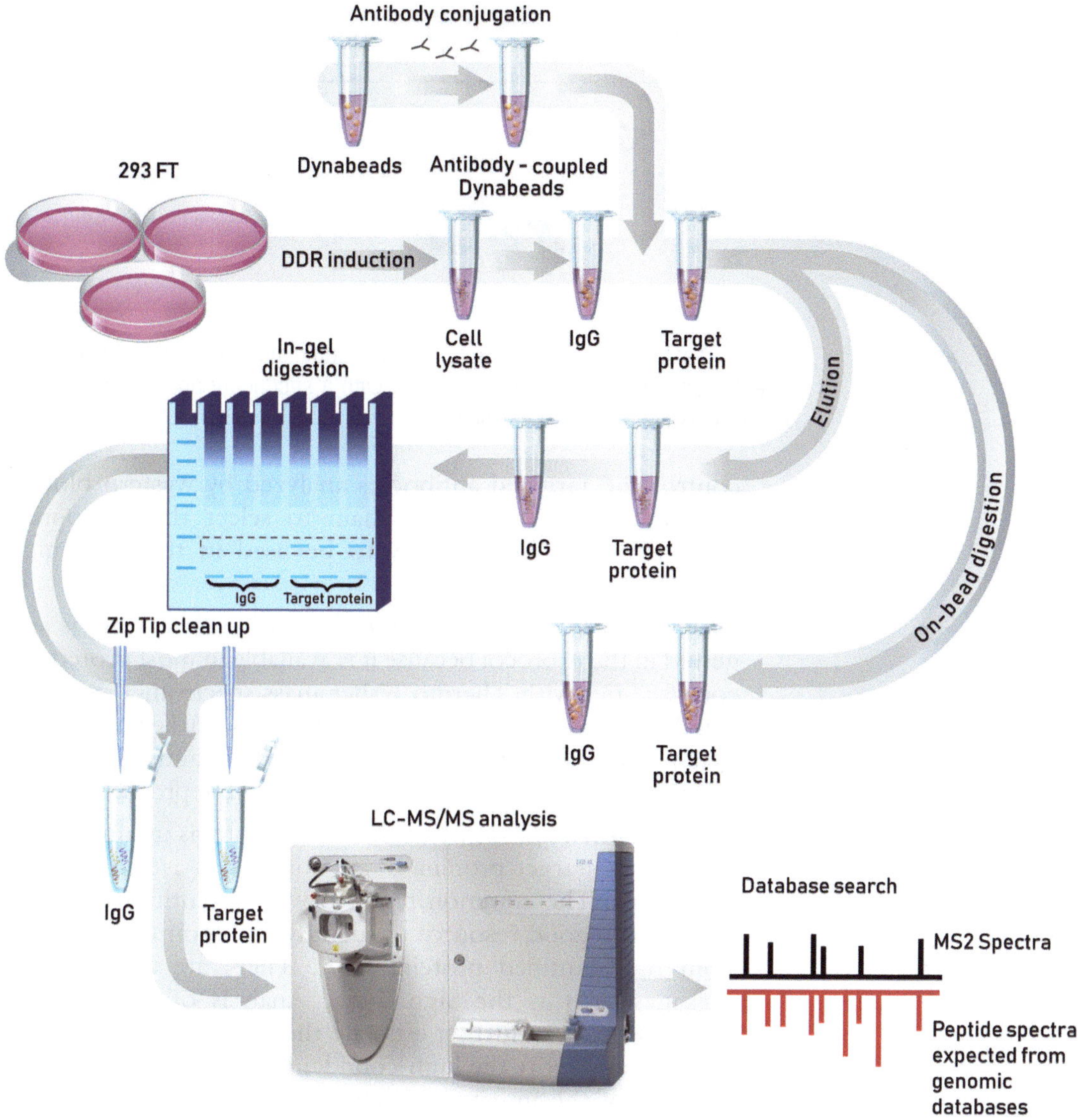

Fig. 1 Overview of the interactome profiling workflow

be used to introduce DNA damage response to the cells. Different types of DNA damage can induce specific signaling networks and PPI, and some PPI can be early, mid-, or late events in the repair process. Therefore, the DNA damage conditions used for each study need to be carefully determined and optimized by the end user (Fig. 1).

3. *Titration of the control antibody*: In our experience, the concentration of antibodies often deviate from that reported by the vendors. In many instances, the normal IgG antibodies used as negative controls contain much more antibody than those targeting specific proteins. It is preferable that the amount of antibody used as a negative control is comparable to that of the targeted antibody. We routinely perform a serial dilution of the

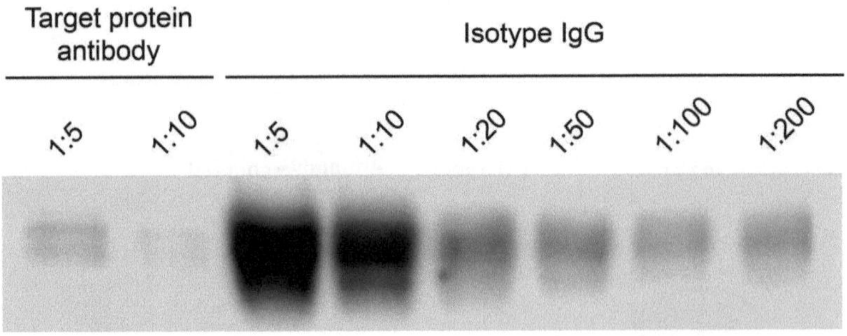

Fig. 2 Western blot image of control antibody titration. The antibodies are directly loaded onto the gel and probed with a secondary antibody to evaluate the abundance of the heavy chain

control and targeted antibodies analyzed by western blot for the intensity of the heavy chain to select the appropriate amount of control antibody to be used in the experiment (Fig. 2).

4. *Choice of database search engine*: Proteome Discoverer is suggested in the protocol because it is available in most proteomics core facilities with Thermo Fisher mass spectrometers. Other search engines, such as Mascot, Peptide Shaker, and MaxQuant, can be used as substitutes. The MS and the MS/MS tolerance will be set the same as shown in the protocol.
5. *Determination of the interactome*: The proteins uniquely identified in the target protein IP are considered as the potential interactors. The common repository of adventitious proteins (cRAP) is a good resource to identify common contaminants from the identified proteins [10]. There are also software packages such as the Significance Analysis of INTeractome (SAINT) that can further estimate the confidence of the interaction and identify the high-confidence interactors [11]. Visualization of the interaction networks and additional pathway enrichment analyses and differential networks dependent upon DNA damage stimulus can be achieved with Cytoscape and the suite of applications available (cytoscape.org).

References

1. Mesquita RD, Woods NT, Seabra-Junior ES, Monteiro AN (2010) Tandem BRCT domains: DNA's praetorian guard. Genes Cancer 1: 1140–1146. https://doi.org/10.1177/1947601910392988
2. Gerloff DL, Woods NT, Farago AA, Monteiro AN (2012) BRCT domains: a little more than kin, and less than kind. FEBS Lett 586:2711–2716. https://doi.org/10.1016/j.febslet.2012.05.005
3. Li J et al (2002) Structural and functional versatility of the FHA domain in DNA-damage signaling by the tumor suppressor kinase Chk2. Mol Cell 9:1045–1054. https://doi.org/10.1016/s1097-2765(02)00527-0

4. Durocher D, Jackson SP (2002) The FHA domain. FEBS Lett 513:58–66. https://doi.org/10.1016/s0014-5793(01)03294-x
5. Sharp MF, Bythell-Douglas R, Deans AJ, Crismani W (2021) The Fanconi anemia ubiquitin E3 ligase complex as an anti-cancer target. Mol Cell 81:2278–2289. https://doi.org/10.1016/j.molcel.2021.04.023
6. Tarsounas M, Sung P (2020) The antitumorigenic roles of BRCA1-BARD1 in DNA repair and replication. Nat Rev Mol Cell Biol 21:284–299. https://doi.org/10.1038/s41580-020-0218-z
7. Woods NT et al (2012) Charting the landscape of tandem BRCT domain-mediated protein interactions. Sci Signal 5:rs6. https://doi.org/10.1126/scisignal.2002255
8. Hu WF et al (2019) CTDP1 regulates breast cancer survival and DNA repair through BRCT-specific interactions with FANCI. Cell Death Discov 5:105. https://doi.org/10.1038/s41420-019-0185-3
9. Lagundzin D et al (2019) Delineating the role of FANCA in glucose-stimulated insulin secretion in beta cells through its protein interactome. PLoS One 14:e0220568. https://doi.org/10.1371/journal.pone.0220568
10. Mellacheruvu D et al (2013) The CRAPome: a contaminant repository for affinity purification-mass spectrometry data. Nat Methods 10:730–736. https://doi.org/10.1038/nmeth.2557
11. Choi H et al (2012) Analyzing protein-protein interactions from affinity purification-mass spectrometry data with SAINT. *Curr Protoc Bioinformatics* 39:8–15. https://doi.org/10.1002/0471250953.bi0815s39

Chapter 13

Using Affinity Pulldown Assays to Study Protein–Protein Interactions of Human NEIL1 Glycosylase and the Checkpoint Protein RAD9–RAD1–HUS1 (9-1-1) Complex

Drew T. McDonald, Pam S. Wang, Jennifer Moitoza Johnson, and Miaw-Sheue Tsai

Abstract

Affinity pulldown is a powerful technique to discover novel interaction partners and verify a predicted physical association between two or more proteins. Pulldown assays capture a target protein fused with an affinity tag and analyze the complexed proteins. Here, we detail methods of pulldown assays for two high-affinity peptide fusion tags, Flag tag (DYKDDDDK) and hexahistidine tag (6xHis), to study protein–protein interactions of human NEIL1 glycosylase and the checkpoint protein complex RAD9–RAD1–HUS1 (9-1-1). We uncover unique interactions between 9-1-1 and NEIL1, which suggest a possible inhibitory role of the disordered, phosphorylated C-terminal region of RAD9 in regulating NEIL1 activity in base excision repair through lack of physical association of 9-1-1 and NEIL1.

Key words Protein-protein interactions, Affinity pulldown, Flag pulldown, Ni pulldown, Human NEIL1, Human RAD9-RAD1-HUS1

1 Introduction

Approximately tens of thousands of oxidative/alkylated base lesions occur daily in the human genome. Most oxidative damage is efficiently repaired by base excision repair (BER) and single-stranded break repair (SSBR) to prevent mutations and maintain the genomic integrity. DNA repair pathways, including BER, are complicated by sub-pathways and alternative pathways, as well as their cross talks with other DNA processes such as replication and transcription. These DNA transactions involve multiprotein complexes and dynamic macromolecular assemblies of both canonical proteins and noncanonical accessory factors [1]. These megadalton assemblies are highly coordinated and regulated in a temporal and

Kishor K. Bhakat and Tapas K. Hazra (eds.), *Base Excision Repair Pathway: Methods and Protocols*, Methods in Molecular Biology, vol. 2701, https://doi.org/10.1007/978-1-0716-3373-1_13,

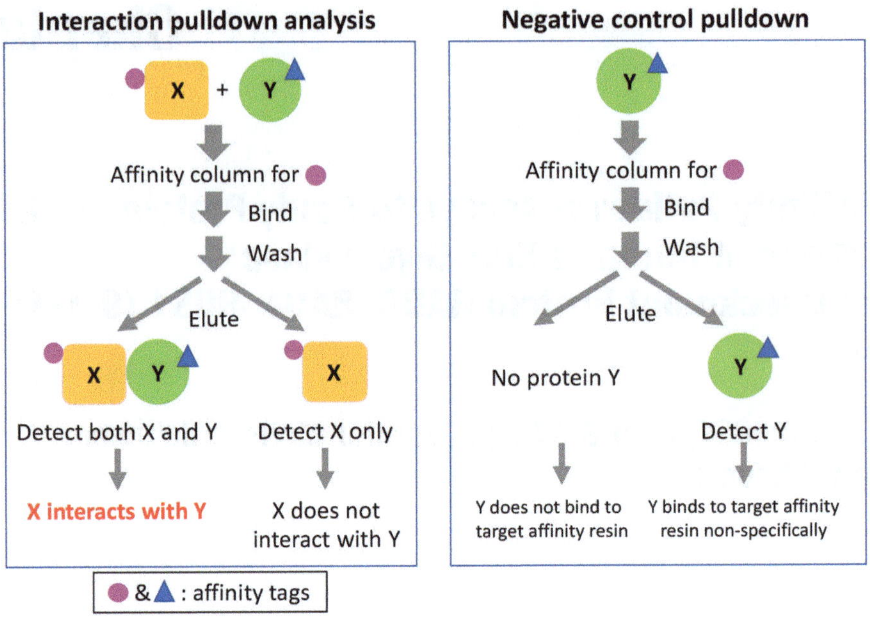

Fig. 1 Schematic of affinity pulldown for direct protein–protein interactions. In the left panel, proteins of interest X and Y are incubated together and bound to an affinity column specific to the affinity tag on protein X (e.g., the bait protein). If both proteins X and Y are detected in the eluted fractions after affinity pulldown, protein Y interacts with protein X. If only protein X is present in the eluted fractions, the two proteins either do not physically interact with each other or the interaction of the two proteins is not stable to be detected by this method. The right panel depicts a negative control pulldown that should always be performed to ensure the prey protein (e.g., protein Y) does not bind to the affinity column nonspecifically to prevent any false-positive interaction with the bait protein (e.g., protein X)

spatial manner, often driven by specific protein–protein interactions. To gain insights on how DNA repair machines work, it is necessary to understand the architecture and functional interactions of proteins in these machines.

Recent development in cutting-edge techniques with increased throughput has enabled qualitative and quantitative analysis of protein–protein interactions and established protein interactome networks. The emerging findings help define the role of the interactions in complex biological processes, shape the molecular basis for human diseases that may arise from dysregulated protein–protein interactions, and facilitate the mechanistic understanding and potential therapeutic discovery to human diseases. An array of genetic, cell biological, biochemical, and biophysical methodologies are available to validate, characterize, and confirm protein–protein interactions [2]. Among those methods, affinity pulldown or affinity co-purification (Fig. 1) is a classical, low-to-medium throughput, but powerful technique to identify direct protein–protein interactions and functional associations of proteins by providing a qualitative analysis for both stable and transient protein complexes. Specifically, affinity pulldown are (1) robust to set up in vitro by mixing purified proteins of interest or using cell extracts containing co-expressed proteins from recombinant hosts, such as *E. coli* or insect cells; (2) readily adaptable for multi-protein interaction studies, in addition to binary interactions; (3) useful for mapping interaction domains and screening for interaction-defective mutations; (4) easy to adjust assay stringency to increase specificity or detect weak interactions; and (5) affordable by using common reagents and inexpensive laboratory equipment. To increase throughput of affinity pulldown, spin columns, magnetic beads, and magnetic microplates are applicable options.

The checkpoint sliding clamp RAD9–RAD1–HUS1 (9-1-1) interacts with and stimulates enzymes involved in base excision repair (BER), such as NEIL1, MYH, TDG, FEN-1, and DNA LIGASE I, thus linking BER activities to checkpoint coordination [3, 4]. Here we describe a general method of pulldown assays to confirm physical associations between NEIL1 glycosylase and the heterotrimeric 9-1-1 complex. We found that NEIL1 interacts with full-length 9-1-1 and a truncated 9-1-1 (9Δ-1-1) that lacks a C-terminal tail of RAD9, when 9-1-1 proteins were expressed and purified from *E. coli* (Fig. 2a, b), consistent with previous reports. When 9-1-1 was expressed and purified from insect cells, surprisingly, NEIL1 does not interact with full-length 9-1-1 (Fig. 2c) while retaining interaction with truncated 9Δ-1-1 (Fig. 2d). Although the C-terminal region of RAD9 is dispensable for interacting with NEIL1, the absence of the C-terminal RAD9 leads to hyperactivation of NEIL1 in vivo [5]. Our finding suggests that the disordered and phosphorylated C-terminal region of RAD9 might regulate NEIL1 activity in human cells by destabilizing or inhibiting interaction of 9-1-1 with NEIL1.

2 Materials

Prepare all reagents using deionized ultrapure water and molecular biology grade reagents. Store all reagents at 4 °C, unless indicated otherwise. Filter sterilize buffers and reagents for long-term usage. Follow all waste disposal and safety regulations when disposing waste materials.

2.1 Purified Proteins

Avoid repeated freezing and thawing of purified proteins. All purified proteins are stored in the presence of 5–10% glycerol in small aliquots at −80 °C. Proteins tagged with an affinity tag are highlighted in bold text.

1. Untagged human NEIL1: Purified from *E. coli* as described in [6].
2. Flag-tagged full-length human 9-1-1 (**Flag-RAD9**–RAD1–HUS1): Expressed and purified from insect cells as described in [7, 8].

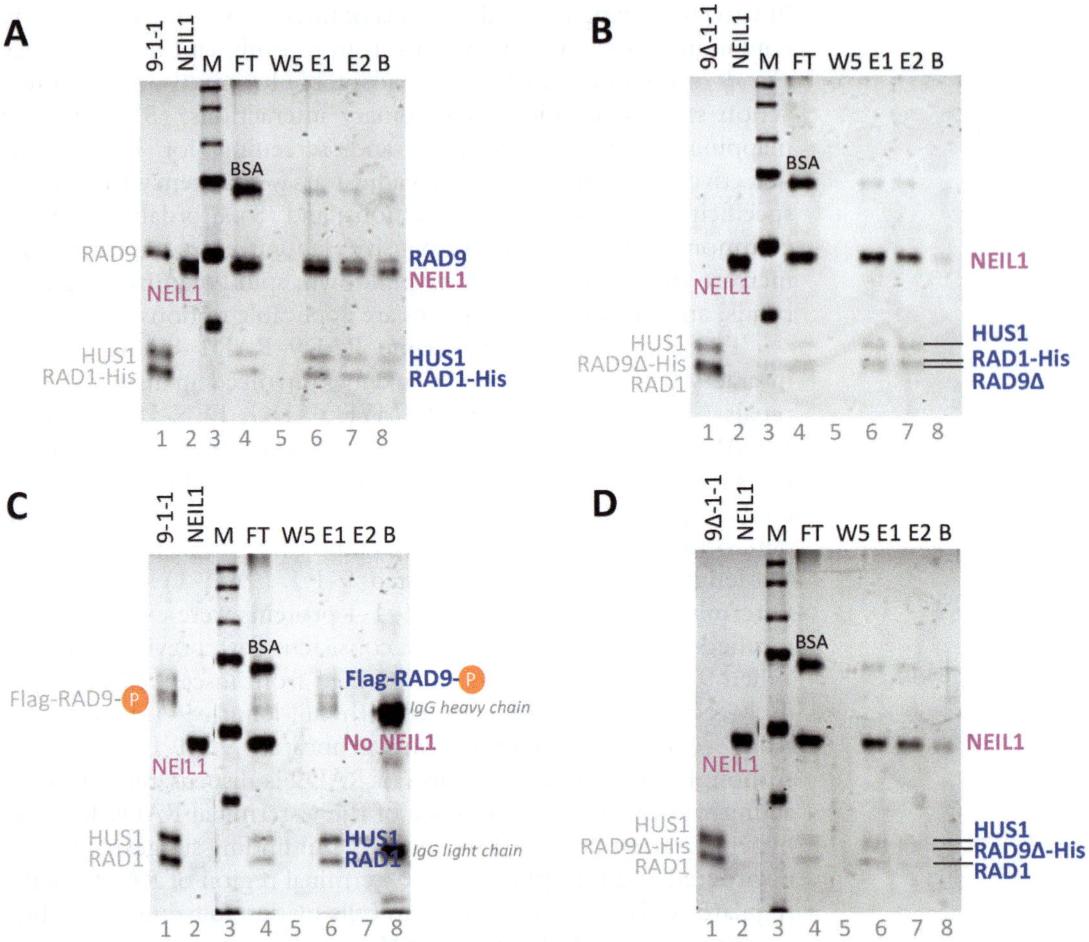

Fig. 2 Physical interaction studies of NEIL1 glycosylase and the 9-1-1 sliding clamp by affinity pulldown. Selected affinity resins, either anti-Flag M2 resin or Ni-NTA resin, were used to pull down Flag- or 6xHis-tagged 9-1-1 proteins and detect if NEIL1 was co-purified with 9-1-1. NEIL1 was pulled down by Ni-NTA resin with 6xHis-tagged (**a**) full-length 9-1-1 and (**b**) truncated 9Δ-1-1, when the 9-1-1 proteins were purified from *E. coli*. NEIL1 was not pulled down with full-length, Flag-tagged 9-1-1 by anti-Flag M2 resin, when the 9-1-1 complex was purified from insect cells (**c**), but was co-purified by Ni-NTA resin with 6xHis-tagged truncated 9Δ-1-1 that was purified from insect cells (**d**). The affinity pulldown fractions were analyzed by SDS-PAGE with Coomassie staining, as follows: in lane 1, 9-1-1 or 9Δ-1-1: protein load for full-length or truncated 9-1-1; lane 2, NEIL1, protein load for NEIL1; lane 3, M, protein molecular weight markers, from top to bottom, including 250-, 150-, 100-, 75-, 50-, and 37-kDa standard markers; lane 4, FT, flow-through fractions; lane 5, W5, last wash fractions; lane 6, E1, eluted fraction 1; lane 7, E2, eluted fraction 2; and lane 8, B, leftover resin fractions. RAD9, RAD1, HUS1, and NEIL1 with or without an affinity tag are indicated on the Coomassie-stained polyacrylamide gels

3. 6xHis-tagged truncated human 9Δ-1-1 lacking the C-terminal region of RAD9 (**RAD9ΔC-6xHis**–RAD1–HUS1): Expressed and purified from insect cells as described in [7, 8].
4. Human 6xHis-tagged full-length 9-1-1 (RAD9–**RAD1-6xHis**–HUS1) and truncated 9Δ-1-1 complex (RAD9ΔC–**RAD1-6xHis**–HUS1): Expressed and purified from *E. coli* as described in [9].

2.2 Buffers and Affinity Chromatography Resins

1. Column binding buffer: 50 mM Tris–HCl, pH 7.5, 300 mM NaCl, 10% glycerol, 50 μg/mL BSA, and 1 mM DTT.
2. Column washing buffer: 50 mM Tris–HCl, pH 7.5, 300 mM NaCl, 10% glycerol, and 1 mM DTT.
3. Ni elution buffer: 300 mM imidazole in the column washing buffer.
4. Flag elution buffer: 200 μg/mL 3× Flag peptide in the column washing buffer.
5. Glycine buffer: 100 mM glycine, pH 3.5.
6. Anti-Flag M2 resin (Sigma Millipore).
7. Ni-NTA Fast Flow resin (Qiagen).
8. 5× lane marker reducing sample buffer: 0.3 M Tris–HCl, 5% SDS, 50% glycerol, 100 mM DTT, and proprietary pink tracking dye (Thermo Fisher).
9. 1× Tris–glycine SDS running buffer for protein gel electrophoresis: 25 mM Tris, 192 mM glycine, and 0.1% SDS, pH 8.3.
10. 12% Tris–glycine SDS polyacrylamide gels.
11. SimplyBlue SafeStain solution (Thermo Fisher).

2.3 Equipment

1. Rotator.
2. High-speed microcentrifuge.
3. Heat block.
4. Protein electrophoresis apparatus.

3 Methods

Carry out all steps of affinity pulldown at 4 °C, unless otherwise specified.

3.1 Column Equilibration

3.1.1 Column Equilibration for Anti-Flag M2 Column

1. Gently mix and fully resuspend the anti-Flag M2 resin. Pipette 25 μL column volume (CV) of anti-Flag M2 resin into a 1.5-mL microtube (*see* **Note 1**).
2. Add 1 mL of column binding buffer to the microtube to wash the resin and get rid of 50% glycerol in the resin storage medium. Use low-speed centrifugation at 500 × *g* for 1 min to collect the resin by packing the resin at the bottom of the tube. Remove and discard the supernatant using a micropipette (*see* **Note 2**).
3. Wash one more time with column binding buffer as described in **step 2.**

4. Add 1 mL of glycine buffer to the microtube to resuspend and wash the resin. Repeat glycine buffer wash two more times (*see* **Note 3**).
5. Then wash three times with 1 mL of column binding buffer to equilibrate the resin.

3.1.2 Column Equilibration for Ni-NTA Column

1. Gently swirl the Ni-NTA Fast Flow bottle and thoroughly resuspend the resin before use. Pipette 25 μL column volume (CV) of Ni-NTA Fast Flow resin into a 1.5 mL microtube.
2. Wash the Ni-NTA resin two times with 1 mL of distilled water to remove 20% ethanol in the resin storage medium.
3. Wash the resin three times with 1 mL of column binding buffer to equilibrate the resin.

3.2 Column Batch Binding

1. Thaw aliquots of purified proteins on ice. Sample protein load fractions (L) for each protein and save for SDS-PAGE analysis.
2. Mix purified full-length 9-1-1 or 9Δ-1-1 proteins (~10 μg) with purified NEIL1 (~12 μg) at a 1:3 molar ratio in column binding buffer to final 200 μL. Incubate mixed proteins on ice for 1 h and gently tap the tube every 15 min to mix.
3. Transfer mixed proteins into the microtube containing the equilibrated column resin and tap the tube to resuspend the resin (*see* **Note 4**). Incubate for 1 h while mixing gently on a rotator at 4 °C (*see* **Note 5**).
4. Centrifuge at 4 °C for 1 min at 500 × *g* to collect the resin. Remove and save the supernatant as column flow-through fractions (FT) for SDS-PAGE analysis.

3.3 Column Batch Washing

1. Add 0.5 mL of column wash buffer to the microtube to resuspend and wash the resin. Pack the resin by centrifugation and discard the supernatant (wash buffer).
2. Wash the resin four more times as in **step 1**, and sample the last wash (W5) and save for SDS-PAGE analysis (*see* **Note 6**).

3.4 Column Batch Elution

1. Add 1 column volume (CV) of elution buffer to the resin, and incubate on ice for 15 min for anti-Flag M2 resin and 5 min for Ni-NTA resin. Pack the resin by centrifugation and collect the supernatant as elution 1, E1.
2. Repeat **step 1** one more time and collect the supernatant as elution 2, E2 (*see* **Note 7**).

3.5 SDS-PAGE Analysis

1. Sample 20 μL from the affinity pulldown fractions as described in Subheadings 3.2, 3.3, and 3.4, including load (L), flow-through (FT), last wash (W5), and eluted fractions 1 and 2 (E1 and E2). Add 5 μL of 5× lane marker reducing sample buffer to each sample. Add 1 CV of 5× lane marker reducing sample buffer to the leftover column beads (B) (*see* **Note 8**). Denature all samples for 5 min at 95 °C.

2. Load 20 μL of each denatured sample to 12% Tris–glycine SDS polyacrylamide gels and electrophorese in 1× Tris–glycine SDS running buffer at 200 V for 1 h and 15 min.
3. After electrophoresis, rinse polyacrylamide gels three times for 5 min with 100 mL of distilled water with gentle shaking. Add sufficient SimplyBlue SafeStain solution (~20 mL) to cover the polyacrylamide gel, and then incubate at room temperature for 1 h with gentle shaking (*see* **Note 9**).
4. Discard the staining solution. Wash with 100 mL of distilled water for 1–3 h to destain the polyacrylamide gel. Perform more washes until the background of gels is clear.

4 Notes

1. Cutting the end of the micropipette tip is recommended if pipette tip opening is too narrow. Ni-NTA and anti-Flag M2-Flag resins are stored as 50% slurry in the storage medium. Pipette double the desired column volume of the resin slurry to get the working column volume of resin. For example, pipette 50 μL of resin slurry to have the column volume of 25 μL resin for experiments. Equilibration may be done at room temperature.
2. When removing buffer from the resin after centrifugation, use a micropipette to carefully remove the supernatant and leave a small volume of buffer (15–20 μL) above the resin to minimize loss of the resin. Add buffer with force to the microtube to help dislodge and resuspend the packed resin. Then gently invert the tube several times to wash the resin. Remove the last wash completely from the resin without leaving any residual buffer.
3. Glycine washes remove protein contaminants from antibody-conjugated affinity resins, such as anti-Flag M2 resin. Skip this step if the affinity resin is not antibody-based. The anti-Flag M2 resin can be regenerated and reused by glycine washes as well. Make sure that there is no residual glycine buffer in the final wash, because leftover glycine buffer may denature proteins due to its low pH.
4. Always include a negative control pulldown to ensure that the complexed protein does not bind nonspecifically to the targeting affinity resin that may result in false-positive interactions. When feasible, perform a reciprocal pulldown using an affinity resin targeting the other protein(s) to confirm interactions.
5. Batch binding is carried out with constant mixing of the resin and proteins in the microtube. Fasten the tube to a rotator and

mix gently in a 4 °C cold room. Make sure that the resin and proteins are fully mixed on the rotator before beginning the 1-h binding.

6. Remove the final wash buffer from the resin completely before adding the elution buffer. Leftover wash buffer dilutes the elution buffer and results in inefficient elution. Tap the tube to mix every 5 min during each elution step to resuspend the resin to ensure complete elution.
7. Most published pulldown assays do not elute proteins from the affinity resin, but interpret proteins bound or not bound on the resin as positive or negative interactions. We routinely carry out two runs of elution to ensure complexed proteins, when expressed and purified recombinantly, are properly folded, soluble, non-aggregated proteins to minimize false-positive interactions. In addition, the ability of eluting complexed proteins from the affinity resin warrants subsequent biochemical, physical, and structural analysis.
8. The 5x lane marker reducing sample buffer contains DTT as a reducing agent and is stored at −20 °C for long-term stability, but SDS in the sample buffer tends to precipitate at low temperatures. Warm up the 5× sample buffer at 37 °C for 5–10 min to dissolve SDS before use. Do not heat the sample buffer for more than 15 min, because DTT may get oxidized.
9. A water-based, nonhazardous SimplyBlue SafeStain is used to visualize protein bands on polyacrylamide gels. It is important to have thorough water rinse to remove SDS and buffer salt; otherwise, they interfere with binding of the Coomassie dye to proteins and reduce staining sensitivity. Stain polyacrylamide gels in SimplyBlue SafeStain for up to 3 h, but sensitivity decreases after 3 h. Add 2 mL of 20% NaCl (w/v) for every 20 mL of stain, when leaving polyacrylamide gels overnight in the stain.

Acknowledgements

We thank Dr. Priscilla K. Cooper (Lawrence Berkeley National Laboratory, Berkeley, California) for her scientific guidance and the Expression and Molecular Biology Core staff Khanh K. Nguyen, Aye Su Hlaing, Phyo S. Aung, and Ahmed M. Akbar for their technical assistance in preparing both full-length and C-terminally truncated human 9-1-1 trimeric proteins from bacteria and insect cells. We thank Dr. Vinay Kumar Singh from the Zongchao Jia's lab (Queen's University, Kinston, Ontario, Canada) for providing the *E. coli* tricistronic expression vector for full-length human 9-1-1 and Dr. Vladimir P. Bermudez from the Hurwitz lab

(Memorial Sloan-Kettering Cancer Institute, New York, New York) for their generous gifts of the baculovirus transfer vectors, baculovirus stocks, and in-house rabbit polyclonal antibodies for RAD9, RAD1, and HUS1 and Drs. Muralidhar L. Hegde and Sankar Mitra (Houston Methodist Cancer Center, Houston Methodist Research Institute, Huston, Texas) for providing purified NEIL1 protein. This work was supported by the National Cancer Institute/ National Institutes of Health (NCI/NIH) program project grant P01 CA092584 for the Structural Cell Biology of DNA Repair Machines (SBDR) to all authors DTM, PSW, JMJ, and MST.

References

1. Dutta A, Yang C, Sengupta S, Mitra S, Hegde ML (2015) New paradigms in the repair of oxidative damage in human genome: mechanisms ensuring repair of mutagenic base lesions during replication and involvement of accessory proteins. Cell Mol Life Sci 72:1679–1698
2. Meyerkord CL, Fu H (eds) (2015) Protein-proteininteractions. In Methods and applications. Methods in molecular biology
3. Guan X, Bai H, Shi G, Theriot CA, Hazra TK, Mitra S et al (2007) The human checkpoint sensor Rad9-Rad1-Hus1 interacts with and stimulates NEIL1 glycosylase. Nucleic Acids Res 35:2463–2472
4. Balakrishnan L, Brandt PD, Lindsey-Boltz LA, Sancar A, Bambara RA (2009) Long patch base excision repair proceeds via coordinated stimulation of the multienzyme DNA repair complex. J Biol Chem 284:15158–15172
5. Panigrahi SK, Hopkins KM, Lieberman HB (2015) Regulation of NEIL1 protein abundance by RAD9 is important for efficient base excision repair. Nucleic Acids Res 43:4531–4546
6. Hazra TK, Mitra S (2006) Purification and characterization of NEIL1 and NEIL2, members of a distinct family of mammalian DNA glycosylases for repair of oxidized bases. Methods Enzymol 408:33–48
7. Bermudez VP, Lindsey-Boltz LA, Cesare AJ, Maniwa Y, Griffith JD, Hurwitz J et al (2003) Loading of the human 9-1-1 checkpoint complex onto DNA by the checkpoint clamp loader hRad17-replication factor C complex in vitro. Proc Natl Acad Sci U S A 100:1633–1638
8. Querol-Audi J, Yan C, Xu X, Tsutakawa SE, Tsai MS, Tainer JA et al (2012) Repair complexes of FEN1 endonuclease, DNA, and Rad9-Hus1-Rad1 are distinguished from their PCNA counterparts by functionally important stability. Proc Natl Acad Sci U S A 109:8528–8533
9. Singh VK, Nurmohamed S, Davey SK, Jia Z (2007) Tri-cistronic cloning, overexpression and purification of human Rad9, Rad1, Hus1 protein complex. Protein Expr Purif 54:204–211

Chapter 14

Tandem Affinity Purification and Mass-Spectrometric Analysis of FACT and Associated Proteins

Amala Kaja, Priyanka Barman, Shalini Guha, and Sukesh R. Bhaumik

Abstract

Isolation of a protein/complex is important for its biochemical and structural characterization with mechanistic insights. TAP (tandem affinity purification) strategy allows rapid isolation of cellular proteins/complexes with a high level of purity. This methodology involves an immuno-affinity-based purification followed by a conformation-based isolation to obtain a highly homogeneous protein/complex. Here, we describe the TAP-mediated isolation of endogenous FACT (facilitates chromatin transcription; a heterodimer), an essential histone chaperone associated with BER (base excision repair). However, it is not clearly understood how FACT regulates BER. Such knowledge would advance our understanding of BER with implications in disease pathogenesis, since BER is an evolutionarily conserved process that is linked to various diseases including ageing, neurodegenerative disorders, and cancers. Using isolated FACT by TAP methodology, one can study the mechanisms of action of FACT in BER. Further, isolated FACT can be used for studies in other DNA transactions such as transcription and replication, as FACT is involved in these processes. Furthermore, TAP-mediated isolation strategy can be combined with mass spectrometry to identify the protein interaction partners of FACT.

Key words TAP, FACT, Pob3

1 Introduction

Proteins play essential roles in cellular processes via their interactions with other molecules including proteins [1]. Thus, the study of the protein interactions and their biochemical and structural characterization with mechanistic insights becomes crucial for delineating their roles in the maintenance of cellular health. Such studies require purified proteins/complexes. A tandem affinity purification (TAP) approach allows a rapid two-step affinity-based isolation of proteins/complexes with a high degree of purity [2]. In this methodology, the protein of interest is tagged with calmodulin-binding protein (CBP) and protein A that flank a tobacco etch virus (TEV) cleavage site. Cells expressing such TAP-tagged protein of interest are grown, harvested, and lysed to

Kishor K. Bhakat and Tapas K. Hazra (eds.), *Base Excision Repair Pathway: Methods and Protocols*, Methods in Molecular Biology, vol. 2701, https://doi.org/10.1007/978-1-0716-3373-1_14,

obtain the whole cell extract (WCE), which is incubated with matrix-bound IgG (immunoglobulin G) to retain the TAP-tagged protein of interest via the interaction of protein A with IgG. Subsequently, the matrix is washed and incubated with a buffer containing TEV protease to elute the protein of interest. Eluted protein containing the CBP tag is then incubated with calmodulin-attached resin in the presence of calcium to allow the binding of CBP with calmodulin. Following washing of the resin, the protein of interest is released in the presence of EGTA [ethylene glycol-bis (β-aminoethyl ether)-N,N,N′,N′-tetraacetic acid] that chelates calcium for altering the conformation of calmodulin to impair calmodulin–CBP interaction. Such TAP-mediated isolation methodology and other TAP variants can be used for numerous proteins and multi-protein complexes [e.g., 3–9]. Here, we describe how this methodology can be used for isolation of FACT (facilitates chromatin transcription; a heterodimer) and identification of its interaction partners in deciphering cellular functions.

FACT is the first identified histone chaperone involved in chromatin regulation associated with transcription [10–18]. Later on, it was also found to regulate other DNA transactions [13, 19–25]. Recently, FACT is shown to be involved in base excision repair (BER) [26, 27]. BER deals with the DNA base lesions resulting from base oxidation, deamination, methylation, etc. caused by intrinsic (e.g., reactive oxygen species) as well as extrinsic (e.g., alkylating, methylating, and deaminating agents) factors. BER begins with the recognition and excision of the damaged base from the ribose sugar moiety to generate abasic (apurinic/apyrimidinic [AP]) site via the catalytic action of DNA glycosylase [28–30]. The AP site, thus generated, is subsequently recognized by AP endonuclease (APE) that specifically cuts (or hydrolyzes) the phospho-ester bond 5′ to the AP site, generating a nick with 3′-OH and 5′-deoxyribose phosphate. Such nick recruits DNA polymerase that copies information from the other strand incorporating the correct nucleotide at the place of the damaged base. This is followed by the removal of the displaced DNA strand containing the AP site by flap endonuclease with subsequent sealing of the ends by DNA ligase. BER can also occur without flap endonuclease [30]. Malfunction or misregulation of the BER machinery impairs BER, which leads to genomic alteration and is associated with premature ageing, neurodegenerative disorders, and cancers [28–30].

FACT has been recently implicated in BER at the levels of recognition/excision of damaged bases and nick generation at the DNA backbone via DNA glycosylase and APE1, respectively [26, 27]. However, it is not clearly understood how FACT controls BER. Therefore, it is important to isolate FACT for biochemical studies to understand its role in BER and identify its interaction partners in response to DNA base lesions. We describe below the detailed protocol for TAP-mediated isolation of FACT from *Saccharomyces cerevisiae*, which can be used to study biochemical mechanisms of FACT in BER. We have used this protocol in our laboratory to isolate FACT (using TAP-tagged Pob3 component of FACT) from *S. cerevisiae* to study FACT in transcription [18]. Our methodology is based on different protocols (e.g., 5, 8, 9, 31, 32) used for the TAP-mediated isolation of FACT and other proteins/complexes. This TAP-mediated isolation strategy can also be used to identify the protein interaction partners of FACT following the induction of the base damages (or under different conditions), which would provide functional and mechanistic insights regarding FACT's role in BER. This can be achieved by carrying out the above TAP-mediated FACT isolation under different washing conditions followed by mass-spectrometric analysis of FACT-interacting proteins. In addition, FACT interaction partners in response to the induction of the base damages can be trapped in living cells by formaldehyde-based in vivo crosslinking followed by TAP-mediated isolation of FACT and interacting proteins under high-stringency washing conditions with subsequent mass-spectrometric analysis [33]. The FACT interaction partners, thus identified, can be further validated/verified genetically, biochemically, and physiologically/cytologically. This approach can be generally used for other proteins/complexes/processes.

2 Materials

2.1 Equipment

- BeadBeater (BioSpec; Cat no. 1107900-101).
- BeadBeater chamber (BioSpec; Cat no. 110803-50SS): 50 mL.
- Benchtop orbital shaker (Bellco Biotechnology).
- Cell culture roller drum (New Brunswick Scientific TC-7).
- Centrifuge bottles (Thermo Fisher Scientific™; Cat no. 75006443): 750 mL.
- Freezer (−80 °C).
- Incubator shaker (New Brunswick Scientific C25 incubator shaker).
- Low-temperature incubator (Thermo Fisher Scientific).
- Microcentrifuge (Eppendorf 5415C and 5424).
- Microcentrifuge tube stands.
- Microscope.
- Micropipettor/Pipetman.
- Pipet-Aid (VWR; Cat no. 75856458).
- Refrigerated centrifuge machines (Sorvall® Legend RT and Beckman J-6B).

- Rotisserie shaker (Barnstead Labquake).
- Spectrophotometer (Hitachi U-2000).

2.2 Reagents and Buffers

- Acetic acid, reagent grade (Fisher; Cat no. BP1185).
- Buffer E: Lysis buffer (see below), 1 mM benzamidine, 25 μg/mL tosyl-L-lysyl-chloromethyl ketone (TLCK), 50 μg/mL tosyl phenylalanyl chloromethyl ketone (TPCK), 20 μg/mL antipain, 10 μg/mL aprotinin, 1 μg/mL leupeptin, and 1 μg/mL pepstatin.
- Calmodulin-Sepharose^R 4B/resin (Sigma; Cat no. GE17052901).
- Calmodulin buffer for binding: 10 mM Tris-HCl (pH 8.0), 150 mM NaCl, 1 mM magnesium acetate, 1 mM imidazole, 2 mM calcium chloride, and 10 mM β-mercaptoethanol (β-ME). *Note: Store buffer without β-ME (which is added before use).*
- Calmodulin buffer for elution: 10 mM Tris–HCl (pH 8.0), 150 mM NaCl, 1 mM magnesium acetate, 1 mM imidazole, 10 mM EGTA, 0.02% NP-40, and 10 mM β-ME. (*see* **Note 1**).
- Coomassie (Bradford) protein assay reagent (Fisher; P123200).
- IgG-Sepharose^R 6 Fast Flow (GE Healthcare; Cat no. 17096901).
- Lysis buffer: 40 mM HEPES (pH 7.5), 150 mM NaCl, 0.1% Tween-20, and 10% glycerol.
- Methanol, reagent grade (Fisher; Cat no. A433P).
- PMSF (phenylmethylsulfonyl fluoride): 250 mM.
- TEV cleavage buffer: 10 mM Tris–HCl (pH 8.0), 150 mM NaCl, 0.5 mM EDTA, 0.1% NP-40, and 1 mM dithiothreitol (DTT). *Note: Store buffer without DTT (which is added before use).*
- TEV protease (Sigma; Cat no. T4455).
- Tris-buffered saline: 10 mM Tris–HCl (pH 8.0), 1 mM EDTA, and 150 mM NaCl.
- YPD (yeast extract and peptone plus dextrose): 1% yeast extract, 2% Bacto-peptone, and 2% dextrose.

2.3 Consumables and Other Items

- Acid-washed glass beads (425–600 μm) (Sigma; Cat no. G8772).
- Amicon® ultra-0.5 centrifugal filter device kit (Sigma/Millipore, Cat no. UFC500308).
- Cell culture flasks.
- Cell culture tubes.

- Column stand.
- Cryogenic dewar flask.
- Disposable cuvettes (Fisher; Cat no. 14955127).
- Duct tape.
- Falcon tubes (50 mL).
- Falcon tube stands.
- Glass beakers.
- Glass slides.
- Ice buckets.
- Liquid nitrogen.
- Microcentrifuge tubes (1.5 mL).
- Micropipettor/Pipetman tips.
- Paper towels.
- Screw.
- Stand clamps.
- Sterile needles: 21G × 1 (Fisher; Cat no. REF305165).
- Sterile pipets.
- Poly-Prep® chromatography columns (Bio-Rad; Cat no. 7311550).
- Silver staining kit (Bio-Rad; Cat no. 1610449).

3 Methods

3.1 Growing Culture

1. Inoculate a single colony of yeast cells expressing TAP-tagged Pob3 in 5 mL YPD medium in a cell culture tube and put in a rotating drum (with a step of 56 rpm) in an incubator (30 °C) overnight.
2. Next day, transfer the whole 5 mL yeast culture (in log phase) into a 500 mL cell culture flask containing 95 mL YPD medium, and put in an incubator shaker (New Brunswick Scientific C25) with a speed of 196 rpm at 30 °C overnight.
3. Next morning, transfer 10 mL of the 100 mL culture (with an OD_{600} of ~1.5–2.0) to 990 mL YPD growth medium in a cell culture flask with a minimum capacity of 4 L. Transfer another 10 mL of the 100 mL culture into the second 4-L cell culture flask containing 990 mL YPD medium. Put these two 4-L flasks in the incubator shaker with a speed of 150 rpm at 30 °C overnight.
4. On the next day, measure the OD_{600} of the yeast culture in the 4-L flask in a spectrophotometer and harvest the cells when the OD_{600} is about 2–2.5 (*see* **Note 2**).

3.2 Harvesting Cells

All the steps are carried out on ice and/or in cold temperature (4 °C).

1. Transfer the 2 L cell culture into four pre-chilled 750 mL centrifuge bottles, and then balance these bottles by weighing in a lab weighing machine prior to centrifugation.
2. Spin down cells in a pre-chilled centrifugation machine (Sorvall) at 3500 rpm for 20 min at 4 °C. (*see* **Note 3**).
3. Immediately put the four centrifugation bottles with cell pellets in an ice bucket, and cover with ice till the neck.
4. Carefully decant supernatant from each bottle (by slowly tilting) and leave about 15 mL supernatant (medium) to suspend the cell pellet in the next step.
5. Suspend the cell pellet by the residual medium in each of the four centrifugation bottles (by slowly pipetting in and out using a 10 mL pipet and Pipet-Aid), and transfer the cell suspension of each bottle to a 50 mL pre-chilled Falcon tube.
6. Spin down cell suspension of four 50 mL Falcon tubes in the pre-chilled centrifugation machine (Sorvall) at 3500 rpm for 5 min at 4 °C.
7. Immediately transfer the Falcon tubes in an ice bucket. Decant the supernatant, and then put the Falcon tubes back on ice.
8. Add 25 mL pre-chilled Tris-buffered saline (TBS) to each of the four Falcon tubes, and resuspend cells using a pre-chilled 25 mL pipet (by slowly pipetting in and out with the help of pipet and Pipet-Aid).

   Falcon *tube with cell pellet is kept mostly covered with ice during resuspension.*
9. Spin down cell suspension in four Falcon tubes in the pre-chilled centrifugation machine (Sorvall) at 3500 rpm for 5 min at 4 °C.
10. Transfer the Falcon tubes to an ice bucket, decant supernatant, and put the Falcon tubes back inside the ice.
11. Repeat the above **steps 8**, **9**, and **10** to wash the cells one more time (*see* **Note 4**).
12. Close the Falcon tubes and immerse in liquid nitrogen (about 2 min) to quickly freeze cells. Store these Falcon tubes with frozen cell pellets in −80 °C freezer.

3.3 WCE Preparation

All steps for WCE preparation are performed in the cold room.

1. Take out the four Falcon tubes containing cell pellets from the −80 °C freezer, and put in an ice bucket. Completely cover the Falcon tubes with ice, and let the cell pellets thaw.

2. While the cell pellets are thawing, put BeadBeater and accessories, glass- and plasticwares, glass beads, and buffers in the cold room to chill. Also, prepare buffer E (for WCE preparation) by adding all protease inhibitors except PMSF to the lysis buffer (*see* **Note 5**).

   When cell pellet is thawed in the Falcon tube, add 15 mL buffer E each to two of the four Falcon tubes with thawed cell pellet. Suspend the cells gently with a pre-chilled 25 mL pipet until they are homogenous suspensions (by slowly pipetting in and out with the help of pipet and Pipet-Aid). Transfer each of the homogenous cell suspensions to one other Falcon tube with thawed cell pellet and continue to suspend cells. These homogenously suspended cells are pooled into one of the four Falcon tubes.

3. Wash the other three Falcon tubes thoroughly with 2 mL of buffer E each, and transfer to the fourth Falcon tube containing the pooled cell suspension. (*see* **Note 6**).

4. Add 45 mL-equivalent glass beads (measured using a 50 mL Falcon tube) to the 50 mL BeadBeater chamber followed by the cell suspension. There may be ~13 mL space left in the BeadBeater chamber after adding glass beads and cell suspension. Fill it up by adding 10 mL buffer E. Then, keep adding 1 mL buffer E several times (~3 times), and check whether the chamber is full (feel by lightly touching the surface of the cell suspension in the chamber with the white plastic lid that fits under the black cover of the chamber). A few drops of cell suspension will be or about to be spilled out if the chamber is full (*see* **Note 7**).

5. Add 400 μL of 250 mM PMSF to the cell suspension in the BeadBeater chamber. Pipet in and out a few times to mix PMSF with the cell suspension, and immediately close the BeadBeater chamber.

6. Place the filled 50 mL BeadBeater chamber upright inside the ice chamber and then put on the rotor of the BeadBeater. Make sure the grooves of the rotor and chamber fit well. Then, fill the ice chamber with ice (completely cover the BeadBeater chamber by ice).

7. Connect the BeadBeater to the power outlet, using an adaptor (15A and 250 V). Hold tight the cap of the BeadBeater chamber (pressing downward) on the rotor of the BeadBeater, and then turn the BeadBeater *on* for 15 s (*see* **Note 8**).

8. Wait for 30 s. Repeat another 11 pulses (15 s each) with a pause of at least 30 s in between pulses. Perform a total of 12 pulses (15 s each) of bead-beating (i.e., a total of 3-min bead-beating). At the end of 12 pulses, cell lysis can be monitored by transferring 10 μL cell extract from the BeadBeater chamber to

a glass slide and visualizing under the microscope (lysed cells appear dark in comparison to refractile intact cells) (*see* **Note 9**).

9. Transfer the WCE and glass bead suspension from the BeadBeater chamber to two 50 mL Falcon tubes (approximately equal volume) using a 25 mL pipet. The little leftover wet glass beads in the BeadBeater chamber can be transferred to these two Falcon tubes using a clean spatula.
10. Close the Falcon tubes and invert them on a flat surface to allow the cell suspension and glass beads to move towards the cap/lid. Then make three equidistant dents at the bottom of each Falcon tube using a screw or nail to facilitate easy punching of holes by a needle in the next step.
11. Using a 21G × 1 needle, make the three holes over the dents made in the above step.
12. Place a new Falcon tube (without its cap/lid) over the inverted tube. Firmly join them using duct tape at and around the junction of these two tubes so that they can be placed vertically upright in the pre-chilled centrifuge machine (Beckman) for quick spin to collect WCE. Put this assembly in the upright position in the centrifuge machine.
13. Spin down at 1000 rpm for 1 min at 4 °C to collect the WCE in the bottom Falcon tube.
14. Take out the bottom Falcon tube in upright position (by carefully peeling off the duct tape), close it with its lid, and put on ice. WCE, thus collected, is in two 50 mL Falcon tubes.
15. Put the Falcon tubes containing WCE in the pre-chilled centrifuge machine (Sorvall) at 3500 rpm for 5 min at 4 °C.
16. Carefully collect supernatant or WCE (without disturbing the cell debris at the bottom) into two new 50 mL Falcon tubes using a 25 mL pipet.
17. Spin these two Falcon tubes in the centrifuge machine (Sorvall) at 3500 rpm for 5 min at 4 °C to remove the last traces of cell debris. Carefully collect the supernatant (WCE) into one new 50 mL Falcon tube and put on ice for use in the next step (C3.4). The total volume of the WCE is about 50 mL (*see* **Note 10**).

3.4 Incubation of WCE with IgG-Sepharose Beads

1. Transfer 500 μL of IgG-Sepharose suspension to a sterile 1.5 mL Eppendorf tube using a Pipetman (P-1000) and 1000 μL Pipetman tip (which is cut at the tip by clean scissors), and spin it down at 3000 rpm for 2 min at room temperature in a microcentrifuge (Eppendorf 5415C). Carefully discard the supernatant (without disturbing the beads) using a P-200

Pipetman. Add around 200 μL of lysis buffer using a P-200 Pipetman to the IgG-Sepharose beads in the Eppendorf tube, and put in the rotor (rotisserie shaker) for 2 min at room temperature for equilibration of the IgG-Sepharose beads with lysis buffer. Then spin down the IgG-Sepharose beads in a microcentrifuge (Eppendorf 5415C) at 3000 rpm for 2 min. Carefully discard supernatant (without disturbing the beads) by P-200 Pipetman (*see* **Note 11**).

2. Transfer the equilibrated IgG-Sepharose beads to the 50 mL Falcon tube containing the WCE (Use ~400 μL of WCE to suspend IgG-Sepharose beads in the Eppendorf tube prior to transferring to the 50 mL Falcon tube containing WCE). Close the Falcon tube and wrap it with parafilm. Then, rotate it on the rotisserie shaker for 3 h in a cold room to allow the binding of TAP-tagged protein with the IgG-Sepharose beads.

3.5 Washing of the IgG-Sepharose Beads

1. Break the snap-off seal (by rotating anticlockwise) at the bottom of the Poly-Prep column, and clamp it to a column stand. Place a beaker under the Poly-Prep column for the collection of flow-through.
2. Transfer the incubated suspension of IgG-Sepharose beads and WCE (from **step C3.4.2**) into the Poly-Prep column, using a 5 mL pipet (*see* **Note 12**).
3. Completely transfer the suspension of IgG-Sepharose beads and WCE to the Poly-Prep column and collect the flow-through in the beaker at the bottom of the column. Let IgG-Sepharose beads settle down on the polyethylene bed in the Poly-Prep column. Allow the last trace of flow-through to come out at the bottom of the Poly-Prep column prior to washing.
4. Wash the IgG-Sepharose beads in the Poly-Prep column three times with 10 mL of lysis buffer each time. Wait until the last drop of lysis buffer comes out at the end of the first wash before proceeding to the second wash, but also ensure that the beads are not completely dried out (follow the same at the end of the second wash). A small portion of the third wash can be stored in a –80 °C freezer for WB analysis, if needed (*see* **Note 13**).

   Proceed to equilibrate the IgG-Sepharose beads in the Poly-Prep column with TEV cleavage buffer (TCB) by passing 10 ml TCB through the column.
5. After the TCB buffer is completely dripped out from the bottom of the column, close the column bottom with the yellow cap (provided with the Poly-Prep column by the vendor), wrap it with parafilm, and then proceed to the next step (C3.6) for TEV cleavage.

3.6 TEV Cleavage

1. Add 1 mL TCB to the IgG-Sepharose beads in the Poly-Prep column followed by 400 units (U) TEV protease (40 μL of 10 U/μL TEV protease stock provided by the vendor).
2. Close the top of the column with the flat cap (provided with the Poly-Prep column by the vendor) and wrap parafilm around the cap.
3. Take out the Poly-Prep column from the column stand and put on the Labquake Rotisserie shaker to rotate for 1.5 h at 16 °C (using low-temperature incubator), and then for 8 h in the cold room.
4. After the above 9.5-h incubation, remove the parafilm from the bottom of the above Poly-Prep column, carefully remove the yellow cap in an inverted position of the Poly-Prep column and put a 1.5 mL Eppendorf tube over the outlet of the inverted column, and then carefully attach the column in an upright position to the column stand with Eppendorf tube at the bottom on an Eppendorf tube stand (all these are carried out in the cold room).
5. Remove the top flat cap from the above Poly-Prep column and carefully collect the flow-through from the column into the Eppendorf tube. After the last drop of the flow-through comes out, close the Eppendorf tube and put it on ice. Label this Eppendorf tube as E1-TEV (this contains eluted proteins).
6. Close the bottom of the Poly-Prep column with a yellow cap and add 1 mL TCB to the IgG-Sepharose beads in the column. Close the top of the Poly-Prep column. Parafilm both ends of the column securely. Then, put the Poly-Prep column on the Labquake Rotisserie shaker to rotate for 3 min in the cold room. Collect the flow-through as done in **step 5** above into a new 1.5 mL Eppendorf tube and label it E2-TEV. Put this Eppendorf tube on ice.
7. Measure the volumes of the above two eluates (E1-TEV and E2-TEV) before proceeding to the next step (C3.7) for incubation with calmodulin-Sepharose beads.

3.7 Incubation of TEV Eluate with Calmodulin-Sepharose Beads

1. Aliquot 300 μl of calmodulin-Sepharose bead suspension into a 1.5 mL Eppendorf tube, using a Pipetman (P-1000) and 1000 μL Pipetman tip (which is cut at the tip by clean scissors), and spin it down at 3000 rpm for 2 min at room temperature in a microcentrifuge (Eppendorf 5415C). Carefully discard the supernatant (without disturbing the beads) using a P-200 Pipetman. Add 1 mL calmodulin buffer for binding (CB-B) using a P-1000 to the calmodulin-Sepharose beads in the Eppendorf tube (use this buffer without β-ME for this equilibration step), and put in the rotator (Labquake) for 1 min at room temperature. Then spin down calmodulin-Sepharose

beads in microcentrifuge (Eppendorf 5415C) at 3000 rpm for 2 min. Carefully discard supernatant (without disturbing the beads) by P-200 Pipetman (*see* **Note 14**).

Repeat the above equilibration step using 1 mL CB-B, and then, put the Eppendorf tube containing calmodulin-Sepharose beads on ice until use in the next step (**step 3**).

2. Attach a new Poly-Prep column to the column stand in the cold room (do not break the snap-off seal at the bottom). Suspend calmodulin-Sepharose beads (in the Eppendorf tube of **step 2** above) in E1-TEV and transfer to the Poly-Prep column (using a 1000 μL pipet tip that is cut at the tip by clean scissors). Add E2-TEV to the same Eppendorf tube to resuspend the leftover calmodulin-Sepharose beads, and then transfer to the Poly-Prep column.
3. Add CB-B-containing β-ME (three times the total volume of E1-TEV and E2-TEV) to the Poly-Prep column (e.g., for 2 mL TEV eluate [E1-TEV + E2-TEV], use 6 mL CB-B containing β-ME).
4. Add 1 M $CaCl_2$ ([three times the total volume of E1-TEV and E2-TEV]/1000) to the Poly-Prep column (e.g., for 2 mL TEV eluate [E1-TEV + E2-TEV], use 6 μL of 1 M $CaCl_2$).
5. Close the top of the column with the flat cap, wrap it with parafilm, and then put the column on the Labquake Rotisserie shaker to rotate for 3 h in the cold room.

3.8 Protein Elution from Calmodulin-Sepharose Beads

All steps during elution are performed in the cold room.

1. After a 3-h incubation in the above step (i.e., C3.7.5), remove the parafilm and break the snap-off seal from the bottom of the above Poly-Prep column. Then, attach the Poly-Prep column to the column stand, and remove the top flat cap of the column to drain the flow-through into a beaker. Wait until the last drop comes out from the bottom of the column.
2. Wash the calmodulin-Sepharose beads in the Poly-Prep column three times with 10 mL CB-B (containing β-ME) each time. A portion of the third wash can be stored in a –80 °C freezer for WB analysis, if needed.
3. Elute the protein from the calmodulin-Sepharose beads into ten fractions (E1–E10) using calmodulin buffer for elution (CB-E). Close the bottom of the Poly-Prep column with the yellow cap (with parafilm wrap), add 500 μL of CB-E (which contains 10 mM EGTA) to the column, close the top of the column with the flat cap (and wrap it with parafilm), and rotate on the Labquake Rotisserie shaker for 10 min (for the first five elutions) and 5 min (for last five elutions). After each incubation, remove the yellow cap in an inverted position, put a 1.5-

mL Eppendorf tube over the outlet of the inverted column, carefully attach to the column stand in an upright position with the Eppendorf tube at the bottom, remove the top flat cap, and collect flow-through in the Eppendorf tube. Put the Eppendorf tube containing the eluate (~500 μL) on ice. Take out the 10 μL eluted protein for measuring protein quantity by Bradford assay (C3.9). The remaining ~490 μL eluate in the Eppendorf tube is put in liquid nitrogen for quick freezing and then transferred to a –80 °C freezer for storage.

3.9 Protein Quantification by Bradford Assay

1. Label ten 1.5 mL Eppendorf tubes as E1-PQ through E10-PQ and another one as "Blank." Arrange these tubes in the Eppendorf tube rack.
2. Add 790 μL of MilliQ water to all 11 Eppendorf tubes. Subsequently, add 10 μL of purified protein (from the above ten elutions from **step C3.8.3**) to first ten Eppendorf tubes (E1-PQ to E10-PQ) and 10 μL CB-E to the Eppendorf tube labeled as "Blank." Then, add 200 μL of Coomassie (Bradford) protein assay reagent to all 11 Eppendorf tubes.
3. Invert the tubes several times and incubate in the dark for 5 min.
4. Set the spectrophotometer to 595 nm, blank it with the solution in the Eppendorf tube labeled as "Blank." Measure the absorbances of the solutions in the Eppendorf tubes labeled as E1-PQ through E10-PQ.
5. Use the absorbance values to calculate the protein concentration with the help of a standard curve generated in the next **step 6**.
6. Standard curve: Using the BSA (bovine serum albumin) stock provided with Coomassie (Bradford) protein assay reagent, prepare six BSA solutions with concentrations between 1 and 6 μg/mL (in 2 ml or above). Mix 800 μL of BSA solution with 200 μL Coomassie (Bradford) protein assay reagent, incubate in the dark for 5 min, transfer to cuvettes to record absorbance at 595 nm. The plot of absorbance vs BSA concentration will generate the standard curve (*see* **Note 15**).

3.10 Concentrating Eluted Proteins from the Calmodulin-Sepharose Beads

1. Thaw frozen protein eluate on ice.
2. For concentrating protein solution, use two collection tubes and one filter device that are provided with Amicon® ultra-0.5 centrifugal filter device kit with 3 K NMWL (nominal molecular weight limit). Transfer the protein solution to the filter device, and then, place it in collection tube 1 and close the filter's mouth with the lid.

Note: *The lower NMWL is used for retention of smaller proteins, while the higher NMWL will ensure filtering out the smaller proteins.*

3. Put collection tube 1 and filter the device assembly in the microcentrifuge machine (Eppendorf 5424) in the cold room, and spin at 14000 *g* for 30 min.
4. After the centrifugation, the concentrated protein is retained in the filter device and the filtrate in collection tube 1. Place the filter upside down into collection tube 2, and spin down at 1000 *g* for 2 min in Eppendorf 5424 microcentrifuge machine in the cold room. The concentrated protein solution from collection tube 2 is transferred to a 1.5 mL Eppendorf tube for SDS-PAGE (sodium dodecyl sulfate–polyacrylamide gel electrophoresis), mass spectrometry, and other analyses. The remaining concentrated protein solution is put in liquid nitrogen for a minute to freeze quickly prior to storage in a –80 °C freezer.

3.11 SDS-PAGE

1. Put about 5 μg concentrated protein solution (that was originally eluted from above calmodulin-Sepharose beads in **step C3.8.3**) in a 1.5 mL Eppendorf tube on ice. Add equal volume of 2× SDS-PAGE dye and put in the heat block at 95 °C for 5 min.
2. After heating the protein sample for 5 min, vortex it and spin down at 13000 rpm for 2 min in a microcentrifuge machine (5415C). Then, load the protein sample for SDS-PAGE (12%) and run at a constant volt of 120 till the dye front reaches the buffer at the bottom of the gel.

3.12 Silver Staining

The Silver Stain Plus™ kit (Bio-Rad) is used for silver staining with the following four steps.

I. **Fixative Step.**

1. Prepare fixative enhancer solution (400 mL).
   (a) Measure 120 mL deionized distilled water in a cylinder and put in a glass tray.
   (b) Measure 200 mL methanol in a cylinder and put in the same tray.
   (c) Add 40 mL of reagent-grade acetic acid (using 25 mL pipet) to the same tray.
   (d) Add 40 mL of fixative enhancer concentrate of the Silver Stain Plus™ kit (using a 25 mL pipet) to the same tray and mix well with the same pipet (as well as by swirling).

2. Place the gel in the fixative enhancer solution carefully.
3. Put the tray on a platform shaker keeping at a low speed (benchtop orbital shaker, 2.8 speed) for 20 min.

II. **Gel Rinsing Step.**

1. After 20 min of the fixation step, carefully decant the fixative enhancer solution from one corner of the tray into the methanol waste container in the chemical hood.
2. Rinse the gel in 400 mL deionized distilled water for a minute.
3. Decant water, and transfer the gel to a new glass tray. Add 400 mL deionized distilled water and put it on the shaker (benchtop orbital shaker, 2.8 speed) for 10 min. Repeat this step two more times (a total of three washes) (*see* **Note 16**).
4. Staining solution:
   (a) Put 35 mL deionized distilled water in a beaker (250/300 mL) and stir with a stir bar (PTFE coated). Add the following reagents to the beaker in the same order:
   - 5 mL silver complex solution of the Silver Stain Plus™ kit
   - 5 mL reduction moderator solution of the Silver Stain Plus™ kit
   - 5 mL of image development reagent of the Silver Stain Plus™ kit

   (b) Keep stirring for 5–10 min.
5. Stop solution: Put 95 mL of deionized distilled water and 5 mL of reagent-grade acetic acid in a 100 mL cylinder. Close the cylinder with multiple parafilm layers (about three), and invert the cylinder several times (by putting one hand on the parafilm).
6. Developing solution:
   (a) Measure 2.5 g of the development accelerator powder of the Silver Stain Plus™ kit in a 50 mL Falcon tube and add water to make up 50 mL volume. Close the Falcon tube, cover with aluminum foil (as the solution is light sensitive), and vortex to mix.

   (b) When the third wash of the gel is complete, quickly add 50 mL of the above development accelerator solution to the beaker stirring the staining solution. Stir well for 2 min. The solution will start turning cloudy. Then, put this solution on the glass tray with gel in the next step for staining and development (*see* **Note 17**).

III. **Staining and Developing Step.**

1. Decant the water from the glass tray after the third wash, and carefully transfer the gel to a new glass tray.
2. Immediately add the content of the above beaker (i.e., developing solution from **step 6b** of the above gel rinsing) to the glass tray with gel. Put the tray on the shaker (bench-top orbital shaker, 2.8 speed) for ~15–20 min until the desired staining intensity is reached. It may take around 10–15 min for bands to be visible. When the gel starts getting blackish, stop gel staining and development in the next step (stop step) (*see* **Note 18**).

IV. **Stop Step.**

- Decant the staining and developing solution from the glass tray into the waste container and place the gel in the stop solution in a new glass tray for a minimum of 15 min. After stopping the reaction, rinse the gel in deionized distilled water for 5 min. The gel is then ready to be photographed in the gel documentation system (iBright FL100). Put the gel on white screen of the gel documentation system, and capture image.

3.13 Mass-Spectrometric Analysis of the Interacting Proteins

TAP-mediated isolation of a protein/complex may pull down its interacting protein partner(s), depending on the stringency of washing conditions (e.g., NaCl concentration). These interactions can be weak, strong, or nonspecific. To identify protein interaction partner(s), one can carry out the above TAP-mediated isolation of a protein/complex at the physiological and higher concentrations of NaCl during washing of the IgG-Sepharose and calmodulin-Sepharose beads, and then perform mass-spectrometric analysis to identify the protein(s) associated with the TAP-tagged protein/complex. Mass spectrometry is not described here, as it is performed with analysis at the instrumentation facility/center. A strong protein interactor(s) is likely to be found in TAP with washing at the physiological as well as higher concentrations of NaCl. Weak protein interactor(s) would be found in TAP with washing at the physiological concentration of NaCl, but not at higher concentration. The weak protein interactor(s), thus identified, could be nonspecific. Therefore, the above TAP and mass-spectrometric analysis can identify strong protein interactor(s), but the identified weak protein interactor(s) can be dubious. Further, the identified strong protein interactor(s) may not be physiologically relevant, but rather shows the propensity to interact with the TAP-tagged protein/complex in the WCE. Further, physiologically relevant protein interactor(s) may not be identified by the combined TAP and mass-spectroscopic analyses, as the above TAP-mediated pull-down identifies the interaction of a protein/complex with other protein(s)/complex(es) in the WCE, but not in the living cells in real time. Moreover, the identification of the protein interactor(s) from the above combined methods of TAP and mass spectrometry can be varied by changing the stringency of washing of IgG and calmodulin-Sepharose beads in TAP-mediated pull-down from the WCE, leading to dubious results/conclusions.

To overcome the above limitations in identifying protein interaction partner(s) of a protein/complex, one can trap the protein interactions in living cell in real time (or during biological processes) by formaldehyde-based in vivo crosslinking, prepare WCE, and then perform the TAP-mediated isolation of a protein/complex under high-stringency washing conditions. Such high-stringency washing would eliminate noncovalent interactor(s), but not the interactor(s) that got covalently linked to the protein/complex of interest by formaldehyde during biological processes in living cells. The in vivo protein interactor(s) will be identified by mass-spectrometric analysis of TAP-mediated pull-down of a protein/complex following formaldehyde-based in vivo crosslinking, but all nonspecific interactors will be washed out during TAP. The protein interactor(s), thus identified, would be physiologically/functionally relevant, which can be validated/verified by various experiments including the analysis of the above identified interaction(s) in a purified system as well as in vivo [1] and genetic/mutational studies. Using this experimental strategy, one can identify FACT-interacting protein(s) with functional analysis.

4 Notes

1. Store buffer without β-ME (which is added before use).
2. Dilute the yeast culture two- to tenfold by the YPD growth medium to have an OD_{600} between 0.1 and 0.5, which is then multiplied by the dilution factor to obtain the final OD_{600} of the culture.
3. Around 80% of the cells are pelleted under the above centrifugation conditions. However, longer spinning would help to recover more cells.
4. After decanting the supernatant, the tubes are gently inverted over a bunch of paper towels in a cold room to get rid of the last trace of TBS.
5. PMSF has a short half-life in the aqueous solution, and therefore, PMSF is added to the yeast cell suspension just before bead-beating.
6. The total volume of cell suspension is ~43.5 mL.
7. Make sure the chamber is full. Otherwise, the presence of air bubble will generate froth which can denature proteins/complexes.

8. Make sure the blade in the 50 mL BeadBeater chamber (plastic) is moving (when the BeadBeater is *on*). If not, appropriately fit the grooves of the rotor and chamber, and run again.
9. As the BeadBeater generates heat that denatures proteins, it is critical that the outer chamber be filled completely with ice and the *on* time be short. The *off* time between pulses can be extended beyond 30 sec.
10. While slowly collecting supernatant (WCE) by pipet in an inclined position, leave ~2 mL supernatant at the bottom to avoid cell debris, if any (usually, cell debris is not visible in this step). This remaining supernatant can be stored in a −80 °C freezer for WCE analysis.
11. Mix IgG-Sepharose beads by inverting the IgG bead container multiple times before use to ensure homogenous suspension.
12. Even though the use of a 5 mL pipet would require multiple rounds to transfer the suspension of IgG-Sepharose beads and WCE in comparison to using a 25 mL pipet, it will have lesser surface area in the inner wall for the beads to stick, thus ensuring minimal loss of the beads during the transfer.
13. For the first wash, the same 5 mL pipet from **step C3.5.2** can be used, as this would help retrieve the beads stuck to the inner wall of the pipet. The same 5 mL pipet can also be used in the second and third washes using 20 mL lysis buffer aliquoted in a Falcon tube. Each wash takes about 25 min.
14. Invert gently the calmodulin-Sepharose container multiple times before use to ensure homogenous suspension.
15. Dilute protein to get absorbance at 595 below 0.5.
16. During **step 3**, prepare solutions for staining, developing, and stop steps.
17. Prepare the developing solution just before the staining and developing step of the gel.
18. The staining time depends on the protein and quantity. It takes about 10–15 min for 5 μg of protein.

Acknowledgements

We thank Zhiguo Zhang for the yeast strain (TAP-tagged Pob3). The work in the Bhaumik laboratory was supported by the grants from the National Institutes of Health (GM088798-03). AK and SG were supported by the predoctoral fellowship of the American Heart Association and the doctoral fellowship of Southern Illinois University, respectively.

References

1. Bhaumik SR (2021) Fluorescence resonance energy transfer in revealing protein-protein interactions in living cells. Emerg Top Life Sci 5:49–59
2. Rigaut G, Shevchenko A, Rutz B, Wilm M, Mann M, Séraphin B (1999) A generic protein purification method for protein complex characterization and proteome exploration. Nat Biotechnol 17:1030–1032
3. Puig O, Caspary F, Rigaut G, Rutz B, Bouveret E, Bragado-Nilsson E, Wilm M, Séraphin B (2001) The tandem affinity purification (TAP) method: a general procedure of protein complex purification. Methods 24: 218–229
4. Miller T, Krogan NJ, Dover J, Erdjument-Bromage H, Tempst P, Johnston M, Greenblatt JF, Shilatifard A (2001) COMPASS: a complex of proteins associated with a trithorax-related SET domain protein. Proc Natl Acad Sci U S A 98:12902–12907
5. Krogan NJ, Kim M, Ahn SH, Zhong G, Kobor MS, Cagney G, Emili A, Shilatifard A, Buratowski S, Greenblatt JF (2002) RNA polymerase II elongation factors of Saccharomyces cerevisiae: a targeted proteomics approach. Mol Cell Biol 22:6979–6992
6. Cheeseman IM, Desai A (2005) A combined approach for the localization and tandem affinity purification of protein complexes from metazoans. Sci STKE 2005:pl1
7. Gregan J, Riedel CG, Petronczki M, Cipak L, Rumpf C, Poser I, Buchholz F, Mechtler K, Nasmyth K (2007) Tandem affinity purification of functional TAP-tagged proteins from human cells. Nat Protoc 2:1145–1151
8. Gould KL, Ren L, Feoktistova AS, Jennings JL, Link AJ (2004) Tandem affinity purification and identification of protein complex components. Methods 33:239–244
9. Pemsel A, Rumpf S, Roemer K, Heyne K, Vogt T, Reichrath J (2018) Tandem affinity purification and nano HPLC-ESI-MS/MS reveal binding of vitamin D receptor to p53 and other new interaction partners in HEK 293T cells. Anticancer Res 38:1209–1216
10. Orphanides G, LeRoy G, Chang CH, Luse DS, Reinberg D (1998) FACT, a factor that facilitates transcript elongation through nucleosomes. Cell 92:105–116
11. Belotserkovskaya R, Oh S, Bondarenko VA, Orphanides G, Studitsky VM, Reinberg D (2003) FACT facilitates transcription-dependent nucleosome alteration. Science 301:1090–1093
12. Mason PB, Struhl K (2003) The FACT complex travels with elongating RNA polymerase II and is important for the fidelity of transcriptional initiation in vivo. Mol Cell Biol 23: 8323–8333
13. Sen R, Ferdoush J, Kaja A, Bhaumik SR (2016) Fine-tuning of FACT by the ubiquitin proteasome system in regulation of transcriptional elongation. Mol Cell Biol 36:1691–1703
14. Sen R, Kaja A, Ferdoush J, Lahudkar S, Barman P, Bhaumik SR (2017) An mRNA capping enzyme targets FACT to the active gene to enhance the engagement of RNA polymerase II into transcriptional elongation. Mol Cell Biol 37:e00029–e00017
15. Kaja A, Adhikari A, Karmakar S, Zhang W, Rothschild G, Basu U, Batra SK, Davie JK, Bhaumik SR (2021) Proteasomal regulation of mammalian SPT16 in controlling transcription. Mol Cell Biol 41:e00452–e00420
16. Winkler DD, Luger K (2011) The histone chaperone FACT: structural insights and mechanisms for nucleosome reorganization. J Biol Chem 286:18369–18374
17. Formosa T (2013) The role of FACT in making and breaking nucleosomes. Biochim Biophys Acta 1819:247–255
18. Barman P, Sen R, Kaja A, Ferdoush J, Guha S, Govind CK, Bhaumik SR (2022) Genome-wide regulations of the preinitiation complex formation and elongating RNA polymerase II by an E3 ubiquitin ligase, San1. Mol Cell Biol 42:e0036821
19. Keller DM, Zeng X, Wang Y, Zhang QH, Kapoor M, Shu H, Goodman R, Lozano G, Zhao Y, Lu H (2001) A DNA damage-induced p53 serine 392 kinase complex contains CK2, hSpt16, and SSRP1. Mol Cell 7:283–292
20. Zhou Y, Wang TS (2004) A coordinated temporal interplay of nucleosome reorganization factor, sister chromatin cohesion factor, and DNA polymerase alpha facilitates DNA replication. Mol Cell Biol 24:9568–9579
21. Tan BC, Chien CT, Hirose S, Lee SC (2006) Functional cooperation between FACT and MCM helicase facilitates initiation of chromatin DNA replication. EMBO J 25:3975–3985
22. Heo K, Kim H, Choi SH, Choi J, Kim K, Gu J, Lieber MR, Yang AS, An W (2008) FACT-mediated exchange of histone variant H2AX regulated by phosphorylation of H2AX and ADP-ribosylation of Spt16. Mol Cell 30:86–97
23. Kumari A, Mazina OM, Shinde U, Mazin AV, Lu H (2009) A role for SSRP1 in recombination-mediated DNA damage response. J Cell Biochem 108:508–518

24. Dinant C, Ampatziadis-Michailidis G, Lans H, Tresini M, Lagarou A, Grosbart M, Theil AF, van Cappellen WA, Kimura H, Bartek J, Fousteri M, Houtsmuller AB, Vermeulen W, Marteijn JA (2013) Enhanced chromatin dynamics by FACT promotes transcriptional restart after UV-induced DNA damage. Mol Cell 51:469–479
25. Sand-Dejmek J, Adelmant G, Sobhian B, Calkins AS, Marto J, Iglehart DJ, Lazaro JB (2011) Concordant and opposite roles of DNA-PK and the "facilitator of chromatin transcription" (FACT) in DNA repair, apoptosis and necrosis after cisplatin. Mol Cancer 10:74
26. Charles Richard JL, Shukla MS, Menoni H, Ouararhni K, Lone IN, Roulland Y, Papin C, Ben Simon E, Kundu T, Hamiche A, Angelov D, Dimitrov S (2016) FACT assists base excision repair by boosting the remodeling activity of RSC. PLoS Genet 12:e1006221
27. Song H, Zeng J, Lele S, LaGrange CA, Bhakat KK (2021) APE1 and SSRP1 is overexpressed in muscle invasive bladder cancer and associated with poor survival. Heliyon 7:e06756
28. Xu G, Herzig M, Rotrek V, Walter CA (2008) Base excision repair, aging and health span. Mech Ageing Dev 129:366–382
29. Dianov GL, Hübscher U (2013) Mammalian base excision repair: the forgotten archangel. Nucleic Acids Res 41:3483–3490
30. Hegde ML, Hazra TK, Mitra S (2008) Early steps in the DNA base excision/single-strand interruption repair pathway in mammalian cells. Cell Res 18:27–47
31. Zheng S, Crickard JB, Srikanth A, Reese JC (2014) A highly conserved region within H2B is important for FACT to act on nucleosomes. Mol Cell Biol 34:303–314
32. Gerace E, Moazed D (2015) Affinity purification of protein complexes using TAP tags. Methods Enzymol 559:37–52
33. Rosenbaum JC, Fredrickson EK, Oeser ML, Garrett-Engele CM, Locke MN, Richardson LA, Nelson ZW, Hetrick ED, Milac TI, Gottschling DE, Gardner RG (2011) Disorder targets misorder in nuclear quality control degradation: a disordered ubiquitin ligase directly recognizes its misfolded substrates. Mol Cell 41:93–106

Part IV

Analysis of Genome-Wide Binding of DNA Repair Proteins and Copy Number Variations of DNA Damage Response Gene in Tumor

Chapter 15

Analysis of Copy Number Variation of DNA Repair/Damage Response Genes in Tumor Tissues

Tadahide Izumi

Abstract

Cells experience increased genome instability through the course of disease development including cancer initiation and progression. Point mutations, insertion/deletions, translocations, and amplifications of both coding and noncoding regions all contribute to cancer phenotypes. Copy number variation (CNV), i.e., changes of the number of copies of nuclear DNA, occurs in the genome of even normal somatic cells. Studies to understand the effects of CNV on tumor development, especially aspects concerning tumor aggressiveness and the influence on outcomes of therapeutic modalities, have been reignited by the breakthrough technologies of the molecular genomics. This section discusses the significance of analyzing CNVs that cause simultaneous increase/decrease of clusters of genes, using the expression profile of XRCC1 with its neighbor genes LIG1, PNKP, and POLD1 as an example. Methods for CNV assay at the individual gene level on formalin-fixed, paraffin-embedded (FFPE) tissues using the NanoString nCounter technology will then be described.

Key words Copy number variation (CNV), DNA base excision repair, XRCC1, Formalin-fixed paraffin embedded tissues (FFPE), NanoString nCounter

1 Introduction

1.1 Significance of Examining Copy Number Variation of Individual Genes

Before describing protocols for copy number variation (CNV) assay, it is worth discussing why evaluating CNV at the individual gene level is important to understand resilience of tumors against the therapeutic use of DNA-damaging reagents. Genome stability is continuously threatened not only by endogenously and exogenously generated DNA damage, but also by irregular reactions of DNA replication and recombination [1, 2]. The genomic alteration hence includes not only point mutations and insertion/deletions, but structural changes such as chromosomal translocations and copy number variations (gene amplification) [1, 3]. Evidently, point mutations profoundly contribute to cellular transformation and tumorigenesis, with missense mutations in the tumor suppressor TP53 gene as one of the most extensively studied examples

Kishor K. Bhakat and Tapas K. Hazra (eds.), *Base Excision Repair Pathway: Methods and Protocols*, Methods in Molecular Biology, vol. 2701, https://doi.org/10.1007/978-1-0716-3373-1_15,

Table 1
Relative locations of the XRCC1, LIG1, PNKP, and POLD1 genes based on the GRC38 genome coordinate and expressional correlations. Relative distance and expression correlations of LIG1, PNKP, and POLD1 genes are compared with those of XRCC1.

Gene symbol	Start position (GRC38)	Relative distance (Mbp)	Expression correlation in normal tissues	Expression correlation in tumor tissues
XRCC1	43,543,311	0	–	–
LIG1	48,115,445	4.59	0.29	0.81*
PNKP	49,859,882	6.30	0.31	0.51*
POLD1	50,384,204	6.84	0.17	0.72*

Asterisks (*) show $p < 2.2e\text{-}16$ based on the evaluations using the cor.test function in R [7]. Number of head and neck cancer tissues for analyzing RNA expressions and their correlations (TCGA): 44 (normal tissues) and 522 (tumor tissues)

[4]. On the other hand, copy number variation (CNV) may not necessarily lead to functional alterations of affected genes, but can change their expressions due to the altered gene dosage. Because CNVs occur in regions large enough to contain multiple genes, there is likely a similar pattern of expressional alterations of genes in the neighborhood sharing the same CNV.

Let us explore the human XRCC1 gene locus as an example. XRCC1 functions in DNA base excision repair (BER) as well as in DNA single-strand break repair (SSBR) [5]. XRCC1 has no known enzymatic activities; it interacts with APE1 (AP-endonuclease), PNKP (polynucleotide kinase/phosphatase), POLB (DNA polymerase beta), and LIG3 (DNA ligase III) and facilitates the "short-patch" BER [5, 6]. The XRCC1 gene is 37 kbp long at chr19: 43,543,311-43,580,473, according to the Genome Reference Consortium genome assembly GRCh38 (Table 1). Interestingly, the LIG1 (DNA ligase I), PNKP (polynucleotide kinase/phosphatase), and POLD1 (DNA polymerase δ subunit 1) genes are in the vicinity, distancing from the XRCC1 gene by 4.6, 6.3, and 6.8 Mbp, respectively. PNKP is a key partner of XRCC1 in the SSBR and also functions in the repair of tyrosine-derived DNA adducts generated by immature reaction attenuation of DNA topoisomerases [8]. POLD and LIG1 are key components of "long-patch" BER, which functions separately from the "short-patch" BER [6, 9]. Therefore, the XRCC1 locus contains genes that are critical for the two main BER pathways.

Tumor-associated expressional correlations of XRCC1 with these genes were analyzed using RNAseq data of head and neck squamous cell carcinoma tissues based on the data from the Cancer Genome Atlas (TCGA) (Fig. 1a). It is apparent that expression of XRCC1 is highly correlated with those of PNKP, LIG1, and POLD1 (Fig. 1a and Table 1). The SNP array data set of TCGA was used to examine CNV at the transcription initiation sites of

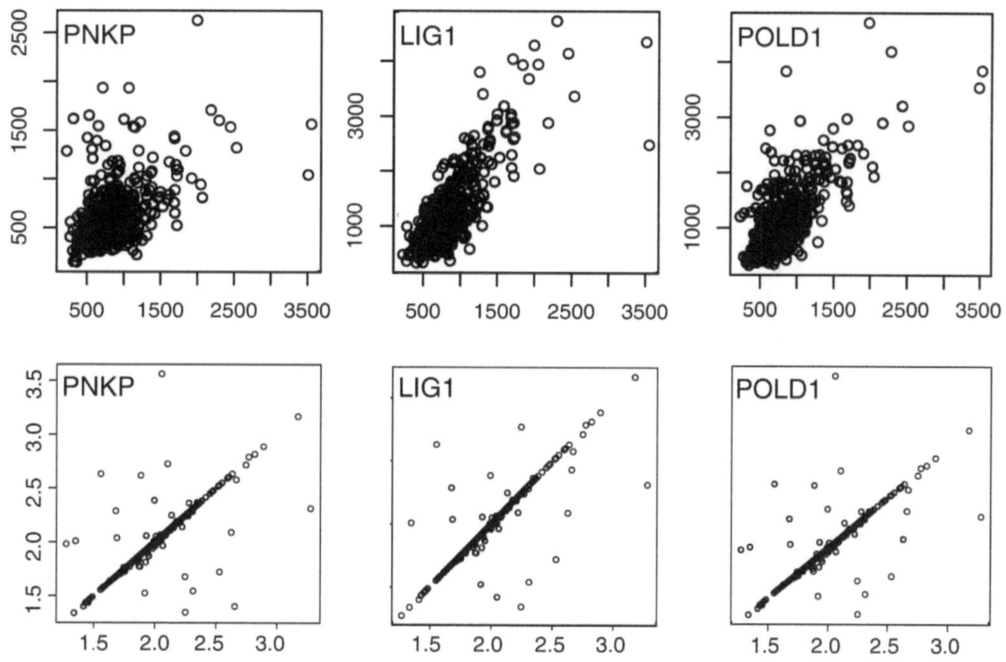

Fig. 1 Correlation of expression and CNV of XRCC1 with PNKP, LIG1, and POLD1 genes. (**a**) Gene expression correlation of XRCC1 with PNKP, LIG1, and POLD1. Normalized RNAseq intensities are blotted for XRCC1 (X-axis) vs. PNKP1 (left), LIG1 (middle), and POLD1 (right) on the Y-axis. (**b**) Correlation of CNV at the XRCC1 gene (X-axis) with PNKP, LIG1,and POLD1. RNAseq (**a**) and SNP array (**b**) data of 521 head and neck squamous cell carcinoma tissues from TCGA were used

these genes (Fig. 1b). The plots clearly indicate that the CNV values are almost identical among these genes, which is predicted from the fact that these genes are localized closely to one another. In addition, RNA expression levels of these genes are highly correlated with the extents of CNV (Fig. 2). Together, these data indicate that expressions of XRCC1, PNKP, LIG1, and POLD1 are all co-regulated in tumor tissues as a result of almost identical CNV in the genomic region.

It is known for a long time that XRCC1 can facilitate the overall BER/SSBR reactions in vitro and in cell culture studies [5, 10]. In vivo, high expression of XRCC1 has been associated with the poor outcome of chemoradiotherapies in patients with head and neck cancer [11]. Similar results were reported for LIG1 [12, 13]. The CNV/transcript profiles above suggest that simultaneous expressional alterations of these genes caused by CNV may synergistically affect the therapeutic outcome. Thus, accurate evaluation of CNV at the individual gene level in cancer tissues should illuminate the mechanism of resistance of tumors against therapeutic reagents and should help in improving diagnosis accuracy and treatment supported by modern genomics.

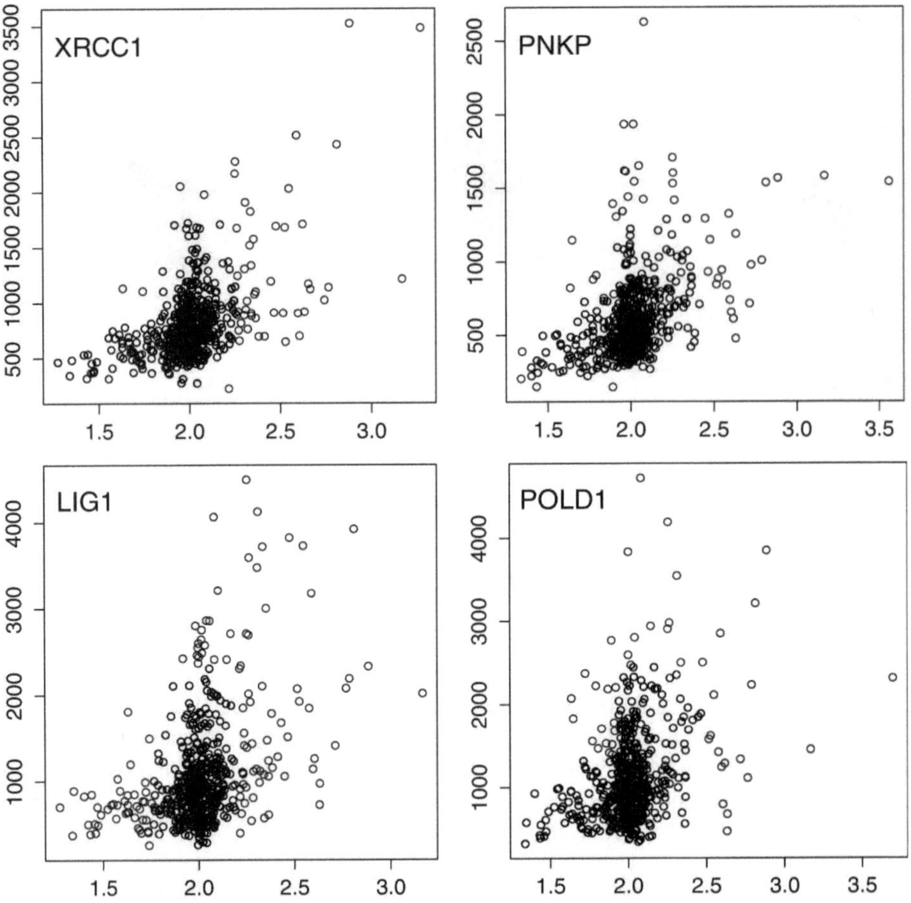

Fig. 2 Correlation between the extent of CNV (X-axis) and RNAseq (Y-axis) of the XRCC1, PNKP, LIG1, and POLD1 genes

2 Materials

2.1 NanoString Probe Code Set

First described in 2008 as a tool to digitally quantify RNA in biological samples [14], NanoString extended its capability to CNV determination [15, 16]. NanoString nCounter technology has advantage in that (1) the technique does not involve amplification of the genome DNA by PCR. (2) The assay is highly sensitive and can detect a subfemtomolar range of nucleic acids, to quantify the CNV of small amounts of DNA prepared from formalin-fixed, paraffin-embedded (FFPE) tissues. It provides digital values with high quality and reproducibility, realizing a replicate correlation of 0.999 [14]. (3) It has a lower cost compared to the massively parallel sequencing (aka next-generation sequencing). The principle of the analysis, which utilizes fluorescence-based bar codes, was described by Geiss et al. [14]. Each investigator needs to have access to a facility equipped with a NanoString nCounter system (either Sprint, Max, or Flex). There are facilities that accept assay request outside of their institutes, such as *Transcriptomics and Deep Sequencing Core* at the Johns Hopkins University. The PanCancer CNV panel, available from NanoString, is a preset of probe codes that covers 770 genes prone to CNV in carcinogenesis [17, 18]. However, investigators need to generate their own probe code sets for their own research interests with the technical assistance by NanoString. Three independent probes for a gene should be generated to avoid misleading results due to a bias of one probe.

2.2 Genome DNA Preparation from FFPE Slides

Pinpoint slide DNA isolation system (D3001, Zymo Research, Irvine, CA).

Pinpoint solution (blue solution).

Extraction buffer.

Binding buffer.

Washing buffer.

Spin columns.

Proteinase K.

5 or 10 μm FFPE slides.

Xylenes (laboratory grade, e.g., Thermo Fisher X4-4).

Ethanol (molecular biology grade, 99.5%, e.g., Sigma Aldrich E7023).

2.3 AluI Digestion of the Genomic DNA

Genomic DNA extracted from FFPE tissues: 320 ng.

Reagents provided by NanoString.

*Alu*I.

*Alu*I digestion buffer.

×10 CNV control probe (for *Alu*I digestion efficiency).

Genome DNA.

Up to 11 genome DNA extracted from FFPE.

A batch of human diploid genomic DNA as a control.

3 Methods: Analysis of CNV Using NanoString Sprint

3.1 Determination of Region of Interest (ROI) in FFPE Tissues

FFPE tissues are mixtures showing a variety of pathological features. These include normal, low to severe dysplasia, carcinoma in situ, and invasive carcinomas [19]. Obtaining genome DNA from these regions separately provides opportunities to examine pathological stage-specific CNV development. Although advanced algorithms have enabled effective use of artificial intelligence [20, 21] or simply computer-associated tissue analyses [19], experienced

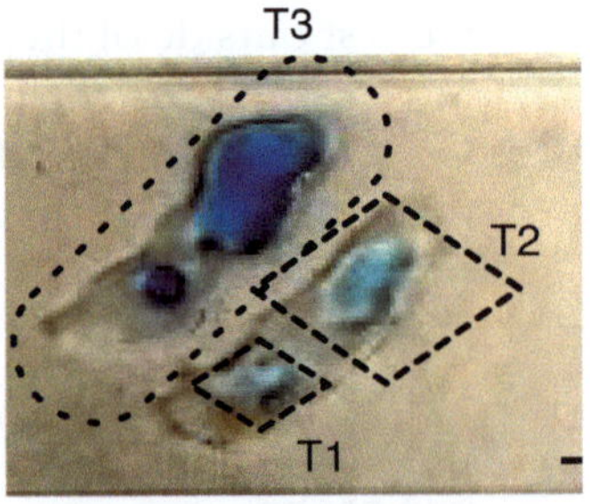

Fig. 3 An example of the FFPE tissue slide. The pinpoint solution (blue paste) was applied directly on the surface of tumor tissues with the ROIs marked (T1, T2, and T3)

pathologists dedicated to the specific cancer types is the most rigorous approach to determine ROI in the FFPE, e.g., benign epithelia vs. invasive carcinoma. With a pathologist with a long-time experience on specific type of cancers, CNV analysis can be expanded to precancerous regions such as carcinoma in situ, moderate to severe dysplasia, and difference between stroma and carcinoma tissues.

There are many commercially available kits for extracting genomic DNA from FFPE tissues. Snow et al. reported a quality control study comparing QIAamp FFPE (Qiagen, Valencia, CA) and Pinpoint FFPE isolation kit (Zymo Research) and found that there was not a significant difference between the preparations using the two kits (with the former performing a slightly better yield) [22]. We tested both kits in our hand and found that there was not much difference in the quality and the quantity of the purified genome DNA (data not shown). However, we find the pinpoint isolation kit helpful to determine the region of interest (ROI) in FFPE tissues on slides. FFPE tissues are always in mix of the ROI and nontarget areas (e.g., benign vs. invasive carcinoma regions). The blue viscous pinpoint solution makes tissue regions thin filmlike materials after being dried, which are easy to peel off from the slides with a scalpel or blade (Fig. 3). The following protocol is based on the pinpoint isolation Kit.

3.2 Genome DNA Preparation from FFPE Slides

3.2.1 Removal of Paraffin from FFPE Tissues

- Fill a glass staining jar (such as Thermo Fisher 08-816) with xylene, and dip FFPE slides. Incubate for 30 min at room temperature.
- Place the slides in a jar containing 90%, 70%, and 50% ethanol sequentially each for 10 min.
- Wash the slides in H2O.
- Dry the slides completely. The slides can be left at room temperature overnight in a drawer.

3.2.2 Marking ROIs on the Tissue Slides

- Mark ROIs on the back of the FFPE slides (Fig. 3). H&E slides with ROI marked by pathologists, such as invasive carcinoma, should be used as references.
- Put the pinpoint (blue) solution on the FFPE slides. The solution is quite viscous and difficult to pipet in/out. Instead, directly scoop the solution using P200 tips to paste it on to the marked regions.
- Dry the solution completely by leaving them for 3 h to overnight at room temperature.
- The blue solution dries to a thin film. Peel off the blue region from the slides using a blade. If only a portion of the tissue should be obtained, first make cutting lines along the borders with the blade first, to peel off only the necessary region easily.
- Put the pieces of filmlike tissue into microtubes. Handle the materials carefully as they are easily lost in the air from tubes due to electrostatic force. Having the lysis solution (below) in the microtubes prior to this step helps to put the tissue safely.

3.2.3 Lysis

- The following protocol is based on the vendor's instruction manual.
- Prepare the extraction buffer of the kit by mixing the 50 μL pinpoint extraction buffer with 5 μL of proteinase K solution provided in the kit. Alternatively, the generic proteinase K at 10 mg/mL concentration can also lyse the tissue sufficiently.
- Incubate the tissues in the extraction buffer at 55 °C for 3 h. Alternatively, the tubes can be incubated in a 55 °C chamber overnight.
- Confirm the tissues are lysed thoroughly. Should the thin film-like tissues be still visible, dissolve by a few strokes of pipetting.
- Add the 100 μL pinpoint binding solution in the lysed tissues, and apply on to the spin columns.
- Centrifuge the columns for 30 s at 12,000 rpm in a microcentrifuge. Discard the flow-through, and replace the columns.
- Wash the columns with 150 μL pinpoint wash buffer. Repeat the washing step once.
- Remove any residual solution by centrifuging for 1 min at 10,000 rpm.
- Replace the columns on new microtubes, and put elution 20 μL buffer.
- Elute the DNA by spinning briefly (30 s at 12,000 rpm in a microcentrifuge).

3.2.4 Confirmation of DNA

- Determine DNA concentration by measuring absorbance at 260 nm and 280 nm in a UV spectrophotometer (NanoDrop).

3.3 *Alu*I Digestion of Extracted DNA

The genomic DNA needs to be fragmented to the average approximate size of 200–300 bp. A controlled DNA sheering process with the Covaris AFA system (e.g., Covaris E220) has advantage in both precision and flexibility of the size adjustment, while the ideal condition has to be preliminarily determined. A simple and effective method for fragmentation is to digest the genomic DNA with *Alu*I restriction enzyme. The *Alu*I enzyme and the reaction buffer, provided by the NanoString, generate reproducible complete digestion, generating DNA fragments of approximately 500 bp that is appropriate for the CNV assay. A CNV assay can process 12 samples at a time. It is strongly recommended to include a diploid genomic DNA as a control (e.g., human genomic DNA G1521/G1471, Promega). Therefore, consider preparing up to 11 test samples (and a diploid control) for a CNV assay.

- Mix DNA (320 ng) in 7.5 μL H_2O in microtubes.
- Add 1 μL *Alu*I buffer.
- Add 1 μL ×10 CNV control probe.
- Add 1 μL *Alu*I.
- Incubate the samples at 37 °C for 2 h, and then place the tubes on ice or at 4 °C. It is convenient to use a temperature controller such as a PCR thermocycler.
- Run 0.5 μL of reactants in 1–2% agarose gel to confirm the fragmentation.
- Store the reactants in −20 °C.

3.4 Assay Using NanoString Sprint and the Interpretation of Results

CNV assays are typically done at a core facility equipped with a NanoString nCounter system (Sprint, Max, or Flex). The protocol of the CNV assay can be found in the manufacturer's website (https://www.nanostring.com/support/support-documentation/). Here, interpretation of the probe signals is described.

Data: In addition to data for the DNA to be examined, there are presets of probes for quality controls as followings.

C1: Controls for assessing *Alu*I digestion efficiency.

A set of four probes are always analyzed to check the efficiency of *Alu*I reaction. Probes A and B detect genome regions containing an *Alu*I site, and thus, signals from these probes should be as low as the baseline (C2) when *Alu*I digestion was successful. The C and D probes do not contain *Alu*I sites, and so the signals at C and D probes should be much higher than those of A and B.

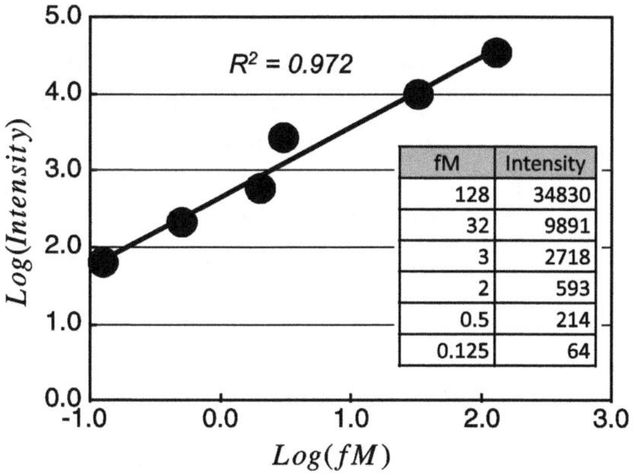

Fig. 4 An example of intensities of control probes determined with NanoString nCounter. The concentration and the corresponding intensity values were plotted on the X- and Y-axis, respectively. The correlation coefficient (R2 = 0.972) by the linear regression evaluation assures the high quality of the NanoString assay run

C2: Negative control probes.

There are eight probes always analyzed as negative controls that do not hybridize to human genomic DNA, and thus, neither of these probes should generate baseline signals near zero.

C3: Positive control probes.

Each CNV assay contains six positive control probes that are at different concentrations (0.125, 0.5, 2.0, 8.0, 32.0, and 128.0 femtomolar). Plotting the signal intensities of these probes over the concentration will provide the coefficient of determination R^2 that should be higher than 0.95 (Fig. 4).

C4: Invariant controls.

Each code set contains ten probes that do not contain genome regions prone to common CNV. The signals from these probes will be used to normalize values of the probes to be examined.

C5: Diploid genomic DNA.

Unlike C1–C4, investigators need to include this control, i.e., genomic DNA from normal human cells (see the *Alu*I digestion step). C5 is used to provide signal intensities for diploid genomic DNA.

3.5 Quality Control

Basic quality control examination should be performed using the control probes C1 and C2, which are described above. Evaluate correlation curves from the C3 positive controls, which must be done for all the samples individually. A low coefficient ($R^2 < 0.95$) of a particular sample implies issues in the sample preparation (low concentration, impurity, etc.). Low R^2 of all samples, particularly with the control genomic DNA C5, may imply a systematic failure of the assay. Check whether the right barcode file for the probes is used during the evaluation. Each NanoString code set needs a specific barcode file, and using incorrect probe files generates unanalyzable results.

For normalization, the diploid genome DNA control (C5) is used to evaluate the copy numbers into biologically meaningful units, i.e., making values from the normal diploid genome DNA 2.0 by the following workflow.

1. Obtain mean probe intensities of invariant controls (M_{C4}) for all samples.
2. Normalize each probe intensities by M_{C4}.
3. Multiply M_{C4} by the corresponding M_{C4} of C5.
4. Multiply by 2.0.

4 Notes

1. RNA contamination will cause overestimation of the concentrations of the genomic DNA from FFPE. The quality of DNA can be monitored by Abs260/Abs280 that should be in the range of 1.7 to 1.9. Alternatively, fluorescence probes specific to duplex DNA, e.g., PicoGreen (Invitrogen) and QuantiFluor (Promega) [23, 24], can be used to quantify DNA. Aliquots of samples can also be treated with/without RNase A and analyzed in an agarose gel, to examine whether the samples mainly contain DNA and not RNA.
2. There are other methods for determination of the genome CNV. These include traditional in situ hybridization [25], Taqman Copy Number Assay [26], OncoScan [27], array-comparative genomic hybridization (aCGH), Affymetrix SNP array [28], or deep sequencing results [29]. Merits using NanoString technology are described above.
3. Transcriptome data such as RNAseq or qRT-PCR panels, if available, can directly assess the expression profiles of a cluster of genes [30] such as XRCC1/LIG1/PNKP/POLD1 for studying prognostic markers. The CNV assay described here has advantage over RNA analysis for genomic DNA is more stable than RNA particularly those from archived FFPE slides. CNV determination may also be necessary to study the mechanism of alteration of transcript expressions of a cluster of gene which may be determined by RNA analyses.

Acknowledgements

This work was supported by NIH grant R03CA249111 (T.I.).

References

1. Hastings PJ, Lupski JR, Rosenberg SM, Ira G (2009) Mechanisms of change in gene copy number. Nat Rev Genet 10:551–564
2. Gerasymchuk M (2021) Genomic instability and aging: causes and consequences. In: Kovalchuk I, Kovalchuk O (eds) Genome stability: from virus to human application, 2nd edn. Academic Press, New York, pp 533–553
3. Moore L, Cagan A, Coorens THH, Neville MDC, Sanghvi R, Sanders MA, Oliver TRW, Leongamornlert D, Ellis P, Noorani A, Mitchell TJ, Butler TM, Hooks Y, Warren AY, Jorgensen M, Dawson KJ, Menzies A, O'Neill L, Latimer C, Teng M, van Boxtel R, Iacobuzio-Donahue CA, Martincorena I, Heer R, Campbell PJ, Fitzgerald RC, Stratton MR, Rahbari R (2021) The mutational landscape of human somatic and germline cells. Nature 597:381–386
4. Yamamoto S, Iwakuma T (2018) Regulators of oncogenic mutant TP53 gain of function. Cancers (Basel) 11:4
5. Caldecott KW (2019) XRCC1 protein; Form and function. DNA Repair (Amst) 81:102664
6. Hegde ML, Izumi T, Mitra S (2012) Oxidized base damage and single-strand break repair in mammalian genomes: role of disordered regions and posttranslational modifications in early enzymes. Prog Mol Biol Transl Sci 110: 123–153
7. Best DJ, Roberts DE (1975) Algorithm AS 89: the upper tail probabilities of Spearman's rho. J R Statist Soci SerC (Appl Statist) 24:377–379
8. Mellon I, Izumi T (2021) Base excision repair and nucleotide excision repair. In: Kovalchuk I, Kovalchuk O (eds) Genome stability: from virus to human application, 2nd edn. Academic Press, New York, pp 293–322
9. Klungland A, Lindahl T (1997) Second pathway for completion of human DNA base excision-repair: reconstitution with purified proteins and requirement for DNase IV (FEN1). EMBO J 16:3341–3348
10. Vidal AE, Boiteux S, Hickson ID, Radicella JP (2001) XRCC1 coordinates the initial and late stages of DNA abasic site repair through protein-protein interactions. EMBO J 20: 6530–6539
11. Ang MK, Patel MR, Yin XY, Sundaram S, Fritchie K, Zhao N, Liu Y, Freemerman AJ, Wilkerson MD, Walter V, Weissler MC, Shockley WW, Couch ME, Zanation AM, Hackman T, Chera BS, Harris SL, Miller CR, Thorne LB, Hayward MC, Funkhouser WK, Olshan AF, Shores CG, Makowski L, Hayes DN (2011) High XRCC1 protein expression is associated with poorer survival in patients with head and neck squamous cell carcinoma. Clin Cancer Res 17:6542–6552
12. Martinez-Terroba E, Ezponda T, Bertolo C, Sainz C, Remirez A, Agorreta J, Garmendia I, Behrens C, Pio R, Wistuba II, Montuenga LM, Pajares MJ (2018) The oncogenic RNA-binding protein SRSF1 regulates LIG1 in non-small cell lung cancer. Lab Investig 98: 1562–1574
13. Hussain MK, Singh DK, Singh A, Asad M, Ansari MI, Shameem M, Krishna S, Valicherla GR, Makadia V, Meena S, Deshmukh AL, Gayen JR, Imran Siddiqi M, Datta D, Hajela K, Banerjee D (2017) A novel Benzocoumarin-stilbene hybrid as a DNA ligase I inhibitor with in vitro and in vivo antitumor activity in breast cancer models. Sci Rep 7:10715
14. Geiss GK, Bumgarner RE, Birditt B, Dahl T, Dowidar N, Dunaway DL, Fell HP, Ferree S, George RD, Grogan T, James JJ, Maysuria M, Mitton JD, Oliveri P, Osborn JL, Peng T, Ratcliffe AL, Webster PJ, Davidson EH, Hood L, Dimitrov K (2008) Direct multiplexed measurement of gene expression with color-coded probe pairs. Nat Biotechnol 26:317–325
15. Ruderfer DM, Chambert K, Moran J, Talkowski M, Chen ES, Gigek C, Gusella JF, Blackwood DH, Corvin A, Gurling HM, Hultman CM, Kirov G, Magnusson P, O'Donovan MC, Owen MJ, Pato C, St Clair D, Sullivan PF, Purcell SM, Sklar P, Ernst C (2013) Mosaic copy number variation in schizophrenia. Eur J Hum Genet 21:1007–1011
16. Brahmachary M, Guilmatre A, Quilez J, Hasson D, Borel C, Warburton P, Sharp AJ (2014) Digital genotyping of macrosatellites and multicopy genes reveals novel biological functions associated with copy number variation of large tandem repeats. PLoS Genet 10: e1004418
17. McIntyre JB, Rambau PF, Chan A, Yap S, Morris D, Nelson GS, Kobel M (2017)

Molecular alterations in indolent, aggressive and recurrent ovarian low-grade serous carcinoma. Histopathology 70:347–358

18. Kutasovic JR, McCart Reed AE, Males R, Sim S, Saunus JM, Dalley A, McEvoy CR, Dedina L, Miller G, Peyton S, Reid L, Lal S, Niland C, Ferguson K, Fellowes AP, Al-Ejeh F, Lakhani SR, Cummings MC, Simpson PT (2019) Breast cancer metastasis to gynaecological organs: a clinico-pathological and molecular profiling study. J Pathol Clin Res 5:25–39
19. Wicker CA, Takiar V, Suganya R, Arnold SM, Brill YM, Chen L, Horbinski CM, Napier D, Valentino J, Kudrimoti MR, Yu G, Izumi T (2020) Evaluation of antioxidant network proteins as novel prognostic biomarkers for head and neck cancer patients. Oral Oncol 111: 104949
20. Srinidhi CL, Ciga O, Martel AL (2021) Deep neural network models for computational histopathology: a survey. Med Image Anal 67: 101813
21. Bera K, Schalper KA, Rimm DL, Velcheti V, Madabhushi A (2019) Artificial intelligence in digital pathology – new tools for diagnosis and precision oncology. Nat Rev Clin Oncol 16: 703–715
22. Snow AN, Stence AA, Pruessner JA, Bossler AD, Ma D (2014) A simple and cost-effective method of DNA extraction from small formalin-fixed paraffin-embedded tissue for molecular oncologic testing. BMC Clin Pathol 14:30
23. Suganya, R., Chakraborty, A., Miriyala, S., Hazra, T. K., and Izumi, T. (2015) Suppression of oxidative phosphorylation in mouse embryonic fibroblast cells deficient in apurinic/apyrimidinic endonuclease. DNA Repair (Amst) 27, 40–48 (PMID 25645679)
24. Kremling, K. A. G., Chen, S. Y., Su, M. H., Lepak, N. K., Romay, M. C., Swarts, K. L., Lu, F., Lorant, A., Bradbury, P. J., and Buckler, E. S. (2018) Dysregulation of expression correlates with rare-allele burden and fitness loss in maize. Nature 555, 520–523 (PMID 29539638)
25. Chia NL, Slater HR, Potter JM (2015) Detection of segmental chromosome copy number gains by improved fluorescence in situ hybridization techniques. J Assoc Genet Technol 41: 5–11
26. Mayo P, Hartshorne T, Li K, McMunn-Gibson C, Spencer K, Schnetz-Boutaud N (2010) CNV analysis using TaqMan copy number assays. *Curr Protoc Hum Genet* 2(Unit2): 13
27. Wang CI, Kao HK, Chen TW, Huang Y, Cheng HW, Yi JS, Hung SY, Wu CS, Lee YS, Chang KP (2019) Characterization of copy number variations in oral cavity squamous cell carcinoma reveals a novel role for MLLT3 in cell invasiveness. Oncologist 24:e1388–e1400
28. Haraksingh RR, Abyzov A, Urban AE (2017) Comprehensive performance comparison of high-resolution array platforms for genome-wide copy number variation (CNV) analysis in humans. BMC Genomics 18:321
29. Chen J (2021) Statistical considerations on NGS data for inferring copy number variations. Methods Mol Biol 2243:27–58
30. Wicker, C. A., and Izumi, T. (2016) Analysis of RNA expression of normal and cancer tissues reveals high correlation of COP9 gene expression with respiratory chain complex components. BMC Genomics 17, 983 (PMID 27903243)

Chapter 16

Genome-Wide Binding Analysis of DNA Repair Protein APE1 in Tumor Cells by ChIP-Seq

Mason Tarpley, Yingling Chen, and Kishor K. Bhakat

Abstract

The base excision repair (BER) is the primary damage repair pathway for repairing most of the endogenous DNA damage including oxidative base lesions, apurinic/apyrimidinic (AP) sites, and single-strand breaks (SSBs) in the genome. Repair of these damages in cells relies on sequential recruitment and coordinated actions of multiple DNA repair enzymes, which include DNA glycosylases (such as OGG1), AP-endonucleases (APE1), DNA polymerases, and DNA ligases. APE1 plays a key role in the BER pathway by repairing the AP sites and SSBs in the genome. Several methods have been developed to generate a map of endogenous AP sites or SSBs in the genome and the binding of DNA repair proteins. In this chapter, we describe detailed approaches to map genome-wide occupancy or enrichment of APE1 in human cells using chromatin immunoprecipitation followed by next-generation sequencing (ChIP-seq). Further, we discuss standard bioinformatics approaches for analyzing ChIP-seq data to identify APE1 enrichment or binding peaks in the genome.

Key words Chromatin immunoprecipitation, Cross-linking, APE1, Next-generation sequencing

1 Introduction

Cellular DNA is under constant attack from endogenous sources of DNA-damaging agents. The primary source of endogenous DNA damage is reactive oxygen species (ROS) which are generated from normal cellular metabolism [1]. The apurinic/apyrimidinic sites, also known as AP sites, are the most prevalent type of endogenous DNA damage in cells [2, 3]. AP sites are primarily generated either by spontaneous hydrolysis of nucleobases or after removal of oxidized or alkylated damaged bases from DNA by glycosylases in the BER pathway [2, 4–7]. Thousands of such AP sites are generated in every cell per day [2]. Repair of AP sites is essential because these are mutagenic and block transcription [3, 8]. Methods to detect formation and location of AP sites in the genome by tagging AP sites in vitro and in live cells (AP-seq) using aldehyde-reactive probe (ARP) have been described [9, 10].

Kishor K. Bhakat and Tapas K. Hazra (eds.), *Base Excision Repair Pathway: Methods and Protocols*, Methods in Molecular Biology, vol. 2701, https://doi.org/10.1007/978-1-0716-3373-1_16,

APE1 is the primary enzyme responsible for repairing the endogenous AP sites in the genome [4, 5]. APE1 recognizes the AP site and cleaves the DNA backbone [4, 5]. This results in the formation of one nucleotide gap flanked by the 3′-hydroxyl and 5′-deoxyribose-phosphate (5′-dRP) end, which is recognized by the downstream enzyme DNA polymerase β (pol β) [11–13]. Besides BER, APE1 also regulates transcription [14–16]. Studies by us and others show that APE1 itself can act as a trans-acting factor, binding to promoters and forming regulatory complexes to modulate gene expression including many genes [17–20]. We discovered earlier that chromatin-bound APE1 can be acetylated (AcAPE1) at multiple lysine residues in cells [21]. We have recently mapped genome-wide binding or occupancy of APE1 or AcAPE1 in the genome of multiple cell lines by immunoprecipitation of chromatin-bound AcAPE1 with specific antibodies followed by next-generation sequencing [22]. Genome-wide mapping studies revealed that AcAPE1 is highly enriched in gene promoters and putative enhancer regions [22, 23]. In this chapter, we provide a detailed methodology for mapping genome-wide binding or occupancy of APE1 in cells by ChIP-seq analysis [24]. This method can also be used to compare the formation of DNA damage and binding of other BER enzymes after induction of DNA damages. Further, we discuss about the standard bioinformatics approaches for analyzing the ChIP-seq data to identify enrichment or binding peaks in the genome [25–32].

2 Materials

2.1 Laboratory Equipment

1. Bioruptor sonicator.
2. Magnetic stand.
3. qPCR machine.
4. Fluorometer or Qubit.
5. Microcentrifuge.
6. Heat block and 37 °C water bath.
7. Rocker at 4 °C.
8. Rocker at room temperature.
9. Agarose gel running apparatus.

2.2 Reagents

1. Hydrogen peroxide (H_2O_2).
2. Methylmethane sulphonate (MMS).
3. Formaldehyde.
4. Glycine.
5. PBS.

6. Protein A or protein G magnetic beads.
7. SPRI beads.
8. 15 mg/mL proteinase K.
9. 20 mg/mL RNA-grade glycogen.
10. 3 M sodium acetate (pH 5.2).
11. 10 mg/mL RNase A.
12. PIPES, pH 8, 1 M.
13. 2 M KCl.
14. Nonidet P40 (NP40), 20%.
15. Protease inhibitors.
16. 1 M Tris–HCl, pH 8.1.
17. 0.5 M EDTA.
18. 20% SDS.
19. Triton X-100.
20. 5 M NaCl.
21. 4 M LiCl.
22. Sodium deoxycholate (SDC) or deoxycholic acid (DOC).
23. 1 M $NaHCO_3$.
24. Phenol/chloroform/isoamyl alcohol (PCI) (25:24:1).
25. 80% ethanol.
26. 70% ethanol.
27. 100% ethanol.
28. SYBR Green.
29. EB buffer.
30. 10× NEB end repair reaction buffer.
31. NEB end repair enzyme mix.
32. 10× NEB buffer 2.
33. 1 mM dATP.
34. NEB Klenow 3′-5′ exo minus.
35. 2× NEB Quick ligase reaction buffer.
36. 1 μM universal adapters.
37. NEB T4 ligase 3′-5′ exo minus.
38. 10 μM PrimerU.
39. 10 μM index primers.
40. ChIP antibodies.

2.3 Buffers and Solutions

1. Cell lysis buffer: 5 mM PIPES (pH 8.0), 85 mM KCl, 0.5% NP40, water, protease inhibitors.
2. Nuclear lysis buffer: 50 mM Tris–HCl, 10 mM EDTA, 1% SDS, water, protease inhibitors.
3. IP dilution buffer: 0.01% SDS, 1.1% Triton X-100, 1.2 mM EDTA, 16.7 mM Tris–HCl, 167 mM NaCl, water, protease inhibitors.
4. Low-salt IP wash buffer: 0.1% SDS, 1% Triton X-100, 2 mM EDTA, 20 mM Tris–HCl, 150 mM NaCl, water, protease inhibitors.
5. High-salt IP wash buffer: 0.1% SDS, 1% Triton X-100, 2 mM EDTA, 20 mM Tris–HCl, 500 mM NaCl, water, protease inhibitors.
6. LiCl buffer: 10 mM Tris–HCl, 1 mM EDTA, 0.25 M LiCl, 1% NP40, 1% SDC or DOC, water, protease inhibitors.
7. TE buffer: 10 mM Tris–HCl, 1 mM EDTA, water, protease inhibitors.

   IP elution buffer: 0.1 M $NaHCO_3$, 1% SDS, water.

3 Methods

Our ChIP protocol is modified from Bowman et al. [24].

3.1 Cell Culture and DNA Protein Cross-Linking

1. For ChIP, enough cells were plated the day before starting the procedure to ensure 20 × 10^6 cells are present. The next morning media was removed, and PBS was added. For cells treated with H_2O_2 or MMS, this was done before removing media and replacing with PBS. Cells were treated with 0.5 mM H_2O_2 for 30 min in a 37 °C incubator. MMS treatment was done with 0.5 mM MMS for 30 min in a 37 °C incubator.
2. Cells were scraped and pipetted into suspension and counted. Then 1% formaldehyde was added directly to the suspended cells and rocked for 10 min at room temperature (*see* **Notes 1** and **2**).
3. Following cross-linking reaction, 0.125 M glycine is added to quench the reaction and rocked for 5 min at room temperature.
4. Pellet the cells and remove the supernatant.
5. Wash the pellet two times with cold PBS, spinning down between washes.
6. Resuspend the pellet in 1 mL cell lysis buffer plus protease inhibitors and incubate for 15 min on ice.
7. Collect the nuclei by spinning the cells at 4000 rpm for 8 min at 4 °C.

8. Resuspend the nuclei in 200 μL ice-cold nuclei lysis buffer plus protease inhibitors and incubate on ice for 20 min.
9. Add 100 μL of cold IP dilution buffer plus protease inhibitors.
10. Using a Bioruptor, set the pulsations for 30 s on, 60 s off for 28 cycles on low power to achieve fragmented DNA in the 300–500 bp range (*see* **Note 3**).
11. Centrifuge the fragmented DNA at max speed for 10 min at 4 °C to pellet cellular debris. Transfer supernatant to new, clean tubes.

3.2 Checking Fragmentation Size

1. Remove 15 μL of supernatant to run as quality control of sonication fragment size.
2. Add 185 μL of IP dilution buffer to get a final volume of 200 μL and then add 0.5 μL 10 mg/mL RNase A.
3. Reverse cross-linking by heating at 95 °C for 15 min (or 65 °C overnight) (*see* **Note 4**).
4. Add 4 μL of 10 mg/mL proteinase K to remove proteins and incubate for 2 h at 50 °C.
5. Extract once with phenol/chloroform/isoamyl alcohol (25: 24:1) by adding an equal volume of P:C:I and vortex to mix.
6. Spin for 5 min at top speed in a microcentrifuge and then transfer the top aqueous layer to a new tube.
7. Add 1 μL of 20 mg/mL RNA-grade glycogen and 1/tenth volume of 3 M sodium acetate (pH 5.2, 0.3 M final concentration).
8. Add 2.5 volumes of ice-cold ethanol and incubate the samples at −80 °C for 30 min to overnight.
9. Spin samples for 10 min at 15,000 rpm at 4 °C.
10. Remove supernatant and wash pellet with cold 70% ethanol to remove salts and then let them air-dry.
11. Once dry, dissolve in 30 μL of UltraPure water and run 10 μL on a 2% agarose gel to confirm sonication fragment size is between 300 and 500 bp.

3.3 DNA Immunoprecipitation

1. For the remaining 285 μL, dilute fivefold with IP dilution buffer plus protease inhibitors (final volume of 1425 μL).
2. Remove 75 μL (5%) of DNA and add to a separate tube to be used as input control.
3. Add 5 μg of antibody and incubate overnight at 4 °C while rocking.
4. On the next morning, add 20 μL of protein A or G Dynabeads and rock for 2 h at 4 °C.
5. Collect the beads using a magnetic stand and discard the supernatant.

3.4 Bead Washing

1. Wash the beads three times with 1 mL low-salt buffer and incubate for 5 min on rocker at room temperature. Remember not to disrupt the bead pellet when removing the supernatant.
2. Wash beads two times with 1 mL high-salt buffer, as above, followed by two times LiCl and one time with TE buffer.
3. Elute the DNA from the beads with 200 μL freshly prepared IP elution buffer as above, pipetting the eluant into fresh tubes.
4. Add NaCl (20 μL of 5 M), EDTA (8 μL of 0.5 M), Tris–HCl (16 μL of 1 M), and incubate overnight at 65 °C to reverse cross-linking.
5. To the input DNA, add 369 μL of IP dilution buffer and take through reverse cross-linking as well.

3.5 DNA Cleanup

1. On the next morning, add 5.3 μL of 15 mg/mL proteinase K to all eluants and inputs and incubate at 50 °C for 2 h.
2. Extract with phenol/chloroform/isoamyl alcohol (25:24:1) by adding an equal volume of P:C:I and vortex to mix.
3. Spin for 5 min at top speed in a microcentrifuge and then transfer the top aqueous layer to a new tube.
4. Add 1 μL of 20 mg/mL RNA-grade glycogen and 1/tenth volume of 3 M sodium acetate (pH 5.2, 0.3 M final concentration).
5. Add 2.5 volumes of ice-cold ethanol and incubate the samples at −80 °C for 30 min to overnight.
6. Spin samples for 10 min at 15,000 rpm at 4 °C.
7. Remove supernatant and wash pellet with cold 70% ethanol to remove salts and then let them air-dry.
8. When the samples are completely dry, dissolve in 30 μL of Qiagen EB buffer (10 mM Tris–HCl pH 8.5). DNA concentrations were checked by Quantus Fluorometer before using for either library prep or qPCR.

3.6 ChIP Library Preparation

3.6.1 Removal of Unwanted DNA Fragments

1. Clean up the ChIP DNA by adding two volumes of room temperature SPRI beads and rotate for 5 min at room temperature.
2. Capture beads on magnetic stand and remove supernatant.
3. Wash two times with 200 μL of freshly prepared 80% ethanol, aspirating between washes. Air-dry samples with caps open, making sure there is no ethanol left in the tubes.
4. Elute in 25.5 μL of EB buffer by letting it sit for 1 min at room temperature and then moving supernatant to a new tube.

3.6.2 End Repair

1. Take the 25.5 μL of eluant from the DNA cleanup and add 3 μL of 10× NEB end repair reaction buffer and 1.5 μL NEB end repair enzyme mix and incubate for 30 min at room temperature.
2. Cleanup the DNA with SPRI beads by adding 2 volumes of SPRI beads and following the same steps as above.
3. Elute the DNA in 19 μL of EB buffer and proceed to the A-tailing reaction.

3.6.3 A-Tail Addition

1. Starting with the 19 μL from the end repair, add 3 μL 10× NEB buffer 2, 7 μL 1 mM dATP, and 1 μL NEB Klenow 3′-5′ exo minus.
2. Incubate for 30 min in a 37 °C water bath.
3. Clean up with two volumes of SPRI beads as before and elute in 24 μL of EB buffer.

3.6.4 Adapter Ligation

1. Add 3 μL 2× NEB quick ligase reaction buffer, 1 μL of 1 μM universal adapters, and 2 μL NEB T4 ligase 3′-5′ exo minus.
2. Incubate for 1 h at room temperature (*see* **Note 6**).
3. Conduct SPRI bead cleanup as before with 1.5 volumes of beads, eluting in 22 μL EB buffer.

3.6.5 PCR Amplification

1. Set up 50 μL qPCR reactions by adding 25 μL 2× KAPA SYBR master mix, 1.5 μL 10 μM PrimerU, and 1.5 μL 10 μM primer number with the proper index code.
2. Pipette 25 μL into a separate PCR tube and place in the machine.
3. Set the qPCR machine to do 30 s at 98 °C followed by 18 cycles of 10 s at 98 °C, 20 s at 64 °C, and 45 s at 72 °C.
4. Check that the reaction has worked by looking at the multi-component plot and seeing that the fluorescence is over 100,000 in 10–14 cycles. Inputs may be ran together, separate from the IP DNA, due to differences in starting DNA quantities.
5. Once PCR amplification has been completed and all samples have worked, conduct SPRI cleanup with 1.2 volumes of beads as before and elute in 15 μL of EB buffer to send to sequencing.

3.7 AcAPE1 ChIP-Seq Analysis

1. Sequencing was completed at the UNMC Genomics Core facility using an Illumina NovaSeq6000 (*see* **Note 5**). Adapter sequences were removed and fastq files were received from the sequencing facility.
2. Reads were mapped to hg 19, or mm 10, and then filtered by mapping quality > = 10 using samtools view [26, 27].
3. Duplicate reads were removed from bam files using Picard tools MarkDuplicates and transformed into bigwig files using bedtools genomecov [28, 33], normalized by reads per million mapped reads (RPM) [31].
4. ChIP-seq peaks were called using macs2 callpeak with the broad peak option for AcAPE1 [32].
5. ChIPseeker, in R, was used to determine the genome-wide distribution of peaks [31].
6. List of genes was downloaded from UCSC including all splice variants [29, 30]. All fastq files and bigwigs have been uploaded to Gene Expression Omnibus (GEO) with a later release date.

4 Notes

1. Always use freshly prepared 1% formaldehyde. Do not use very old (>6 months or more) formaldehyde stock solution.
2. Do not allow formaldehyde cross-linking step for more than 10 min at room temperature.
3. After fragmentation by a Bioruptor Plus, run fragmented chromatin in agarose gel and stain with EtBr to make sure the average fragment sizes are between 300 and 500 bp.
4. For reverse cross-linking, incubate at 65 °C for at least more than 6 h.
5. Please use an appropriate adapter and linker for the Illumina website that are compatible for sequencing.
6. Removing the duplicate sequence using the standard tools is important to eliminate the false enrichment.

References

1. Winterbourn CC (2008) Reconciling the chemistry and biology of reactive oxygen species. Nat Chem Biol 4(5):278–286
2. Lindahl T (1993) Instability and decay of the primary structure of DNA. Nature 362(6422): 709–715
3. Loeb LA, Preston BD (1986) Mutagenesis by apurinic/apyrimidinic sites. Annu Rev Genet 20:201–230
4. Dianov GL, Sleeth KM, Dianova II, Allinson SL (2003) Repair of abasic sites in DNA. Mutat Res 531(1–2):157–163
5. Li M, Wilson DM 3rd. (2014) Human apurinic/apyrimidinic endonuclease 1. Antioxid Redox Signal 20(4):678–707
6. Wallace SS (2014) Base excision repair: a critical player in many games. DNA Repair (Amst) 19:14–26

7. Scott TL, Rangaswamy S, Wicker CA, Izumi T (2014) Repair of oxidative DNA damage and cancer: recent progress in DNA base excision repair. Antioxid Redox Signal 20(4):708–726
8. Tornaletti S, Maeda LS, Hanawalt PC (2006) Transcription arrest at an abasic site in the transcribed strand of template DNA. Chem Res Toxicol 19(9):1215–1220
9. Poetsch AR, Boulton SJ, Luscombe NM (2018) Genomic landscape of oxidative DNA damage and repair reveals regioselective protection from mutagenesis. Genome Biol 19(1):215
10. Liu ZJ, Martinez Cuesta S, van Delft P, Balasubramanian S (2019) Sequencing abasic sites in DNA at single-nucleotide resolution. Nat Chem 11(7):629–637
11. Mitra S, Izumi T, Boldogh I, Bhakat KK, Hill JW, Hazra TK (2002) Choreography of oxidative damage repair in mammalian genomes. Free Radic Biol Med 33(1):15–28
12. Robertson AB, Klungland A, Rognes T, Leiros I (2009) DNA repair in mammalian cells: base excision repair: the long and short of it. Cell Mol Life Sci 66(6):981–993
13. Wilson SH, Kunkel TA (2000) Passing the baton in base excision repair. Nat Struct Biol 7(3):176–178
14. Bhakat KK, Mantha AK, Mitra S (2009) Transcriptional regulatory functions of mammalian AP-endonuclease (APE1/Ref-1), an essential multifunctional protein. Antioxid Redox Signal 11(3):621–638
15. Georgiadis MM, Luo M, Gaur RK, Delaplane S, Li X, Kelley MR (2008) Evolution of the redox function in mammalian apurinic/apyrimidinic endonuclease. Mutat Res 643(1–2):54–63
16. Xanthoudakis S, Miao G, Wang F, Pan YC, Curran T (1992) Redox activation of Fos-Jun DNA binding activity is mediated by a DNA repair enzyme. EMBO J 11(9):3323–3335
17. Bhakat KK, Izumi T, Yang SH, Hazra TK, Mitra S (2003) Role of acetylated human AP-endonuclease (APE1/Ref-1) in regulation of the parathyroid hormone gene. EMBO J 22(23):6299–6309
18. Sengupta S, Mantha AK, Mitra S, Bhakat KK (2011) Human AP endonuclease (APE1/Ref-1) and its acetylation regulate YB-1-p300 recruitment and RNA polymerase II loading in the drug-induced activation of multidrug resistance gene MDR1. Oncogene 30(4):482–493
19. Ziel KA, Campbell CC, Wilson GL, Gillespie MN (2004) Ref-1/Ape is critical for formation of the hypoxia-inducible transcriptional complex on the hypoxic response element of the rat pulmonary artery endothelial cell VEGF gene. FASEB J 18(9):986–988
20. Antoniali G, Lirussi L, D'Ambrosio C, Dal Piaz F, Vascotto C, Casarano E, Marasco D, Scaloni A, Fogolari F, Tell G (2014) SIRT1 gene expression upon genotoxic damage is regulated by APE1 through nCaRE-promoter elements. Mol Biol Cell 25(4):532–547
21. Roychoudhury S, Nath S, Song H, Hegde ML, Bellot LJ, Mantha AK, Sengupta S, Ray S, Natarajan A, Bhakat KK (2017) Human Apurinic/Apyrimidinic Endonuclease (APE1) is acetylated at DNA damage sites in chromatin, and acetylation modulates its DNA repair activity. Mol Cell Biol 37(6):e00401
22. Roychoudhury S, Pramanik S, Harris HL, Tarpley M, Sarkar A, Spagnol G, Sorgen PL, Chowdhury D, Band V, Klinkebiel D, Bhakat KK (2020) Endogenous oxidized DNA bases and APE1 regulate the formation of G-quadruplex structures in the genome. Proc Natl Acad Sci U S A 117(21):11409–11420
23. Pramanik S, Chen Y, Song H, Khutsishvili I, Marky LA, Ray S, Natarajan A, Singh PK, Bhakat KK (2022) The human AP-endonuclease 1 (APE1) is a DNA G-quadruplex structure binding protein and regulates KRAS expression in pancreatic ductal adenocarcinoma cells. Nucleic Acids Res 50(6):3394–3412
24. Bowman SK, Simon MD, Deaton AM, Tolstorukov M, Borowsky ML, Kingston RE (2013) Multiplexed Illumina sequencing libraries from picogram quantities of DNA. BMC Genomics 14:466
25. Lander ES, Linton LM, Birren B, Nusbaum C, Zody MC, Baldwin J, Devon K, Dewar K, Doyle M, FitzHugh W, Funke R, Gage D, Harris K, Heaford A, Howland J, Kann L et al (2001) Initial sequencing and analysis of the human genome. Nature 409(6822):860–921
26. Li H (2011) A statistical framework for SNP calling, mutation discovery, association mapping and population genetical parameter estimation from sequencing data. Bioinformatics 27(21):2987–2993
27. Li H, Durbin R (2010) Fast and accurate long-read alignment with burrows-wheeler transform. Bioinformatics 26(5):589–595
28. Quinlan AR, Hall IM (2010) BEDTools: a flexible suite of utilities for comparing genomic features. Bioinformatics 26(6):841–842
29. Robinson JT, Thorvaldsdottir H, Winckler W, Guttman M, Lander ES, Getz G, Mesirov JP (2011) Integrative genomics viewer. Nat Biotechnol 29(1):24–26

30. Thorvaldsdottir H, Robinson JT, Mesirov JP (2013) Integrative genomics viewer (IGV): high-performance genomics data visualization and exploration. Brief Bioinform 14(2): 178–192
31. Yu G, Wang LG, He QY (2015) ChIPseeker: an R/Bioconductor package for ChIP peak annotation, comparison and visualization. Bioinformatics 31(14):2382–2383
32. Zhang Y, Liu T, Meyer CA, Eeckhoute J, Johnson DS, Bernstein BE, Nusbaum C, Myers RM, Brown M, Li W, Liu XS (2008) Model-based analysis of ChIP-Seq (MACS). Genome Biol 9(9):R137
33. Amemiya HM, Kundaje A, Boyle AP (2019) The ENCODE blacklist: identification of problematic regions of the genome. Sci Rep 9(1): 9354

Chapter 17

Tumorsphere Formation Assay: A Cancer Stem-Like Cell Characterization in Pediatric Brain Cancer Medulloblastoma

Sutapa Ray

Abstract

Cancer is a heterogeneous disease, comprising of a mixture of different cell populations. Cancer stem cells (CSCs), also known as tumor-initiating cells (TICs), are a subpopulation of multipotent cells within the cancer that has self-renewing capability, tumor-initiating ability, multi-differentiation potential, and an inherent capacity for drug and chemoresistance. Sphere-formation assay is commonly used for enrichment and analysis of CSC properties in vitro and is typically used as a metric for testing the viability of tumor cells to anticancer agents. This model is based on the ability of CSCs to grow under ultralow-attachment conditions in serum-free medium supplemented with growth factors. In contrast to the adherent 2D culture of cancer cells, the 3D culture of tumorsphere assay exploits inherent biologic features of CSCs such as anoikis resistance and self-renewal. We describe here the detailed methodology for the generation and propagation of spheres generated from pediatric brain tumor medulloblastoma (MB) cells. As signal transducer and activator of transcription (STAT3) is known to play an important role in maintaining cancer stem cell properties, we accessed the effect of depleting or inhibiting STAT3 on MB-sphere sizes, numbers, and integrity. This may serve as a promising platform for screening potential anti-CSC agents and small-molecule inhibitors.

Key words Tumorspheres, Cancer stem cells, Tumor-initiating cells, Self-renewal

1 Introduction

Accumulating evidence has suggested that a subpopulation of stem-like cells within tumors, known as CSCs, exhibits characteristics of both stem cells and cancer cells and can form tumors when transplanted into an animal host [1, 2]. Also, this subgroup of cancer cells is suggested to be responsible for drug resistance and cancer relapse due in part to their ability to self-renew themselves and differentiate into heterogeneous lineages of cancer cells [3–5]. Therefore, the development of treatments that target cancer stem cells is an important objective. The first evidence for CSCs

Kishor K. Bhakat and Tapas K. Hazra (eds.), *Base Excision Repair Pathway: Methods and Protocols*, Methods in Molecular Biology, vol. 2701, https://doi.org/10.1007/978-1-0716-3373-1_17,

came from acute myeloid leukemia [6]. There is now increasing evidence for CSCs in a variety of solid tumors, e.g., breast, brain, melanoma, prostate, colon, pancreatic, and lung cancers [7–13]. As CSCs have the ability to grow through self-renewal as nonadherent spherical clusters, it is commonly known as tumorspheres.

Medulloblastoma (MB) is the most common malignant pediatric brain tumor. It is an embryonic tumor that arises from the cerebellar neuronal progenitor cells [14, 15]. The current standard of care for MB patients involves a combination of maximal surgical resection followed by craniospinal radiation and high doses of chemotherapy [16–19]. Although the current treatment regimen in MB kills the bulk of proliferating tumor cells, a subset of remaining CSCs can survive and promote cancer relapse through their self-renewal capability and greater DNA repair ability, displaying resistance to conventional anticancer therapies [4, 20–23]. Therefore, the improvement of therapies targeting CSCs may raise hope for the treatment of the MB patient. Several studies have identified key players involved in the intracellular signal transduction pathways regulating stem cell renewal and proliferation. Leukemia inhibitory factor (LIF)-induced activated STAT3 signaling is known to maintain the pluripotent and self-renewing state of murine embryonic stem (ES) [24]. STAT3 is constitutively activated by tyrosine 705 phosphorylation in many murine and human malignant tumors, including MB, and recent studies reported that inhibition of the STAT3-mediated signaling pathway reduces growth of MB cells and affects MB cell sensitivity to conventional chemotherapeutic agents such as cisplatin [25–30]. Here, we demonstrate that the selective inhibition of STAT3 in the tumorsphere assay can attenuate MB cell growth and self-renewal, suggesting that specific targeting of STAT3 would be valuable to inhibit tumor growth at the level of primitive cells [31–34].

2 Materials

2.1 MB Cell Lines

DAOY (purchased from ATCC, Rockville, MD, USA) and ONS-76 (purchased from Sekisui-XenoTech; Kansas City, KS, USA) MB cells were used for sphere formation.

2.2 Reagents

1. EMEM and RPMI 1640 media.
2. Fetal bovine serum (FBS).
3. Dulbecco's PBS without calcium and magnesium (D-PBS).
4. Sodium pyruvate 100 mM solution.
5. Penicillin–streptomycin.
6. 0.05% trypsin/EDTA.
7. Trypan Blue solution (0.4%).

8. Ultralow-attachment 6-well plates.
9. EGF 20 ng/mL.
10. bFGF 40 ng/mL.
11. Heparin 2 μg/mL.
12. β-Me 0.1 mM.
13. B27 1%.
14. N2 1%.
15. Doxycycline (Dox) was purchased from Sigma Aldrich.
16. WP1066 was purchased from Selleckchem.

2.3 Laboratory Equipment

1. Cell counter (Cellometer Auto 2000).
2. Tabletop centrifuge.
3. EVOS Cell Imaging Systems.

3 Methods

1. Prior to MB-sphere (medullosphere) culture, DAOY cells were grown in a monolayer-adherent condition in EMEM, supplemented with 10% FBS and 1% penicillin–streptomycin and ONS 76 cells harboring doxycycline (Dox)-inducible STAT3 shRNA (ONS-shSTAT3) were grown in RPMI 1640 media, supplemented with 10% FBS, 1 mM sodium pyruvate and 1% penicillin-streptomycin. Both cells were grown in a T-75 flask in a humidified incubator with 5% CO_2 and 95% air atmosphere at 37 °C [31].
2. When cells reach confluence, remove media from the flask and gently rinse cells once with 10 mL PBS.
3. Detach cells from flask with 3 mL of trypsin. Tilt the flask to disperse trypsin evenly all over the cells and remove excess trypsin. Incubate the flask in 37 °C, 5% CO_2 incubator for 5 min.
4. When the cells will start to round up and detach from the flask, add 10 mL EMEM or RPMI 1640 serum-free media. Count cells (with Trypan Blue) in a cell counter and then centrifuge 300 × *g* for 5 min.
5. Add sphere media (serum-free EMEM or RPMI media containing either EGF 20 ng/mL, bFGF 40 ng/mL, heparin 2 μg/mL, β-Me 0.1 mM, B27 1%, N2 1%, and 1% penicillin–streptomycin) to obtain 1×10^5 cells/ml. Add 10 K, 20 K, and 40 K cells per well in a total volume of 2 mL in low-attachment 6-well plates.
6. Incubate the culture for 7–14 days, depending on the cell types. Add 1 mL fresh sphere media (with supplements) every 3–4 days (*see* **Note 1**).

7. Sphere formation should appear within 4–7 days (*see* **Note 2**).
8. Count small-, medium-, and large-size medullosphere numbers using EVOS Cell Imaging System and the 40× magnification lens.
9. Stem cell markers (CD133, Nestin, Sox2, Oct3/4, etc.) in medullospheres can be detected by centrifuging the spheres at 300 g/5 min followed Western Blot [31, 35].
10. Propagation of medullospheres:
    (i) For passaging medullospheres, medium- to large-size spheres from 7 to 10 days' culture were selected.
    (ii) Transfer spheres and the media into 15 mL conical tubes using a sterile 10 mL serological pipet and allow the spheres to settle at the bottom of the tube for 10–15 min.
    (iii) Aspirate media from the top leaving 200–300 μL of media in the tube containing the spheres.
    (iv) Wash the spheres with 10 mL PBS and let the spheres settle at the bottom of the tube for 10–15 min. Aspirate PBS as above leaving 200–300 μL of PBS with the spheres in the conical tube. Repeat washing with PBS one more time.
    (v) Now add 1.5 mL trypsin into the tube containing 200 μL spheres in PBS and incubate it for 5 min at 37 °C incubator to disperse the cells from the sphere cluster. Further, suspend the spheres along with the trypsin by pipetting (with the help of 1 mL pipet tips) up and down once every min for ten times.
    (vi) Add 3 mL serum-free media and determine cell number and viability before centrifuging the cells at 300 g/5 min. Resuspend pellet in sphere media and reseed the cells at the same seeding density as mentioned in #5 (*see* **Note 3**).
11. Effect of STAT3 knockdown with Dox or inhibition with small-molecule inhibitor WP1066 on medullosphere size and number:

    To determine whether STAT3 pathway activity is required for the survival or self-renewal of medullosphere, we knockdown STAT3 level either with Dox-inducible expression of STAT3 shRNA in ONS-shSTAT3 cells or with a small-molecule inhibitor WP1066, which blocks STAT3 phosphorylation in DAOY cells and performed a tumorsphere formation assay in vitro. We found that ONS-shSTAT3 (Fig. 1a) and DAOY (Fig. 1b) cells formed medullospheres of varied numbers and sizes, whereas in the presence of Dox or WP1066, the number and size of medullospheres were reduced significantly [35].

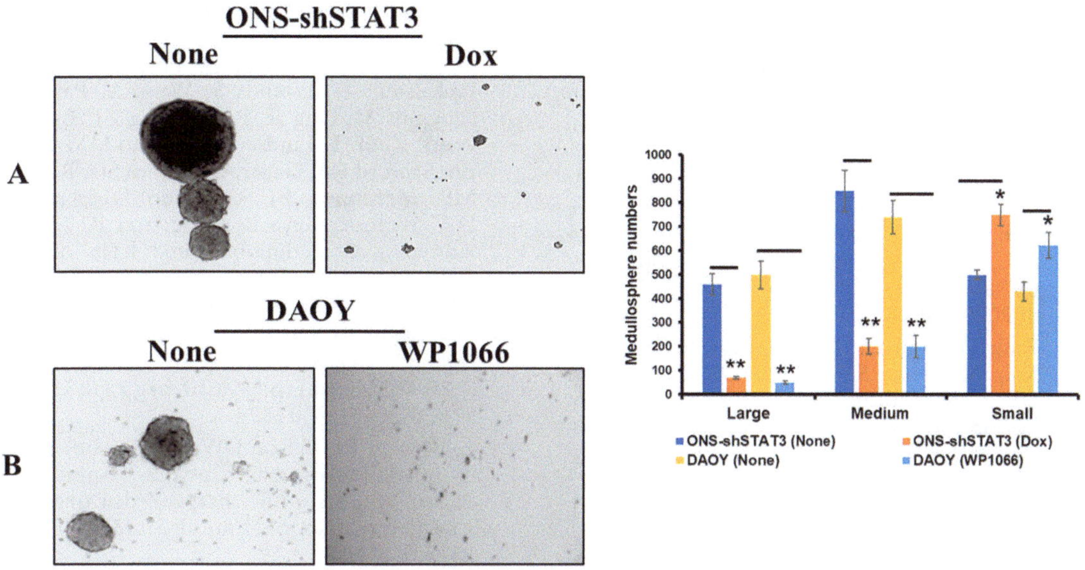

Fig. 1 Effect of STAT3 inhibition on medullosphere formation. Figure 1a and b shows representative morphologies of the medullosphere from ONS-shSTAT3 and DAOY cells, respectively, grown in serum-free media in the presence and absence of either 0.5 μg/mL Dox or 2 μM WP1066. Bar graph on right represents sphere numbers and sizes (small, medium, and large) before and after treatment. *$p < 0.05$, **$p < 0.01$

4 Notes

1. Do not change the medium; otherwise, you will lose the floating spheres; only add the fresh medium.
2. It will depend on the cell seeding density. Do not add more than 40 K cells per well in a 6-well plate or it will form clumps/aggregates.
3. Select spheres before it starts to develop a dark center that represents the hypoxic area.

References

1. Yu Z, Pestell TG, Lisanti MP, Pestell RG (2012) Cancer stem cells. Int J Biochem Cell Biol 44:2144–2151. https://doi.org/10.1016/j.biocel.2012.08.022
2. Clarke MF, Fuller M (2006) Stem cells and cancer: two faces of eve. Cell 124:1111–1115. https://doi.org/10.1016/j.cell.2006.03.011
3. Visvader JE, Lindeman GJ (2008) Cancer stem cells in solid tumours: accumulating evidence and unresolved questions. Nat Rev Cancer 8: 755–768. https://doi.org/10.1038/nrc2499
4. Dean M, Fojo T, Bates S (2005) Tumour stem cells and drug resistance. Nat Rev Cancer 5: 275–284. https://doi.org/10.1038/nrc1590
5. Shlush LI, Mitchell A, Heisler L, Abelson S, Ng SWK, Trotman-Grant A, Medeiros JJF, Rao-Bhatia A, Jaciw-Zurakowsky I, Marke R, McLeod JL, Doedens M, Bader G, Voisin V, Xu C, McPherson JD, Hudson TJ, Wang JCY, Minden MD, Dick JE (2017) Tracing the origins of relapse in acute myeloid leukaemia to stem cells. Nature 547:104–108. https://doi.org/10.1038/nature22993
6. Lapidot T, Sirard C, Vormoor J, Murdoch B, Hoang T, Caceres-Cortes J, Minden M, Paterson B, Caligiuri MA, Dick JE (1994) A cell initiating human acute myeloid leukaemia after transplantation into SCID mice. Nature

367:645–648. https://doi.org/10.1038/367645a0

7. Singh SK, Clarke ID, Terasaki M, Bonn VE, Hawkins C, Squire J, Dirks PB (2003) Identification of a cancer stem cell in human brain tumors. Cancer Res 63:5821–5828
8. Al-Hajj M, Wicha MS, Benito-Hernandez A, Morrison SJ, Clarke MF (2003) Prospective identification of tumorigenic breast cancer cells. Proc Natl Acad Sci U S A 100:3983–3988. https://doi.org/10.1073/pnas.0530291100
9. Fang D, Nguyen TK, Leishear K, Finko R, Kulp AN, Hotz S, Van Belle PA, Xu X, Elder DE, Herlyn M (2005) A tumorigenic subpopulation with stem cell properties in melanomas. Cancer Res 65:9328–9337. https://doi.org/10.1158/0008-5472.CAN-05-1343
10. Collins AT, Berry PA, Hyde C, Stower MJ, Maitland NJ (2005) Prospective identification of tumorigenic prostate cancer stem cells. Cancer Res 65:10946–10951. https://doi.org/10.1158/0008-5472.CAN-05-2018
11. O'Brien CA, Pollett A, Gallinger S, Dick JE (2007) A human colon cancer cell capable of initiating tumour growth in immunodeficient mice. Nature 445:106–110. https://doi.org/10.1038/nature05372
12. Li C, Heidt DG, Dalerba P, Burant CF, Zhang L, Adsay V, Wicha M, Clarke MF, Simeone DM (2007) Identification of pancreatic cancer stem cells. Cancer Res 67:1030–1037. https://doi.org/10.1158/0008-5472.CAN-06-2030
13. Eramo A, Lotti F, Sette G, Pilozzi E, Biffoni M, Di Virgilio A, Conticello C, Ruco L, Peschle C, De Maria R (2008) Identification and expansion of the tumorigenic lung cancer stem cell population. Cell Death Differ 15:504–514. https://doi.org/10.1038/sj.cdd.4402283
14. Ray S, Chaturvedi NK, Bhakat KK, Rizzino A, Mahapatra S (2021) Subgroup-specific diagnostic, prognostic, and predictive markers influencing pediatric Medulloblastoma treatment. Diagnostics (Basel) 12. https://doi.org/10.3390/diagnostics12010061
15. Taylor MD, Northcott PA, Korshunov A, Remke M, Cho YJ, Clifford SC, Eberhart CG, Parsons DW, Rutkowski S, Gajjar A, Ellison DW, Lichter P, Gilbertson RJ, Pomeroy SL, Kool M, Pfister SM (2012) Molecular subgroups of medulloblastoma: the current consensus. Acta Neuropathol 123:465–472. https://doi.org/10.1007/s00401-011-0922-z
16. Gibson P, Tong Y, Robinson G, Thompson MC, Currle DS, Eden C, Kranenburg TA, Hogg T, Poppleton H, Martin J, Finkelstein D, Pounds S, Weiss A, Patay Z, Scoggins M, Ogg R, Pei Y, Yang ZJ, Brun S, Lee Y, Zindy F, Lindsey JC, Taketo MM, Boop FA, Sanford RA, Gajjar A, Clifford SC, Roussel MF, McKinnon PJ, Gutmann DH, Ellison DW, Wechsler-Reya R, Gilbertson RJ (2010) Subtypes of medulloblastoma have distinct developmental origins. Nature 468:1095–1099. https://doi.org/10.1038/nature09587
17. Gilbertson RJ (2004) Medulloblastoma: signalling a change in treatment. Lancet Oncol 5:209–218. https://doi.org/10.1016/S1470-2045(04)01424-X
18. Gilbertson RJ, Ellison DW (2008) The origins of medulloblastoma subtypes. Annu Rev Pathol 3:341–365. https://doi.org/10.1146/annurev.pathmechdis.3.121806.151518
19. Cavalli FMG, Remke M, Rampasek L, Peacock J, Shih DJH, Luu B, Garzia L, Torchia J, Nor C, Morrissy AS, Agnihotri S, Thompson YY, Kuzan-Fischer CM, Farooq H, Isaev K, Daniels C, Cho BK, Kim SK, Wang KC, Lee JY, Grajkowska WA, Perek-Polnik M, Vasiljevic A, Faure-Conter C, Jouvet A, Giannini C, Rao AAN, Li KKW, Ng HK, Eberhart CG, Pollack IF, Hamilton RL, Gillespie GY, Olson JM, Leary S, Weiss WA, Lach B, Chambless LB, Thompson RC, Cooper MK, Vibhakar R, Hauser P, van Veelen MC, Kros JM, French PJ, Ra YS, Kumabe T, Lopez-Aguilar E, Zitterbart K, Sterba J, Finocchiaro G, Massimino M, Van Meir EG, Osuka S, Shofuda T, Klekner A, Zollo M, Leonard JR, Rubin JB, Jabado N, Albrecht S, Mora J, Van Meter TE, Jung S, Moore AS, Hallahan AR, Chan JA, Tirapelli DPC, Carlotti CG, Fouladi M, Pimentel J, Faria CC, Saad AG, Massimi L, Liau LM, Wheeler H, Nakamura H, Elbabaa SK, Perezpena-Diazconti M, de Leon FCP, Robinson S, Zapotocky M, Lassaletta A, Huang A, Hawkins CE, Tabori U, Bouffet E, Bartels U, Dirks PB, Rutka JT, Bader GD, Reimand J, Goldenberg A, Ramaswamy V, Taylor MD (2017) Intertumoral heterogeneity within Medulloblastoma subgroups. Cancer Cell 31:737. https://doi.org/10.1016/j.ccell.2017.05.005
20. Stavrou T, Bromley CM, Nicholson HS, Byrne J, Packer RJ, Goldstein AM, Reaman GH (2001) Prognostic factors and secondary malignancies in childhood medulloblastoma. J Pediatr Hematol Oncol 23:431–436. https://doi.org/10.1097/00043426-200110000-00008

21. Konrad CV, Murali R, Varghese BA, Nair R (2017) The role of cancer stem cells in tumor heterogeneity and resistance to therapy. Can J Physiol Pharmacol 95:1–15. https://doi.org/10.1139/cjpp-2016-0079
22. Pastrana E, Silva-Vargas V, Doetsch F (2011) Eyes wide open: a critical review of sphere-formation as an assay for stem cells. Cell Stem Cell 8:486–498. https://doi.org/10.1016/j.stem.2011.04.007
23. Ayob AZ, Ramasamy TS (2018) Cancer stem cells as key drivers of tumour progression. J Biomed Sci 25:20. https://doi.org/10.1186/s12929-018-0426-4
24. Cartwright P, McLean C, Sheppard A, Rivett D, Jones K, Dalton S (2005) LIF/-STAT3 controls ES cell self-renewal and pluripotency by a Myc-dependent mechanism. Development 132:885–896. https://doi.org/10.1242/dev.01670
25. Bromberg JF, Wrzeszczynska MH, Devgan G, Zhao Y, Pestell RG, Albanese C, Darnell JE Jr (1999) Stat3 as an oncogene. Cell 98:295–303
26. Levy DE, Darnell JE Jr (2002) Stats: transcriptional control and biological impact. Nat Rev Mol Cell Biol 3:651–662. https://doi.org/10.1038/nrm909
27. Darnell JE Jr (1997) STATs and gene regulation. Science 277:1630–1635
28. Xiao H, Bid HK, Jou D, Wu X, Yu W, Li C, Houghton PJ, Lin J (2015) A novel small molecular STAT3 inhibitor, LY5, inhibits cell viability, cell migration, and angiogenesis in medulloblastoma cells. J Biol Chem 290:3418–3429. https://doi.org/10.1074/jbc.M114.616748
29. Sreenivasan L, Wang H, Yap SQ, Leclair P, Tam A, Lim CJ (2020) Autocrine IL-6/STAT3 signaling aids development of acquired drug resistance in group 3 medulloblastoma. Cell Death Dis 11:1035. https://doi.org/10.1038/s41419-020-03241-y
30. Chen X, Pan L, Wei J, Zhang R, Yang X, Song J, Bai RY, Fu S, Pierson CR, Finlay JL, Li C, Lin J (2021) LLL12B, a small molecule STAT3 inhibitor, induces growth arrest, apoptosis, and enhances cisplatin-mediated cytotoxicity in medulloblastoma cells. Sci Rep 11:6517. https://doi.org/10.1038/s41598-021-85888-x
31. Ray S, Coulter DW, Gray SD, Sughroue JA, Roychoudhury S, McIntyre EM, Chaturvedi NK, Bhakat KK, Joshi SS, McGuire TR, Sharp JG (2018) Suppression of STAT3 NH2 -terminal domain chemosensitizes medulloblastoma cells by activation of protein inhibitor of activated STAT3 via de-repression by microRNA-21. Mol Carcinog 57:536–548. https://doi.org/10.1002/mc.22778
32. Zaiter A, Audi ZF, Shawraba F, Saker Z, Bahmad HF, Nabha RH, Harati H, Nabha SM (2022) STAT3 in medulloblastoma: a key transcriptional regulator and potential therapeutic target. Mol Biol Rep 49:10635–10652. https://doi.org/10.1007/s11033-022-07694-6
33. Yang MY, Lee HT, Chen CM, Shen CC, Ma HI (2014) Celecoxib suppresses the phosphorylation of STAT3 protein and can enhance the radiosensitivity of medulloblastoma-derived cancer stem-like cells. Int J Mol Sci 15:11013–11029. https://doi.org/10.3390/ijms150611013
34. Garg N, Bakhshinyan D, Venugopal C, Mahendram S, Rosa DA, Vijayakumar T, Manoranjan B, Hallett R, McFarlane N, Delaney KH, Kwiecien JM, Arpin CC, Lai PS, Gomez-Biagi RF, Ali AM, de Araujo ED, Ajani OA, Hassell JA, Gunning PT, Singh SK (2017) CD133+ brain tumor-initiating cells are dependent on STAT3 signaling to drive medulloblastoma recurrence. Oncogene 36:606–617. https://doi.org/10.1038/onc.2016.235
35. Rohrer KA, Song H, Akbar A, Chen Y, Pramanik S, Wilder PJ, McIntyre EM, Chaturvedi NK, Bhakat KK, Rizzino A, Coulter DW, Ray S (2023) STAT3 inhibition attenuates MYC expression by modulating co-activator recruitment and suppresses Medulloblastoma tumor growth by augmenting cisplatin efficacy in vivo. Cancers 15(8):2239. https://doi.org/10.3390/cancers15082239

Index

Kishor K. Bhakat and Tapas K. Hazra (eds.), *Base Excision Repair Pathway: Methods and Protocols*, Methods in Molecular Biology, vol. 2701, https://doi.org/10.1007/978-1-0716-3373-1,

MIX
Papier aus verantwortungsvollen Quellen
Paper from responsible sources
FSC® C105338
FSC
www.fsc.org

If you have any concerns about our products, you can contact us on
ProductSafety@springernature.com

In case Publisher is established outside the EU, the EU authorized representative is:
Springer Nature Customer Service Center GmbH
Europaplatz 3, 69115 Heidelberg, Germany

Printed by Libri Plureos GmbH
in Hamburg, Germany